Comparing the Geological and Fossil Records: Implications for Biodiversity Studies

Geological Society books refereeing procedures

The Society makes every effort to ensure that the scientific and production quality of its books matches that of its journals. Since 1997, all book proposals have been refereed by specialist reviewers as well as by the Society's Books Editorial Committee. If the referees identify weaknesses in the proposal, these must be addressed before the proposal is accepted.

Once the book is accepted, the Society Book Editors ensure that the volume editors follow strict guidelines on refereeing and quality control. We insist that individual papers can only be accepted after satisfactory review by two independent referees. The questions on the review forms are similar to those for *Journal of the Geological Society*. The referees' forms and comments must be available to the Society's Book Editors on request.

Although many of the books result from meetings, the editors are expected to commission papers that were not presented at the meeting to ensure that the book provides a balanced coverage of the subject. Being accepted for presentation at the meeting does not guarantee inclusion in the book.

More information about submitting a proposal and producing a book for the Society can be found on its web site: www.geolsoc.org.uk.

It is recommended that reference to all or part of this book should be made in one of the following ways:

MCGOWAN, A. J. & SMITH, A. B. (eds) 2011. *Comparing the Geological and Fossil Records: Implications for Biodiversity Studies*. Geological Society, London, Special Publications, **358**.

O'CONNOR, A., MONCRIEFF, C. & WILLS, M. A. 2011. Variation in stratigraphic congruence (GER) through the Phanerozoic and across higher taxa is partially determined by sources of bias. *In*: MCGOWAN, A. J. & SMITH, A. B. (eds) *Comparing the Geological and Fossil Records: Implications for Biodiversity Studies*. Geological Society, London, Special Publications, **358**, 31–52.

GEOLOGICAL SOCIETY SPECIAL PUBLICATION NO. 358

Comparing the Geological and Fossil Records: Implications for Biodiversity Studies

EDITED BY

A. J. McGOWAN
University of Glasgow, Scotland

and

A. B. SMITH
Natural History Museum, London

2011
Published by
The Geological Society
London

THE GEOLOGICAL SOCIETY

The Geological Society of London (GSL) was founded in 1807. It is the oldest national geological society in the world and the largest in Europe. It was incorporated under Royal Charter in 1825 and is Registered Charity 210161.

The Society is the UK national learned and professional society for geology with a worldwide Fellowship (FGS) of over 10 000. The Society has the power to confer Chartered status on suitably qualified Fellows, and about 2000 of the Fellowship carry the title (CGeol). Chartered Geologists may also obtain the equivalent European title, European Geologist (EurGeol). One fifth of the Society's fellowship resides outside the UK. To find out more about the Society, log on to www.geolsoc.org.uk.

The Geological Society Publishing House (Bath, UK) produces the Society's international journals and books, and acts as European distributor for selected publications of the American Association of Petroleum Geologists (AAPG), the Indonesian Petroleum Association (IPA), the Geological Society of America (GSA), the Society for Sedimentary Geology (SEPM) and the Geologists' Association (GA). Joint marketing agreements ensure that GSL Fellows may purchase these societies' publications at a discount. The Society's online bookshop (accessible from www.geolsoc. org.uk) offers secure book purchasing with your credit or debit card.

To find out about joining the Society and benefiting from substantial discounts on publications of GSL and other societies worldwide, consult www.geolsoc.org.uk, or contact the Fellowship Department at: The Geological Society, Burlington House, Piccadilly, London W1J 0BG: Tel. + 44 (0)20 7434 9944; Fax + 44 (0)20 7439 8975; E-mail: enquiries@geolsoc.org.uk.

For information about the Society's meetings, consult *Events* on www.geolsoc.org.uk. To find out more about the Society's Corporate Affiliates Scheme, write to enquiries@geolsoc.org.uk.

Published by The Geological Society from:
The Geological Society Publishing House, Unit 7, Brassmill Enterprise Centre, Brassmill Lane, Bath BA1 3JN, UK

(*Orders*:   Tel. + 44 (0)1225 445046, Fax + 44 (0)1225 442836)
Online bookshop: www.geolsoc.org.uk/bookshop

The publishers make no representation, express or implied, with regard to the accuracy of the information contained in this book and cannot accept any legal responsibility for any errors or omissions that may be made.

British Library Cataloguing in Publication Data

A catalogue record for this book is available from the British Library.
ISBN 978-1-86239-336-3

Distributors
For details of international agents and distributors see:
www.geolsoc.org.uk/agentsdistributors

Typeset by Techset Composition Ltd, Salisbury, UK
Printed by MPG Books Ltd, Bodmin, UK

Contents

The ties linking rock and fossil records and why they are important for palaeobiodiversity studies

ANDREW B. SMITH[1]* & ALISTAIR J. McGOWAN[2]

[1]*Natural History Museum, Cromwell Road, London SW7 5BD, UK*

[2]*School of Geographical and Earth Sciences, University of Glasgow, Glasgow, G12 8QQ, UK*

Corresponding author (e-mail: a.smith@nhm.ac.uk)

Abstract: A correlation exists between the quality of the rock record and the diversity of fossils recorded from that rock record but what drives that correlation, and how consistent that correlation is across different environments, remain to be determined. Palaeontologists wishing to investigate past diversity patterns need to first address issues of geological bias in their data.

The fossil record provides the only empirical evidence of how life has diversified over geological time, but it needs to be interpreted with caution. For many years the history of diversity was estimated simply by summing up the numbers of taxa (species, genera, families) palaeontologists have recorded from successive geological time intervals, or extrapolated from their first and last occurrences in the geological record. This time series approach led to what is now a classic view of how diversity has changed over time (Sepkoski *et al.* 1981; Benton 1995; Sepkoski 1997). Such counts take the fossil record at face value, or assume that biases or errors are randomly distributed in such a way that the overall effect is negligible. We now realize that this is only a first, crude, approximation that may be subject to a number of strong biases that arise because of the nature of the sedimentary rock record.

Time series analysis requires that sampling be carefully controlled for best results. Ideally data should be collected so that sampling from each time interval is uniform, or at least sampled fairly using the 'shareholder quorum' subsampling method (Alroy 2010). Otherwise apparent changes in diversity may arise for spurious reasons, for example, because (a) a time intervals being sampled are of variable duration (longer time intervals = more recorded diversity), or (b) a time interval has been more intensively sampled (more localities/specimens/habitats/formations sampled = more recorded diversity). While for a biological survey it is easy to plan a sampling strategy that will give approximately equal effort and coverage for observations, palaeontologists are faced with a much more difficult task. Firstly, the time bins they work with are irregular and highly variable in their duration. For example, the durations of one of the most widely used time scales has intervals spanning two

orders of magnitude (Gradstein *et al.* 2004). Secondly, and much more critically, palaeontologists are collecting from an already incomplete and highly biased set of rocks, which in turn skew the range of taxa and habitats that can be sampled (**Zuschin *et al.* 2011**). Bluntly put, we cannot sample what is not preserved in the rock record, although phylogenetic and molecular approaches can attempt to compensate (Pol & Norrell 2006; Bininda-Emonds *et al.* 2007; Wills 2007; **O'Connor *et al.* 2011**). So while it may be possible to standardize for sampling effort from the rock record that remains, those rocks might provide a far from uniform sample of the sediments and palaeoenvironments that were originally present.

Palaeontologists cannot therefore assume uniform sampling of the fossil record and must try to assess how the rock record they have to work with has affected their ability to sample evenly, and then develop methods that compensate appropriately for this variation. In recent years therefore palaeobiodiversity studies have become more probabilistic in their approach, concerned with establishing confidence limits around estimates that try to correct for uneven sampling (Alroy *et al.* 2001, 2008) and testing empirical patterns against model predictions (Foote 2001; Smith & McGowan 2007; McGowan & Smith 2008). If we are to improve our estimates of biodiversity over time then sampling parameters need to be better quantified and we need to better understand the complex interrelationship between rock and fossil records. This is no simple task and requires a more systematic approach to recording culture as well as better documentation of the variables (**Benton *et al.* 2011**).

Figure 1 summarizes the problem as we see it. Palaeontologists estimate past diversity from remains preserved in the rock record. As with any

From: McGowan, A. J. & Smith, A. B. (eds) *Comparing the Geological and Fossil Records: Implications for Biodiversity Studies*. Geological Society, London, Special Publications, **358**, 1–7.
DOI: 10.1144/SP358.1   0305-8719/11/$15.00 © The Geological Society of London 2011.

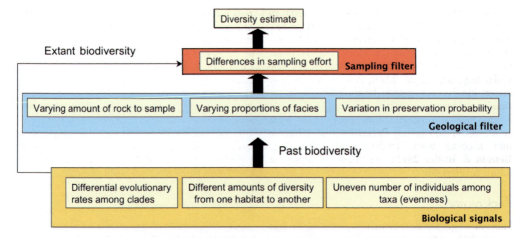

Fig. 1. Schematic flow chart showing how the sampled diversity estimate that palaeontologists have to work with represents a filtered signal of the original biological diversity record from the geological past.

biological survey, variation of sampling effort of the surviving rock record needs to be factored out. But this rock record has already passed through a geological filter that has altered and distorted not only the rock area that survives to sample, but also the proportions of facies and the preservational quality of its fossil record. Were this geological filter to remain approximately constant over time there would be little problem of interpreting the biological signal. However, it is far from uniform with amounts of rock to sample, proportions of facies represented by those rocks and preservation potential all varying from one time interval to the next. To interpret observed changes in sampled biodiversity estimates correctly we must therefore first understand the role of the geological filter.

Variability of the rock and fossil records

Fossil record

All clades rise and fall in diversity over time according to whether speciation or extinction is proceeding faster. Such change can take place rapidly, during mass extinction events and adaptive radiations, or gradually so as to define a long-term rising or falling trend over time. At any one time standing diversity also varies in different habitats and latitudes, and some taxa are much more numerous than others. These, of course, are the biological signals that palaeobiologists wish to isolate. However, superimposed on this comes variation in preservation potential. This varies markedly amongst taxonomic groups but remains relatively invariant within groups over time. Many of the best fossil records belong to microfossil groups

such as coccolithophorids and planktonic foraminifera, where specimens can be collected in abundance against a Milankovitch time scale (20 Ka) (e.g. Dunkley-Jones *et al.* 2008; Ebra *et al.* 2010; **Lazarus 2011**). For many other groups, however, fossil records can be patchy and incomplete at best. Amongst clades with a mineralized skeleton, the worst fossil record must surely be that of birds. Fountaine *et al.* (2005) compiled a total of just 121 specimens recorded from Mesozoic sediments, which encompass 98 species and 70 genera of fossil birds. Thus, over 80% of species and almost 60% of genera are known only from single specimens. Clearly the chances that any new fossil find will represent a new species or new genus are very high, particularly if it comes from a new location or time interval (Benton 2008).

While some groups clearly have better fossil records than others, so long as preservation potential in each group remains approximately similar over time there should be no problem. Changes to preservation potential can occur, as for example with the evolving robustness of the skeleton in echinoids (Smith 2007) or bivalves (Kidwell & Holland 2002; Behrensmeyer *et al.* 2005). However, for the most part, the preservation potential of major higher taxa changes so slowly that it can have little role to play in creating short- and medium-term fluctuations in sampled diversity. It is also unidirectional rather than cyclical. So, while preservation potential is variable amongst taxonomic groups it is unlikely to confound biodiversity studies except in generating simple long-term trends.

Lagerstätten pose a particular problem for the analysis of diversity patterns. The record of groups with low preservational potential or multi-element skeletons that are prone to rapid post-mortem disarticulation, such as insects and vertebrates, may be largely confined to such deposits. As Lagerstätten are non-randomly distributed through time, both at coarse (Allison & Briggs 1993) and fine scales (e.g. Brett *et al.* 2009), this could seriously distort biodiversity counts. Improving our understanding of how Lagerstätten are distributed within a sequence stratigraphic framework (e.g. Brett *et al.* 2009) and taking Lagerstätten distribution into account when analysing diversity (e.g. **Benson & Butler 2011**), are important areas for future development.

Rock record

Palaeontologists collect fossils from the rocks that are available to them at outcrop or from drill cores, and for the most part have done this effectively. There are, never the less, spatial and temporal biases to this sampling (Smith 2001) with the record of Europe and North America dominating global databases. While these are easy to compensate for by standard procedures such as rarefaction and subsampling (e.g. Alroy *et al.* 2008), a deeper, more pernicious problem remains: the rock record being sampled is itself a biased sample of what once existed.

The quality of the sedimentary record varies markedly amongst environmental settings, with some environments, such as cratonic highlands and deep-sea basins, which represent over 50% of the surface area of the planet at present, being much less commonly represented in the geological record than others. However, in marked contrast to the fossil record, the sedimentary rock records from individual settings also show marked temporal variability at all time-scales. At the very largest scale, major plate tectonic cycles of plate accretion and dismemberment generate changes in ocean basin volume that drive sea-level changes on cratons of up to 150 m amplitude (Dewey & Pitman 1998; Miller *et al.* 2005). These sea-level cycles drive major changes to the relative proportions of terrestrial and marine sediments being deposited over the continental blocks, with marine sedimentary rocks dominating the rock record at times of high-stand (Smith 2001; Smith & McGowan 2007). Over much shorter time intervals of 10–100 Ka, changes in land-locked ice drive sea-level oscillations of up to 150 m amplitude by altering the volume of water in the ocean basins (Miller *et al.* 2005). Finally, at intermediate time-scales of 10–50 Ma there is growing evidence that mantle cell cycles create regional sea-level changes of somewhat smaller amplitude (Lovell 2010; Petersen *et al.* 2010) through thermal uplift. All these processes directly affect the quality and nature of the rock record that is laid down by affecting uplift and erosion as well as accommodation space and thus sediment accumulation rates. When sea-level in the past was close to, or below, current levels the resulting rock record available to geologists is dominated by terrestrial sediments deposited in flood plains and continental basin environments. As a consequence there is only a highly restricted set of localities where marine rocks of those time intervals and their fossils can be investigated on land. Conversely, when sea-level stood much higher than present-day levels, as in the early Late Cretaceous, terrestrial and marginal marine deposits are relatively sparse. Note, however, that the link between sea-level change and habitable marine shelf area can be complex. Wyatt (1995) for example showed that, due to hypsometry, a drop in sea-level in the late Ordovician actually resulted in an increased surface area of shallow marine settings.

A further complicating factor is the degree of post-depositional compaction and alteration which sediments have undergone, as this affects how easily and effectively sampling can be carried out (Hendy 2009). The probability of preserving and sampling small and/or delicate fossils (and thus recording higher diversities) is higher in fine-grained, poorly consolidated sediments than in older sequences subjected to tectonic and thermal alteration. Finally, the chemical composition of the skeleton significantly affects the chances of a fossil surviving in the rock record and can seriously bias both land and deep-sea records (e.g. **Cherns & Wright** 2000, **2011**).

The evidence that sedimentary rock and fossil records are intimately linked

Recent efforts to calculate global diversity patterns after standardizing for collecting effort (Alroy *et al.* 2001, 2008) recover a Phanerozoic diversity curve that is different to that using raw sampled diversity. This suggests that the actual and potential collecting effort in different parts of the geological column is a significant factor in shaping our sampled diversity. But it does not tell us whether palaeontologists have unevenly sampled the rocks that are available at outcrop, or whether sampling has been relatively uniform, but those rocks provide a non-random sample of what once existed.

That a positive correlation exists between areal extent of sedimentary rocks on land and sampled diversity has emerged from a number of studies. These include studies that estimate the surface outcrop area of terrestrial or marine sedimentary rocks from geological maps and their accompanying memoirs (Ramp 1976; Smith 2001; Crampton *et al.* 2003; Smith & McGowan 2007;

McGowan & Smith 2008; Barrett *et al.* 2009; **Wall *et al.* 2009, 2011**) or counts of the numbers of named formations (Peters & Foote 2001; Crampton *et al.* 2003; Benson *et al.* 2009; Mannion *et al.* 2010; **Benson & Butler 2011**). In all cases a statistically significant positive correlation has been demonstrated linking the rock and fossil records, both in marine and terrestrial environments.

One drawback of some of these studies is that they match rock and fossil diversity records that are not directly comparable. Most commonly regional rock record compilations have been tested against global diversity estimates (Smith 2001; Peters & Foote 2001; Smith & McGowan 2007; McGowan & Smith 2008; Benson *et al.* 2009). Global rock record outcrop estimates are available (Ronov 1978) but are compiled at a much coarser time-scale compared to diversity estimates (see Wall *et al.* 2009). However, not all studies suffer from this problem. **Crampton *et al.* (2003, 2011)** compare fossil and rock record data from exactly the same geographical region and Peters & Heim (2010) have now combined the Paleobiology Database (a database of taxonomic lists of fossils with accompanying geological and spatial data: http://paleodb.org) with Macrostrat (a database of rock outcrop in North America: http://macrostrat. geology.wisc.edu) to compare North American rock and fossil records directly (**Peters & Heim 2011**). **Lloyd *et al.* (2011)** have a comparable deep-sea rock and fossil database for the central and North Atlantic and adjacent regions, and **Upchurch *et al.* (2011)** have compared regional rock outcrop estimates with regional dinosaur diversities. Where data from rock and fossil records are collected from exactly the same geographical region the evidence for a link between rock and fossil diversity records is generally strengthened for both short and long-term trends.

One additional complication has arisen recently in the use of geological maps to directly estimate the area of exposed rock available to sample for fossils (e.g. Uhen & Pyenson 2007; Wall *et al.* 2009). Dunhill (2011) has shown that, for a series of 50 sites within England and Wales, the area of rock outcropping and the amount of rock exposure (i.e. rock that is not covered by superficial deposits and one could literally stand on) is not well correlated. This finding makes it much more difficult to apply simple species-area reasoning to local diversity fluctuations in the fossil record at small scales. However, such variation becomes negligible for large-scale studies that simply ask what proportion of a large landmass has rocks that yield fossils of a particular age, and present day exposure and historic exposure patterns may be very different.

The conundrum

That rock and fossil records are intimately connected is therefore now well established, for both marine and terrestrial records. However, determining the mechanism that is responsible for generating this linkage remains surprisingly difficult. Climate change, mantle plume cycles and plate tectonic activity all act in concert to create cycles of sea-level change across cratons. This has two important consequences: it changes the relative amount of marine and terrestrial sedimentary rocks that end up being preserved in the geological record and it also changes the surface area of shallow marine and terrestrial habitats where organisms can live. These two factors (regional extent of sedimentary rock and regional extent of original habitat) change in concert: as the outcrop area of marine rock record expands with craton flooding, the surface area of marine habitats also expands, potentially driving speciation and leading to greater standing diversity. Conversely, as sea-level drops, the area over which marine sedimentary rocks are deposited becomes smaller, as does the habitat area for marine organisms driving extinction and leading to smaller standing diversity.

We are therefore faced with two linkage mechanisms acting in parallel. On the one hand the fluctuating quality of the rock record may be controlling sampled diversity directly through altering the potential collecting effort that can be made in each time interval. The more outcrop area and the more environmental heterogeneity that outcrop encompasses, the more biological diversity is likely to be recovered from a simple species/area consideration (Rosenzweig 1995). On the other hand, biological diversity will also be responding directly to sea-level change. The 'common cause' hypothesis (Peters 2005) thus predicts that species diversity and rock record will mirror one another because both macrostratigraphy and biodiversity respond independently and in concert to sea-level cycles.

In truth both sampling and common cause effects must act together to influence the fossil record that we have recovered from the rock record. The key question to be answered then is which, if either, of the two processes dominates? This may turn out to be a far from simple question, as the relative strengths of the two factors may be dependent on the environment, time-scale or geological period being studied (e.g. **Benson & Butler 2011**). For example we may find that sampling effects may dominate in forming small-scale, stage-to-stage, changes in sampled diversity, while common cause effects shape longer-term trends. It could even be possible that sampling effects dominate at certain periods while common cause effects drive biodiversity curves at other time periods.

How we test these ideas remains to be formulated but a start is now being made (Peters 2005; Hannisdal & Peters 2010; **Hannisdal 2011**; Wall *et al.* 2011). Peters (2006) found no correlation between the size of taxonomic loss and the duration of the succeeding sediment hiatus. However, as his null hypothesis assumed uniform diversity over time, all this proves is that there is more to the fossil record than pure sampling bias, something also shown by Smith & McGowan (2007). Peters & Heim (2010) also argued for common cause dominancy based on their demonstration that in North America last occurrences of marine taxa correlate more strongly to marine sediment package terminations than first occurrence rates correlate to sediment package originations. This, however, mirrors the asymmetry of the sediment packages themselves, which provide a relatively complete record of transgressive intervals but whose later parts are artificially truncated by non-deposition and erosion (thereby truncating generic ranges). The weak correlation between taxic and sediment package originations thus argues for a relatively weak sampling bias, while the stronger correlation between sediment trucations and taxic last occurrences argues for strong sampling bias within cycles of deposition.

Where next?

If we are ever to develop a more complete and accurate estimate of Phanerozoic biodiversity patterns then a first important step must be to develop a better understanding of the complex interaction between rock record quality and sampled diversity. This requires better documentation of the way in which the rock record changes over time as well as a more consistent recording strategy that takes account of both sampled diversity and sampling opportunity, comparable to that developed for North American Quaternary land mammals (**Barnosky et al. 2011**). To test these ideas thoroughly we probably need to turn to sedimentary depositional systems where sea-level is not the dominant driving factor of both biological opportunity and rock record quality. Here the deep-sea record offers such an opportunity.

For land-based records the Paleobiology Database Project (PaleoDB) represents a major advance in providing taxonomic occurrence data tied to specific outcrops. This offers the ability to analyse faunas from comparable habitats and to make partial corrections for collection effort. It is not the complete answer, however, as it only provides indirect estimates of alpha diversity (Bush *et al.* 2004) and in calculating regional or global diversity is not able to correct for the biases introduced by the missing rock record and habitat heterogeneity.

Standardizing for sampling effort without also standardizing for the proportion of rocks capturing different palaeoenvironments that are preserved in the geological record, will still produce misleading results. What is really needed is an equivalent database that can be used to measure the diversity and heterogeneity of the geological record. This would need a method to combine the digital data generated by geological surveys summarizing the aerial distribution of rock outcrops, with field evidence about the nature of the sedimentary environments those outcrops encompass. It would also need to be compiled at a temporal resolution equivalent to the PaleoDB. Then it would be truly possible to disentangle the signals coming from biodiversity and rock record.

References

ALLISON, P. A. & BRIGGS, D. E. G. 1993. Exceptional fossil record: distribution of soft-tissue preservation through the Phanerozoic. *Geology*, **21**, 605–608.

ALROY, J. 2010. Geographical, environmental and intrinsic controls on Phanerozoic marine diversification. *Palaeontology*, **53**, 1211–1235.

ALROY, J., MARSHALL, C. R. ET AL. 2001. Effects of sampling standardization on estimates of Phanerozoic marine diversification. *Proceedings of the National Academy of Sciences of the United States of America*, **98**, 6261–6266.

ALROY, J., ABERHAN, M. ET AL. 2008. Phanerozoic trends in the global diversity of marine invertebrates. *Science*, **321**, 97–100.

BARNOSKY, A. D., CARRASCO, M. A. & GRAHAM, R. W. 2011. Collateral mammal diversity loss associated with late Quaternary megafaunal extinctions and implications for the future. *In*: McGOWAN, A. J. & SMITH, A. B. (eds) *Comparing the Geological and Fossil Records: Implications for Biodiversity Studies*. Geological Society, London, Special Publications, **358**, 179–190.

BARRETT, P. M., McGOWAN, A. J. & PAGE, V. 2009. Dinosaur diversity and the rock record. *Proceedings of the Royal Society of London, Series B*, **276**, 2667–2674.

BEHRENSMEYER, A. K., FURSICH, F. T. ET AL. 2005. Are the most durable shelly taxa also the most common in the marine fossil record? *Paleobiology*, **31**, 607–623.

BENSON, R. B. J. & BUTLER, R. J. 2011. Uncovering the diversification history of marine tetrapods: ecology influences the effect of geological sampling bias. *In*: McGOWAN, A. J. & SMITH, A. B. (eds) *Comparing the Geological and Fossil Records: Implications for Biodiversity Studies*. Geological Society, London, Special Publications, **358**, 191–207.

BENSON, R. B. J., BUTLER, R. J., LINDGREN, J. & SMITH, A. S. 2009. Mesozoic marine tetrapod diversity: mass extinctions and temporal heterogeneity in geological megabiases affecting vertebrates. *Proceedings of the Royal Society, B*, **277**, 829–834.

BENTON, M. J. 1995. Diversity and extinction in the history of life. *Science*, **268**, 52–58.

BENTON, M. J. 2008. How to find a dinosaur and the role of synonymy in biodiversity studies. *Paleobiology*, **34**, 516–533.

BENTON, M. J., DUNHILL, A. M., LLOYD, G. T. & MARX, F. G. 2011. Assessing the quality of the fossil record: insights from vertebrates. *In*: McGOWAN, A. J. & SMITH, A. B. (eds) *Comparing the Geological and Fossil Records: Implications for Biodiversity Studies*. Geological Society, London, Special Publications, **358**, 63–94.

BININDA-EMONDS, O. R. P., CARDILLO, M. ET AL. 2007. The delayed rise of present-day mammals. *Nature*, **446**, 507–512.

BRETT, C. E., ALLISON, P. A., DeSANTIS, M. K., LIDDELL, W. D. & KRAMER, A. 2009. Sequence stratigraphy, cyclic facies and lagerstätten in the Middle Cambrian Wheeler and Marjum Formations, Great Basin, USA. *Palaeogeography, Palaeoclimatology, Palaeoecology*, **277**, 9–33.

BUSH, A. M., MARKEY, M. J. & MARSHALL, C. R. 2004. Removing bias from diversity curves: the effects of spatially organized biodiversity on sampling-standardization. *Paleobiology*, **30**, 666–686.

CHERNS, L. & WRIGHT, V. P. 2000. Missing molluscs as evidence of large-scale early skeletal aragonitic dissolution in a Silurian sea. *Geology*, **28**, 791–784.

CHERNS, L. & WRIGHT, V. P. 2011. Skeletal mineralogy and biodiversity of marine invertebrates: size matters more than seawater chemistry. *In*: McGOWAN, A. J. & SMITH, A. B. (eds) *Comparing the Geological and Fossil Records: Implications for Biodiversity Studies*. Geological Society, London, Special Publications, **358**, 9–18.

CRAMPTON, J. S., BEU, A. G., COOPER, R. A., JONES, C. A., MARSHALL, B. & MAXWELL, P. A. 2003. Estimating the rock volume bias in paleobiodiversity studies. *Science*, **301**, 358–360.

CRAMPTON, J. S., FOOTE, M., COOPER, R. A., BEU, A. G. & PETERS, S. E. 2011. The fossil record and spatial structuring of environments and biodiversity in the Cenozoic of New Zealand. *In*: McGOWAN, A. J. & SMITH, A. B. (eds) *Comparing the Geological and Fossil Records: Implications for Biodiversity Studies*. Geological Society, London, Special Publications, **358**, 105–122.

DEWEY, J. F. & PITMAN, W. C. 1998. Sea-level changes: mechanisms, magnitudes and rates. *In*: PINDELL, J. L. & DRAKE, C. L. (eds) *Paleogeographic Evolution and Non-Glacial Eustacy, Northern South America*. SEPM Special Publication, **58**, 1–16, Tulsa.

DUNHILL, A. M. 2011. Using remote sensing and a GIS to quantify rock exposure area in England and Wales: implications for paleodiversity studies. *Geology*, **39**, 111–114.

DUNKLEY-JONES, T., BOWN, P. R., PEARSON, P. N., WADE, B. S. & COXALL, H. K. 2008. Major shifts in calcareous plankton assemblages through the Eocene–Oligocene transition in Tanzania and their implications for low-latitude primary production. *Paleoceanography*, **23**, PA4202, doi: 1029/2008PA001620.

EBRA, E., BOTTINI, C., WEISSERT, H. J. & KELLER, C. E. 2010. Calcareous nannoplankton response to surface water acidification around Anoxic Event 1a. *Science*, **329**, 428–432.

FOOTE, M. 2001. Inferring temporal patterns of preservation, origination, and extinction from taxonomic survivorship analysis. *Paleobiology*, **27**, 602–630.

FOUNTAINE, T. M. R., BENTON, M. J., NUDDS, R. L. & DYKE, G. J. 2005. The quality of the fossil record of Mesozoic birds. *Proceedings of the Royal Society, B*, **272**, 289–294.

GRADSTEIN, F. M., OGG, J. G. & SMITH, A. G. (eds). 2004. *A Geologic Time Scale 2004*. Cambridge University Press, Cambridge, UK.

HANNISDAL, B. 2011. Non-parametric inference of causal interactions from geological records. *American Journal of Science*, in press.

HANNISDAL, B. & PETERS, S. E. 2010. On the relationship between macrostratigraphy and geological processes: quantitative information capture and sampling robustness. *Journal of Geology*, **118**, 111–130.

HENDY, A. J. W. 2009. The influence of lithification on Cenozoic marine biodiversity trends. *Paleobiology*, **35**, 51–62.

KIDWELL, S. M. & HOLLAND, S. M. 2002. The quality of the fossil record: implications for evolutionary analyses. *Annual Review of Ecology and Systematics*, **33**, 561–588.

LAZARUS, D. B. 2011. The deep-sea microfossil record of macroevolutionary change in plankton and its study. *In*: McGOWAN, A. J. & SMITH, A. B. (eds) *Comparing the Geological and Fossil Records: Implications for Biodiversity Studies*. Geological Society, London, Special Publications, **358**, 141–166.

LLOYD, G. T., SMITH, A. B. & YOUNG, J. R. 2011. Quantifying the deep-sea rock and fossil record bias using coccolithophores. *In*: McGOWAN, A. J. & SMITH, A. B. (eds) *Comparing the Geological and Fossil Records: Implications for Biodiversity Studies*. Geological Society, London, Special Publications, **358**, 167–177.

LOVELL, B. 2010. A pulse in the planet. *Journal of the Geological Society, London*, **167**, 1–12.

MANNION, P. D., UPCHURCH, P., CARRANO, M. T. & BARRETT, P. M. 2010. Testing the effect of the rock record on diversity: a multidisciplinary approach to elucidating the generic richness of sauropodomorph dinosaurs through time. *Biological Reviews*, **86**, 157–181, doi: 10.1111/j.1469-185X.2010.00139.x.

McGOWAN, A. J. & SMITH, A. B. 2008. Are global Phanerozoic marine diversity curves truly global? A study of the relationship between regional rock records and global Phanerozoic marine diversity. *Paleobiology*, **34**, 80–103.

MILLER, K. G., KOMINZ, M. A. ET AL. 2005. The Phanerozoic record of global sea-level change. *Science*, **310**, 1293–1297.

O'CONNOR, A., MONCRIEFF, C. & WILLS, M. A. 2011. Variation in stratigraphic congruence (GER) through the Phanerozoic and across higher taxa is partially determined by sources of bias. *In*: McGOWAN, A. J. & SMITH, A. B. (eds) *Comparing the Geological and Fossil Records: Implications for Biodiversity Studies*. Geological Society, London, Special Publications, **358**, 31–52.

PETERS, S. E. 2005. Geological constraints on the macroevolutionary history of marine animals. *Proceedings of the National Academy of Sciences*, **102**, 12326–12331.

PETERS, S. E. 2006. Genus extinction, origination, and the durations of sedimentary hiatuses. *Paleobiology*, **32**, 387–407.

PETERS, S. E. & FOOTE, M. 2001. Biodiversity in the Phanerozoic: a reinterpretation. *Paleobiology*, **27**, 583–601.

PETERS, S. E. & HEIM, N. A. 2010. The geological completeness of paleontological sampling in North America. *Paleobiology*, **36**, 61–79.

PETERS, S. E. & HEIM, N. A. 2011. Macrostratigraphy and macroevolution in marine environments: taking the common cause hypothesis. *In*: MCGOWAN, A. J. & SMITH, A. B. (eds) *Comparing the Geological and Fossil Records: Implications for Biodiversity Studies*. Geological Society, London, Special Publications, **358**, 95–104.

PETERSEN, K. D., NIELSEN, S. B., CLAUSEN, O. R., STEPHENSON, R. & GERYA, T. 2010. Small-scale mantle convection produces stratigraphic sequences in sedimentary basins. *Science*, **329**, 827–830.

POL, D. & NORRELL, M. A. 2006. Uncertainty in the age of fossils and the stratigraphic fit to phylogenies. *Systematic Biology*, **55**, 512–521.

RAUP, D. M. 1976. Species diversity in the Phanerozoic: a tabulation. *Paleobiology*, **2**, 279–288.

RONOV, A. B. 1978. The Earth's sedimentary shell. *International Geology Review*, **24**, 1313–1363.

ROSENZWEIG, M. L. 1995. *Species Diversity in Space and Time*. Cambridge University Press, Cambridge.

SEPKOSKI, J. J., JR. 1997. Biodiversity: past, present, and future. *Journal of Paleontology*, **71**, 533–539.

SEPKOSKI, J. J., JR., BAMBACH, R. K., RAUP, D. M. & VALENTINE, J. W. 1981. Phanerozoic marine diversity and the fossil record. *Nature*, **293**, 435–437.

SMITH, A. B. 2001. Large-scale heterogeneity of the fossil record: implications for Phanerozoic biodiversity studies. *Philosophical Transactions Royal Society of London, Series B*, **356**, 351–367.

SMITH, A. B. 2007. Intrinsic v. extrinsic biases in the fossil record: contrasting the fossil record of echinoids in the Triassic and early Jurassic using sampling data, phylogenetic analysis and molecular clocks. *Paleobiology*, **33**, 310–323.

SMITH, A. B. & MCGOWAN, A. J. 2007. The shape of the Phanerozoic diversity curve. How much can be predicted from the sedimentary rock record of Western Europe? *Palaeontology*, **50**, 765–777.

UHEN, M. D. & PYENSON, N. D. 2007. Diversity estimates, biases, and historiographic effects; resolving cetacean diversity in the Tertiary. *Palaeontologia Electronica*, 10.2.11A.

UPCHURCH, P., MANNION, P. D., BENSON, R. B. J., BUTLER, R. J. & CARRANO, M. T. 2011. Geological and anthropogenic controls on the sampling of the terrestrial fossil record: a case study from the Dinosauria *In*: MCGOWAN, A. J. & SMITH, A. B. (eds) *Comparing the Geological and Fossil Records: Implications for Biodiversity Studies*. Geological Society, London, Special Publications, **358**, 209–240.

WALL, P. D., IVANY, L. C. & WILKINSON, B. H. 2009. Revisiting Raup: exploring the influence of outcrop area on diversity in light of modern sample standardization techniques. *Paleobiology*, **35**, 146–167.

WALL, P. D., IVANY, L. C. & WILKINSON, B. H. 2011. Impact of outcrop area on estimates of Phanerozoic terrestrial biodiversity trends. *In*: MCGOWAN, A. J. & SMITH, A. B. (eds) *Comparing the Geological and Fossil Records: Implications for Biodiversity Studies*. Geological Society, London, Special Publications **358**, 53–62.

WILLS, M. A. 2007. Fossil ghost ranges are most common in some of the oldest and some of the youngest strata. *Proceedings of the Royal Society, B*, **274**, 2421–2427.

WYATT, A. R. 1995. Late Ordovician extinctions and sea-level change. *Journal of the Geological Society, London*, **152**, 899–902.

ZUSCHIN, M., HARZHAUSER, M. & MANDIC, O. 2011. Disentangling palaeodiversity signals from a biased sedimentary record: an example from the Early to Middle Miocene of Central Paratethys Sea. *In*: MCGOWAN, A. J. & SMITH, A. B. (eds) *Comparing the Geological and Fossil Records: Implications for Biodiversity Studies*. Geological Society, London, Special Publications, **358**, 123–140.

Skeletal mineralogy and biodiversity of marine invertebrates: size matters more than seawater chemistry

LESLEY CHERNS[1]* & V. PAUL WRIGHT[2]

[1]School of Earth and Ocean Sciences, Cardiff University, Cardiff CF10 3AT, UK

[2]BG Group, 100 Thames Valley Park, Reading RG6 1PT, UK

*Corresponding author (e-mail: cherns@cardiff.ac.uk)

Abstract: It is now well established that seawater chemistry, as well as influencing non-skeletal marine precipitation ('calcite' and 'aragonite seas'), has affected skeletal mineral secretion in some algal and marine invertebrate groups. Skeletal mineralogy has had a yet more profound consequence on fossil preservation. The realization that the fossil record of marine organisms with an aragonite shell is widely depleted in some shelf settings through early, effectively syn-depositional, dissolution ('missing molluscs' effect) has led to a re-evaluation of the composition, diversity, ecological and trophic structure of marine benthic communities. Comparisons of molluscan lagerstätten from 'calcite' and 'aragonite seas' show a similar pattern of skeletal mineralogical loss, that is, no differences are discernibly linked to changed seawater geochemistry. It is notable that the rare mollusc-rich skeletal lagerstätten faunas in the fossil record include many small individuals. Micromolluscs are quantitatively important among modern shell assemblages, yet small size is a major source of taphonomic and biodiversity loss in the fossil record. In skeletal lagerstätten faunas, micromolluscs contribute variably to mollusc biodiversity but appear particularly significant through at least to Triassic times. They highlight a further 'missing molluscs' effect of taphonomic loss through early dissolution.

There is growing evidence pointing towards early diagenetic dissolution of biogenic carbonate as a widespread process operating in modern and ancient carbonate shelf sea floors (e.g. Morse *et al.* 1985; Walter *et al.* 1993; Hendry *et al.* 1996; Cherns & Wright 2009). In ancient sea floors this process led to remobilization of carbonate to form limestone-marl alternations (Munnecke & Samtleben 1996; Munnecke *et al.* 1997; Westphal *et al.* 2000; Arzani 2006). Preferential destruction of the less stable carbonate polymorphs, aragonite and high-Mg calcite, is increasingly recognized as a significant factor leading to bias in the fossil record, with consequent loss of biodiversity, particularly for molluscs (Cherns & Wright 2000; Wright *et al.* 2003; Bush & Bambach 2004; James *et al.* 2005; Knoerich & Mutti 2006). However, the mollusc fossil record is good, and Kidwell (2005) concluded that original skeletal mineralogy had an insignificant effect on macroevolutionary patterns in bivalves. Since 'skeletal lagerstätten' (*sensu* Cherns & Wright 2000), most often formed through early silicification, are rare, other more frequent processes must have operated to capture the biodiversity of labile carbonate shells in the fossil record. Such processes include storm events, temporary sea floor anoxia, and synsedimentary cementation (Cherns *et al.* 2008).

Through the Phanerozoic there have been secular changes in seawater geochemistry that controlled inorganic carbonate precipitation: 'calcite' and 'aragonite seas' (Sandberg 1983). It is notable that the skeletal lagerstätten case studies that demonstrate bias against originally aragonite shells have come from 'calcite sea intervals' (Ordovician–Early Carboniferous (Mississippian), Jurassic; Cherns & Wright 2009). In this paper we consider whether skeletal lagerstätten faunas from 'aragonite sea' faunas point to comparable depletion of originally aragonitic shells in the fossil record, or whether this is specifically a 'calcite sea' phenomenon. We also consider the effect of shell size on preservation and biodiversity. Small size is a notable characteristic among the unusually mollusc-rich fossil assemblages of case study faunas. What is the role of micromolluscs in fossil biodiversity?

'Calcite' and 'aragonite seas'

Carbonate sea floor precipitation is controlled by the interplay of atmospheric pCO_2 and Mg/Ca ratios in seawater (Hardie 1996). A recent study on Cretaceous seawater indicated that alkalinity is also an important influence on inorganic precipitation (Lee & Morse 2010). Recent oceans have aragonite

From: McGowan, A. J. & Smith, A. B. (eds) *Comparing the Geological and Fossil Records: Implications for Biodiversity Studies.* Geological Society, London, Special Publications, **358**, 9–17.
DOI: 10.1144/SP358.2   0305-8719/11/$15.00 © The Geological Society of London 2011.

and high-Mg calcite as the primary sea floor precipitates in the tropical realm. Through the Phanerozoic, similar 'aragonite seas' characterized oceans back to the end-Cretaceous, while earlier periods of 'aragonite seas' were the late Precambrian–Early Cambrian, and late Mississippian–end-Triassic. These alternated with 'calcite seas', when calcite formed the sea floor precipitate, that is, during the early-mid Palaeozoic, and Jurassic–Cretaceous (Fig. 1).

To date most evidence of aragonite dissolution comes from 'calcite sea' intervals (Cherns & Wright 2009; Fig. 1). 'Calcite sea' intervals are also, in part, greenhouse climates, associated with increased pCO_2 and reduced carbonate saturation state. As well as the skeletal lagerstätten studies mentioned above, sea floor dissolution of aragonite in undersaturated 'calcite seas' was suggested by Palmer et al. (1988) and Palmer & Wilson (2004) for hardgrounds in the Ordovician and Jurassic. Taylor

(2008) noted the concentration of hardground studies during these 'calcite sea' intervals (Ordovician, Jurassic–Cretaceous) as a possible indication of increased sea floor dissolution of aragonitic shells leading to early cementation. However, in an exhaustive study of Middle Jurassic and Ordovician hardgrounds, Kenyon-Roberts (1995) found that evidence for synsedimentary aragonite dissolution was invariably associated with either subaerial exposure or bacterially mediated pore-water modification during very shallow burial. With no evidence for aragonite dissolution in the absence of these effects, he concluded that these calcite sea intervals were not aragonite undersaturated (also, Riding & Liang 2005). From the Upper Cretaceous, dissolutional loss of aragonite shells in a 'calcite sea' interval is apparent from storm bed faunas (Fürsich & Pandey 1999, 2003) and from rudist biostromes (Sanders 1999, 2001). It is however notable that Sanders (2003) in a Phanerozoic review discerned

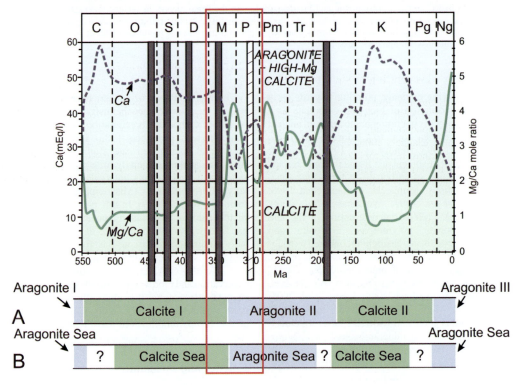

Fig. 1. Distribution of mollusc-rich lagerstätten from the Ordovician–Jurassic against 'aragonite' and 'calcite sea' periods shown by A, secular variation in the Mg/Ca ratio and Ca concentration in seawater at 25°C (Hardie 1996), and B, the temporal distribution of non-skeletal carbonates (Sandberg 1983, 1985; after Stanley & Hardie 1998). Ordovician, Silurian, Devonian, Mississippian and Jurassic faunas (Cherns & Wright 2009; dark boxes) correspond to 'calcite sea' intervals; note that Lower Jurassic ooids and cements are calcitic. Comparison is made herein (red box) between the Pennsylvanian Buckthorn Asphalt fauna (Seuss et al. 2009; striped box), from an 'aragonite sea' interval when marine abiotic precipitates were aragonitic, and Mississippian Compton Martin fauna, when these precipitates were calcitic.

no distinction between 'calcite' and 'aragonite sea' intervals in extent of syndepositional dissolution in neritic carbonates.

Oscillations in inorganic carbonate precipitation correlate broadly with the evolutionary patterns of skeletal calcification in reef building groups such as corals, sponges and rudists, and in sediment producing algae (Stanley & Hardie 1998; Stanley et al. 2010). Experiments show that skeletal mineralogy and growth in scleractinian corals and algae are affected by the ambient seawater geochemistry (Ries et al. 2006; Ries 2006). In cheilostome bryozoans, which though mostly calcitic also secrete bimineralic and aragonitic skeletons, the independent evolutionary acquisition of aragonitic skeletons in several groups increased after the late Palaeogene transition from 'calcite' to 'aragonite seas' (Taylor et al. 2009). Similar control by ambient seawater on biogenic carbonate production was suggested for the evolution of calcitic shelled bivalves during 'calcite sea' intervals (Harper et al. 1997). The implication of control by seawater geochemistry on evolutionary biomineralization is that dissolutional bias affecting biodiversity might be anticipated to increase during 'aragonite sea' intervals.

Diagenetic bias relating to shell mineralogy

The main taphonomic loss of originally aragonitic shells occurs through microbially mediated processes in the taphonomically active zone (TAZ) of shallow burial (Sanders 2003; Cherns et al. 2010; Fig. 2). Those most refractory elements of the skeletal fauna, the originally calcitic shells and shell layers, include the brachiopods and other epifaunal

suspension feeders (stenolaemate bryozoans, rugose and tabulate corals) that dominate Palaeozoic faunas. In Mesozoic shelly faunas the commonly dominant molluscs are largely calcitic or bimineralic pteriomorph bivalves. Preservation of steinkerns of originally aragonitic shells depends largely on the extent of bioturbational reworking of the shallow sediment. Only deep burrowing aragonitic bivalves have an increased preservational potential, either as steinkerns or recrystallized shells in situ in what had been firmground sediment (Cherns & Wright 2009). The most important process is likely aragonite undersaturation caused by oxidation of H_2S (Sanders 2003) in the TAZ; sulphate reduction somewhat deeper in the sediment frees H_2S (Fig. 2). The role of sulphate–sulphide activity in aragonite dissolution is shown by the reduced loss in freshwater (i.e. low sulphate) settings (Wright et al. 2009).

A taphonomic environmental gradient results from the increasing dissolutional drive in low energy, muddy settings of the mid-outer shelf – the limestone-marl zone – and in restricted, lagoonal settings. The former are widely represented in the fossil record (Cherns et al. 2010).

'Aragonite sea' faunas

During 'aragonite sea' intervals (Pennsylvanian–late Triassic and Cenozoic), aragonitic shelly faunas are preserved in storm shell beds across the shelf gradient, as well as in late Palaeozoic mollusc-rich silicified faunas (Cherns et al. 2010). Storm shell beds and coquinas are common but typically thin in the Palaeozoic, but increase in frequency and thickness in the Cenozoic (Kidwell & Brenchley

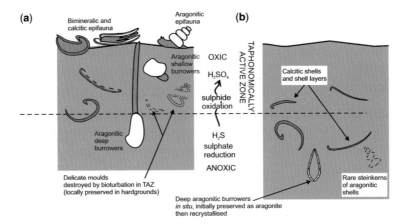

Fig. 2. Calcium carbonate shells in the shallow burial diagenetic environment. (**a**) The processes of aragonite dissolution through sulphide oxidation in the taphonomically active zone (TAZ), where delicate moulds of originally aragonitic (white) shells are vulnerable to destruction through bioturbation. (**b**) Calcitic shells and shell layers (grey) survive early dissolution, with rare steinkerns of originally aragonitic shells, and deep-burrowing bivalves preserved in situ.

1994, 1996), when thick shell beds rich in bivalves are abundant (Hendy *et al.* 2006).

Buckhorn Asphalt fauna

The Pennsylvanian Buckhorn Asphalt Lagerstätte of Oklahoma, USA (Fig. 1) has an exceptionally well preserved and diverse mollusc-rich fauna (Seuss *et al.* 2009). Shell aragonite and original skeletal microstructures became preserved through early impregnation with oil migrating through sediments including storm beds deposited in shallow marine settings. Proximity to the palaeo-shoreline at times is indicated by conglomeratic beds interpreted as lag or mass flow deposits, which include chert pebbles and large logs of wood, as well as marine benthic assemblages with common gastropods and other invertebrates (Seuss *et al.* 2009).

Asphalt impregnation is variable, and in poorly impregnated beds the skeletal material shows diagenetic alteration and loss of aragonite (Brand 1989). Bivalves are mostly distributed in the less impregnated mudstones and sandstones where, except for bimineralic pteriomorphs, they are commonly poorly preserved. Originally aragonitic bivalves are represented in these beds by *in situ* deep burrowing, large anomalodesmatids. However, in the asphaltic shell beds, although bivalves are fairly uncommon they show fine preservation and, by contrast to the pteriomorph domination of assemblages in less impregnated beds, are dominated numerically by originally aragonitic taxa. Highly impregnated beds include the thick to massive 'cephalopod coquina' containing skeletal debris including very large shells of these nektic forms, while other shell beds are rich in gastropods and with varied skeletal benthos.

The biota includes at least 125 genera and *c.* 160 species, of which molluscs form by far the dominant group qualitatively (*c.* 2/3; Seuss *et al.* 2009). Original aragonite retains a 'mother of pearl' appearance

and microstructure, and juvenile shells are preserved. Gastropods are the most abundant and diverse molluscs, common in the shell beds (29 genera >50 species), while bivalves and cephalopods represent the next most diverse groups (Seuss *et al.* 2009, figs 19 & 20). Cephalopods are concentrated into shell beds and coquinas with relatively few benthic molluscs.

The Buckhorn Asphalt Quarry is one of several Pennsylvanian localities in the midcontinent USA (Ohio, Pennsylvania, Kentucky) with skeletal aragonite preservation through oil impregnation (e.g. Kentucky Shale; Batten 1972), and there are other mollusc dominated faunas where aragonite shells are replaced by calcite or silicified (e.g. Kues & Batten 2001). These faunas contrast with the typically brachiopod dominated and calcitic faunas of the Upper Palaeozoic (e.g. Stevens 1971). The preservational bias, as with case studies from 'calcite sea' intervals, suggests that the mollusc fossil record should be under-represented. However, since the various Pennsylvanian mollusc-rich faunas include many of the same genera, the Buckhorn Asphalt fauna has a reduced influence on the biodiversity of the mollusc record. A notable difference between the Buckhorn Asphalt fauna and faunas where the shells are replaced by calcite, however, is that its characteristically very small species are rarely preserved in the latter.

Comparison with 'calcite sea' Mississippian Compton Martin fauna

Both the Buckhorn Asphalt and Compton Martin lagerstätten faunas represent highly diverse and abundant, mollusc dominated assemblages (Mitchell 1987; Cherns & Wright 2009; Seuss *et al.* 2009; Table 1). The benthic assemblages are *c.* 70% molluscs, dominated by gastropods, and are predominantly aragonitic shells with a minor component of bimineralic shells. Other invertebrates in the faunas

Table 1. *Comparison of Carboniferous skeletal lagerstätten benthic faunas from 'calcite' – Compton Martin – and 'aragonite sea' – Buckhorn Asphalt – intervals*

Fauna	Compton Martin, UK Mississippian (Mitchell 1987) No. species = >150 % specific diversity	Buckhorn Asphalt, USA Pennsylvanian (Seuss *et al.* 2009) No. species = 111 % specific diversity
Molluscs	71	70
Other invertebrates	27	18
Vertebrates	<1	3
Algae	<1	9

Note that the mollusc fauna is predominantly aragonitic shells, with a component of bimineralic shells; other invertebrates are calcitic shells. Microfossil groups (foraminiferans and ostracodes) and land plants are omitted from the comparison. Cephalopods as nektobenthos are also omitted. In the Buckhorn Asphalt, cephalopods are concentrated in coquinas and are rarely associated with the benthic shell bed faunas or facies (Seuss *et al.* 2009).

are calcitic shelled groups: brachiopods, bryozoans and echinoderms are common in the Asphalt, while brachiopods and corals are most numerous at Compton Martin. Land plants in the Buckhorn Asphalt fauna indicate proximity to an exposed land surface. Both faunas include many small species, and assemblages have trophic and ecological structures at variance with the epifaunal suspension feeding communities of Sepkoski's (1984) Palaeozoic Fauna (Fig. 3). The trophic shifts resulting from taphonomic loss of originally aragonitic shells appear comparable in the two faunas.

Size bias in the fossil record

Among the skeletal lagerstätten from 'calcite sea' periods, small sized and delicate shells are notably common. The generally poor preservation potential of small and thin shells confirms the unusually early lithification or diagenetic replacement that preserved the originally aragonitic faunas. Such rare faunas in early and late Palaeozoic examples include numerous new taxa that contribute significantly to mollusc biodiversity (Silurian, Liljedahl 1985; Ordovician, Hoare & Pojeta 2006; Lower Carboniferous, Batten 1966). In the Mesozoic case study from the Lower Jurassic, however, the diverse and dominant aragonitic bivalve fauna included no new taxa (Wright et al. 2003; Cherns & Wright 2009). This contrast may indicate a temporal shift in the completeness of the mollusc fossil record, reflecting the increasingly abundant mollusc faunas among the post-Palaeozoic benthos.

Small species are an important feature of the modern shelf benthos dominated by bivalve and gastropod molluscs. In a highly diverse shelf fauna from the tropical Indo-Pacific, where molluscs are

estimated to represent 60% of the marine invertebrate biodiversity, gastropods comprise 80% of species and 64% by abundance, and 34% comprise micromolluscs <4.1 mm (Bouchet et al. 2002). Small species suffer a size bias in fossil assemblages through factors such as fragility, collection failure and lithification (e.g. Kidwell & Bosence 1991; Donovan & Paul 1998; Kidwell 2001; Sessa et al. 2009). Based on Holocene and Cenozoic museum records from New Zealand, Cooper et al. (2006) estimated that only 43% of temperate Holocene molluscs have a fossil record. Micromolluscs <5 mm form the modal size class in both shelf and deeper water Holocene environments (bathyal-abyssal zones; Cooper et al. 2006, fig. 1); they represent >50% of the biodiversity of living faunas, and a 27% loss of total biodiversity into the fossil record. The New Zealand Cenozoic record is c. 32% complete at the species level (Crampton et al. (2006). In a study of Quaternary bivalves from the temperate Pacific coast of South America, where only 45% of living species had a Pleistocene fossil record, size produces a significant bias in preservation potential against micromolluscs <5 mm (Rivadeneira 2010). Valentine et al. (2006) estimated that small size <1 cm accounts for 48% of shelf depth bivalve genera lacking a fossil record.

In bivalves shell mineralogy and life habit are closely linked; aragonitic shells characterize the infaunal groups while calcitic or bimineralic shells characterize most epifaunal groups. Of the 25% of infaunal genera missing from the fossil record, small size (<1 cm) accounts for over half, while relatively few of the epifaunal groups are small-sized (Valentine et al. 2006).

Microgastropod faunas in the fossil record are mostly confined to lagerstätten. The Mississippian Compton Martin microgastropod fauna of 98

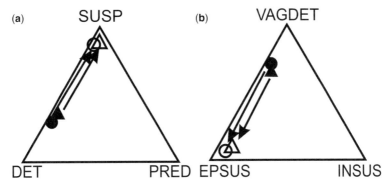

Fig. 3. Trophic ternary diagrams for the 'calcite sea' Compton Martin (filled circle) and 'aragonite sea' Buckhorn Asphalt (filled triangle) faunas (N.B. benthic faunas only, as in Table 1), compared with typical non-lagerstätten faunas (open symbols); trophic shifts resulting from taphonomic loss of aragonitic shells are indicated by arrows. (**a**) feeding habits (DET, detritus-feeding; SUSP, suspension-feeding; PRED, predatory); (**b**) media niches (EPSUS, epifaunal suspension-feeders; INSUS, infaunal suspension feeders; VAGDET, vagrant detritus feeders).

species includes 24 either new (18) or left in open nomenclature (Batten 1966). Two microgastropod lagerstätten faunas represent *c.* 40% of the Triassic gastropod biodiversity (Sinbad Limestone, USA; Gastropod Oolite, Italy; Fraiser & Bottjer 2004; Nützel & Schulbert 2005). Similarly, Permian silicified micromollusc faunas include numerous new taxa (Pan & Erwin 2002). From the Devonian, a silicified microgastropod assemblage of *c.* 40 species includes 9 new (Frýda & Blodgett 2004).

Micromollusc-rich shelly faunas are preserved in lagerstätten also in younger rocks. Black shales of the Gault Clay from Folkestone, UK yield a diverse fauna of molluscs (ammonites, bivalves, gastropods) with original shell aragonite preserved, as well as fish and arthropods, and associated with an extensive, mostly pyritized microfauna (Price 1874; Young *et al.* 2010). The mollusc assemblage includes >50 species of bivalves and >100 species of gastropods. Microgastropods including the delicate larval shells are abundant in some beds. In one bulk sample processed recently, a large (>1000 specimens) micromollusc assemblage comprised >99% gastropods <1mm, many of which retained larval shells (S. Thomas, Cardiff University, pers. comm. 2010). This microgastropod assemblage may represent planktic larvae carried into an oxygen-depleted mud environment inhospitable for benthos (cf. Mississippian Ruddle Shale; Nützel & Mapes 2001; Mapes & Nützel 2009).

Bias into the fossil record is increased by rarity. In spite of time-averaging (Kidwell & Flessa 1995), rare species still have low probability (preservation potential, outcrop, collection) of representation in the fossil record. Bouchet *et al.* (2002) noted that 48% of species in their highly diverse, micromollusc-dominated fauna (2738 species) were represented by <5 individuals (and 20% by only one). The Compton Martin fauna has 26% of species represented by <5 specimens, and 11% by only one specimen (Batten 1966). The lower proportions of poorly represented species make probable an under-representation of the biodiversity of this Mississippian fauna.

Discussion

Skeletal lagerstätten from 'aragonite sea' intervals indicate preservational bias against originally aragonitic shells comparable to that reported from 'calcite sea' intervals. Increased dissolutional bias against originally aragonitic shells arising in 'aragonite sea' intervals through seawater geochemistry controlling evolutionary biomineralization (e.g. Stanley & Hardie 1998), or in 'calcite sea' intervals due to sea floor undersaturation (Palmer *et al.* 1988) do not produce discernible differences in

lagerstätten faunas. Aragonite dissolution is driven by microbially mediated decay processes in the taphonomically active zone, and the extent of dissolution appears similar in 'aragonite' and calcite sea' intervals; the same view was also reached by Sanders (2003).

It is apparent, however, that small size – micromolluscs – may play a significant role in fossil mollusc biodiversity. Micromolluscs today are a major component of marine benthos, and represent a large part of the biodiversity. Their high loss back into the Cenozoic fossil record reflects poor preservation potential and the relative rarity of many species (Cooper *et al.* 2006). In the fossil record, micromolluscs and small individuals are notable among skeletal lagerstätten faunas yet otherwise rare in shelly faunas, with collection bias being an obvious contributory factor. Where preserved, for example Pennsylvanian Buckthorn Asphalt fauna, micromolluscs represent a diverse assemblage (Seuss *et al.* 2009). The contribution of each lagerstätte to biodiversity depends on the frequency of penecontemporaneous faunas – several microgastropod faunas are known from the Pennsylvanian of USA (e.g. Batten 1995; Kues & Batten 2001; Bandel *et al.* 2002; Seuss *et al.* 2009), yet in the early Triassic when shelly faunas are less common, a single lagerstätte may represent a high percentage of gastropod biodiversity (Sinbad Limestone; Batten & Stokes 1986; Fraiser & Bottjer 2004; Nützel & Schulbert 2005).

Is there a temporal bias in the contribution to biodiversity of skeletal lagerstätten faunas? Early silicification is an important taphonomic process in preserving originally aragonitic faunas, and appears more common among Palaeozoic faunas (Cherns & Wright 2009) although silicified faunas are reported more commonly from the Palaeozoic (Schubert *et al.* 1997). However, silicification also characterizes several younger mollusc-rich lagerstätten faunas (Triassic gastropod faunas –see above; Wright *et al.* 2003). The post-Palaeozoic rise in bivalve diversity involves particularly the radiation of aragonite-shelled groups (Kidwell 2005). If bivalve biodiversity is not affected by the taphonomic bias likely to have widely diminished faunas from subtidal shelf facies (Wright & Cherns 2004, 2008), this requires a sufficient frequency of beds preserving originally aragonitic faunas. Increasing frequency and thickness of storm shell beds, particularly in Cenozoic times, may reduce the effects of taphonomic bias on the mollusc fossil record (Kidwell & Brenchley 1994, 1996; Fürsich & Pandey 1999, 2003; Cherns *et al.* 2008).

The effects of taphonomic bias (Cherns & Wright 2000; Bush & Bambach 2004; Valentine *et al.* 2006) on the veracity of the Phanerozoic mollusc fossil record are open to debate (e.g. Kidwell 2005;

Cooper *et al.* 2006). Of living species, major losses into the fossil record include a large proportion of micromolluscs and rare species (Cooper *et al.* 2006). In addition, changes to the rock outcrop area/volume, facies and palaeolatitudinal/longitudinal representation, and sedimentation rates all influence biodiversity (Crame 2000; Smith *et al.* 2001; Crampton *et al.* 2003, 2006; Smith 2007; Smith & McGowan 2007), and introduce bias against older rocks.

Conclusions

Early, effectively syn-sedimentary dissolution of aragonite, particularly important for molluscs, is a major process affecting the fossil record. Comparison of skeletal lagerstätten from Carboniferous 'calcite' and 'aragonite seas' shows no discernible differential in taphonomic bias caused by ambient seawater composition. These and other lagerstätten faunas highlight the importance of micromolluscs on biodiversity, as is apparent from studies of Holocene and late Cenozoic faunas. The skeletal lagerstätten preserve diverse micromolluscs with macromolluscs, which indicates that the disappearance of micromolluscs is largely another example of the 'missing molluscs' effect.

References

ARZANI, N. 2006. Primary v. diagenetic bedding in the limestone-marl/shale alternations of the epeiric seas, an example from the Lower Lias (early Jurassic) of SW Britain. *Carbonates and Evaporites*, **21**, 94–109.

BANDEL, K., NÜTZEL, A. & YANCEY, T. E. 2002. Larval shells and shell microstructures of exceptionally well-preserved late Carboniferous gastropods from the Buckhorn asphalt deposit (Oklahoma, USA). *Palaeobiodiversity and Palaeoenvironments*, **82**, 639–689.

BATTEN, R. L. 1966. The Lower Carboniferous gastropod fauna from the Hotwells Limestone of Compton Martin, Somerset. *Palaeontographical Society Monographs*, **509** (119, Part 1) and **513** (120, Part 2), 1–52; 53–109.

BATTEN, R. L. 1972. The ultrastructure of five common Pennsylvanian pleurotomarian gastropod species of eastern United States. *American Museum Novitates*, **2501**, 1–34.

BATTEN, R. L. 1995. Pennsylvanian (Morrowan) gastropods from the Magdalena Formation of the Hueco Mountains, Texas. *American Museum Novitates*, **3122**, 1–46.

BATTEN, R. L. & STOKES, W. L. 1986. Early Triassic gastropods from the Sinbad member of the Moenkopi Formation, San Rafael Swell, Utah. *American Museum Novitates*, **2864**.

BOUCHET, P., LOZOUET, P., MAESTRATI, P. & HEROS, V. 2002. Assessing the magnitude of species richness in tropical marine environments: exceptionally high

numbers of molluscs at a New Caledonia site. *Biological Journal of the Linnean Society*, **75**, 421–436.

BRAND, U. 1989. Aragonite-calcite transformation based on Pennsylvanian molluscs. *Geological Society of America Bulletin*, **101**, 377–390.

BUSH, A. M. & BAMBACH, R. K. 2004. Did alpha diversity increase during the Phanerozoic? Lifting the veils of taphonomic, latitudinal, and environmental biases. *Journal of Geology*, **112**, 625–642.

CHERNS, L. & WRIGHT, V. P. 2000. Missing molluscs as evidence of large-scale, early skeletal aragonite dissolution in a Silurian sea. *Geology*, **28**, 791–794.

CHERNS, L. & WRIGHT, V. P. 2009. Quantifying the impacts of early diagenetic aragonite dissolution on the fossil record. *Palaios*, **24**, 756–771.

CHERNS, L., WHEELEY, J. R. & WRIGHT, V. P. 2008. Taphonomic windows and molluscan preservation. *Palaeogeography, Palaeoclimatology, Palaeoecology*, **270**, 220–229.

CHERNS, L., WHEELEY, J. R. & WRIGHT, V. P. 2010. Taphonomic bias in shelly faunas through time: early aragonitic dissolution and its implications for the fossil record. *In*: ALLISON, P. A. & BOTTJER, D. J. (eds) *Taphonomy: Process and Bias Through Time*. 2nd ed. Topics in Paleobiology, **32**. Springer, New York, 79–105.

COOPER, R. A., MAXWELL, P. A., CRAMPTON, J. S., BEU, A. G., JONES, C. M. & MARSHALL, B. A. 2006. Completeness of the fossil record: estimating losses due to small body size. *Geology*, **34**, 241–244.

CRAME, J. A. 2000. Evolution of taxonomic diversity gradients in the marine realm: evidence from the composition of recent bivalve faunas. *Paleobiology*, **26**, 188–214.

CRAMPTON, J. S., BEU, A. G., COOPER, R. A., JONES, C. M., MARSHALL, B. & MAXWELL, P. A. 2003. Estimating the rock volume bias in paleobiodiversity studies. *Science*, **301**, 358–360.

CRAMPTON, J. S., FOOTE, M. ET AL. 2006. Second-order sequence stratigraphic controls on the quality of the fossil record at an active margin: New Zealand Eocene to Recent shelf molluscs. *Palaios*, **21**, 86–105.

DONOVAN, S. K. & PAUL, C. R. C. 1998. *The Adequacy of the Fossil Record*. John Wiley & Sons Ltd, Chichester.

FRAISER, M. L. & BOTTJER, D. J. 2004. The non-actualistic Early Triassic gastropod fauna: a case study of the Lower Triassic Sinbad Limestone Member. *Palaios*, **19**, 259–275.

FRÝDA, J. & BLODGETT, R. B. 2004. New Emsian (late Early Devonian) gastropods from Limestone Mountain, Medfra B-4 Quadrangle, west-central Alaska (Farewell Terrane), and their paleobiogeographic affinities and evolutionary significance. *Journal of Paleontology*, **78**, 111–132.

FÜRSICH, F. T. & PANDEY, D. K. 1999. Sequence stratigraphic significance of sedimentary cycles and shell concentrations in the Upper Jurassic–Lower Cretaceous of Kachchh, western India. *Palaeogeography Palaeoclimatology Palaeoecology*, **145**, 119–139.

FÜRSICH, F. T. & PANDEY, D. K. 2003. Sequence stratigraphic significance of sedimentary cycles and shell concentrations in the Upper Jurassic–Lower

Cretaceous of Kachchh, western India. *Palaeogeography Palaeoclimatology Palaeoecology*, **193**, 285–309.

HARDIE, L. A. 1996. Secular variation in seawater chemistry: an explanation for the coupled secular variation in the mineralogies of marine limestones and potash evaporites over the past 600 my. *Geology*, **24**, 279–283.

HARPER, E. M., PALMER, T. J. & ALPHEY, J. R. 1997. Evolutionary response by bivalves to changing Phanerozoic sea-water chemistry. *Geological Magazine*, **134**, 403–407.

HENDRY, J. P., TREWIN, N. H. & FALLICK, A. E. 1996. Low Mg-calcite marine cement in Cretaceous turbidites: origin, spatial distribution and relationship to sea-water chemistry. *Sedimentology*, **43**, 877–900.

HENDY, A. J. W., KAMP, P. J. J. & VONK, A. J. 2006. Cool-water shell bed taphofacies from Miocene–Pliocene shelf sequences in New Zealand: utility of taphofacies in sequence stratigraphic analysis. *In*: PEDLEY, H. M. & CARANNANTE, G. (eds) *Cool-water Carbonates*. Geological Society, London, Special Publications, **255**, 283–305.

HOARE, R. D. & POJETA, J., JR. 2006. Ordovician Polyplacophora (Mollusca) from North America. *Journal of Paleontology*, **80** (3, Suppl. II), 1–28.

JAMES, N. P., BONE, Y. & KYSER, T. K. 2005. Where has all the aragonite gone? – Mineralogy of Holocene neritic cool-water carbonates, southern Australia. *Journal of Sedimentary Research*, **75**, 454–463.

KENYON-ROBERTS, S. M. 1995. *The petrography and distribution of some calcite sea hardgrounds*. PhD thesis, University of Reading.

KIDWELL, S. M. 2001. Preservation of species abundance in marine death assemblages. *Science*, **294**, 1091–1094.

KIDWELL, S. M. 2005. Shell composition has no net impact on large-scale evolutionary patterns in mollusks. *Science*, **307**, 914–917.

KIDWELL, S. M. & BOSENCE, D. W. J. 1991. Taphonomy and time-averaging of marine shelly faunas. *In*: ALLISON, P. A. & BRIGGS, D. E. G. (eds) *Taphonomy: Releasing the Data Locked in the Fossil Record*. Plenum Press, New York, 115–209.

KIDWELL, S. M. & BRENCHLEY, P. J. 1994. Patterns in bioclastic accumulation through the Phanerozoic – changes in input or in destruction. *Geology*, **22**, 1139–1143.

KIDWELL, S. M. & BRENCHLEY, P. J. 1996. Evolution of the fossil record; thickness trends in marine skeletal accumulations and their implications. *In*: JABLONSKI, D., ERWIN, D. H. & LIPPS, J. H. (eds) *Evolutionary Paleobiology*. University of Chicago Press, Chicago and London, 290–336.

KIDWELL, S. M. & FLESSA, K. W. 1995. The quality of the fossil record: populations, species, and communities. *Annual Review of Ecology and Systematics*, **26**, 269–299.

KNOERICH, A. C. & MUTTI, M. 2006. Missing aragonitic biota and the diagenetic evolution of heterozoan carbonates: a case study from the Oligo-Miocene of the central Mediterranean. *Journal of Sedimentary Research*, **76**, 871–888.

KUES, B. S. & BATTEN, R. L. 2001. Middle Pennsylvanian gastropods from the Flechado Formation, north-central New Mexico. *Journal of Paleontology*, **75** (Suppl. 54), 1–95.

LEE, J. & MORSE, J. W. 2010. Influences of alkalinity and pCO_2 on $CaCO_3$ nucleation from estimated Cretaceous composition seawater representative of 'calcite seas'. *Geology* **38**, 115–118.

LILJEDAHL, L. 1985. Ecological aspects of a silicified bivalve fauna from the Silurian of Gotland. *Lethaia*, **18**, 53–66.

MAPES, R. H. & NÜTZEL, A. 2009. Late Palaeozoic mollusc reproduction: cephalopod egg-laying behavior and gastropod larval palaeobiology. *Lethaia*, **42**, 341–356.

MITCHELL, M. 1987 [for 1986]. The fossil collection of C.B. Salter from Cliff Quarry, Compton Martin, Mendip Hills. *Geological Curator*, **4**, 487–491.

MORSE, J. W., ZULLIG, J. J., BERNSTEIN, L. D., MILLERO, F. J., MILNE, P., MUCCI, A. & CHOPPIN, G. R. 1985. Chemistry of calcium-rich shallow water sediments in the Bahamas. *American Journal of Science*, **285**, 147–185.

MUNNECKE, A. & SAMTLEBEN, C. 1996. The formation of micritic limestones and the development of limestone-marl alternations in the Silurian of Gotland, Sweden. *Facies*, **34**, 159–176.

MUNNECKE, A., WESTPHAL, H., REIJMER, J. J. G. & SAMTLEBEN, C. 1997. Microspar development during early marine burial diagenesis: a comparison of Pliocene carbonates from the Bahamas with Silurian limestones from Gotland (Sweden). *Sedimentology*, **44**, 977–990.

NÜTZEL, A. & MAPES, R. H. 2001. Larval and juvenile gastropods from a Carboniferous black shale: palaeoecology and implications for the evolution of the Gastropoda. *Lethaia*, **34**, 143–162.

NÜTZEL, A. & SCHULBERT, C. 2005. Facies of two important Early Triassic gastropod lagerstätten: implications for diversity patterns in the aftermath of the end-Permian mass extinction. *Facies*, **51**, 495–515.

PALMER, T. J. & WILSON, M. A. 2004. Calcite precipitation and dissolution of biogenic aragonite in shallow Ordovician calcite seas. *Lethaia*, **37**, 417–427.

PALMER, T. J., HUDSON, J. D. & WILSON, M. A. 1988. Palaeoecological evidence for early aragonite dissolution in ancient calcite seas. *Nature*, **335**, 809–810.

PAN, H.-Z. & ERWIN, D. H. 2002. Gastropods from the Permian of Guangxi and Yunnan provinces, South China. *Journal of Paleontology*, **76** (Suppl. 56), 1–49.

PRICE, F. G. H. 1874. On the Gault of Folkestone. *Quarterly Journal of the Geological Society*, **30**, 342–368.

RIDING, R. & LIANG, L. 2005. Seawater chemistry control of marine limestone accumulation over the past 550 million years. *Revista Española de Micropaleontología*, **37**, 1–11.

RIES, J. B. 2006. Aragonitic algae in calcite seas: effect of seawater Mg/Ca ratio on algal sediment production. *Journal of Sedimentary Research*, **76**, 515–523.

RIES, J. B., STANLEY, S. M. & HARDIE, L. A. 2006. Scleractinian corals produce calcite, and grow more slowly, in artificial Cretaceous seawater. *Geology*, **34**, 525–528.

RIVADENEIRA, M. M. 2010. On the completeness and fidelity of the Quaternary bivalve record from the temperate Pacific coast of South America. *Palaios*, **25**, 40–45.

SANDBERG, P. A. 1983. An oscillating trend in Phanerozoic non-skeletal carbonate mineralogy. *Nature*, **305**, 19–22.

SANDBERG, P. A. 1985. Nonskeletal aragonite and pCO_2 in the Phanerozoic and Proterozoic. *In*: SUNDQUIST, E. T. & BROECKER, W. S. (eds) *The Carbon Cycle and Atmospheric CO₂: Natural Variations Archean to Present*. Geophysical Monograph, **32**, American Geophysical Union, Washington D.C., 585–594.

SANDERS, D. 1999. Shell disintegration and taphonomic loss in radiolitid biostromes. *Lethaia*, **32**, 101–112.

SANDERS, D. 2001. Burrow-mediated carbonate dissolution in rudist biostromes (Aurisina, Italy): implications for taphonomy in tropical, shallow subtidal carbonate environments. *Palaeogeography, Palaeoclimatology, Palaeoecology*, **168**, 39–74.

SANDERS, D. 2003. Syndepositional dissolution of calcium carbonate in neritic carbonate environments: geological recognition, processes, potential significance. *Journal of African Earth Sciences*, **36**, 99–134.

SCHUBERT, J. K., KIDDER, D. L. & ERWIN, D. H. 1997. Silica-replaced fossils through the Phanerozoic. *Geology*, **25**, 1031–1034.

SEPKOSKI, J. J. 1984. A Kinetic-Model of Phanerozoic Taxonomic Diversity 3. Post- Paleozoic Families and Mass Extinctions. *Paleobiology*, **10**, 246–267.

SESSA, J. A., PATZKOWSKY, M. E. & BRALOWER, T. J. 2009. The impact of lithification on the diversity, size distribution, and recovery dynamics of marine invertebrate assemblages. *Geology* **37**, 115–118.

SEUSS, B., NÜTZEL, A., MAPES, R. H. & YANCEY, T. E. 2009. Facies and fauna of the Pennsylvanian Buckhorn Asphalt Quarry deposit: a review and new data on an important Palaeozoic fossil Lagerstätte with aragonite preservation. *Facies*, **55**, 609–645.

SMITH, A. B. 2007. Marine diversity through the Phanerozoic: problems and prospects. *Journal of the Geological Society, London*, **164**, 731–745.

SMITH, A. B., GALE, A. S. & MONKS, N. E. A. 2001. Sea-level change and rock record bias in the Cretaceous: a problem for extinction and biodiversity studies. *Paleobiology*, **27**, 241–253.

SMITH, A. B. & McGOWAN, A. J. 2007. The shape of the Phanerozoic marine palaeodiversity curve: how much can be predicted from the sedimentary rock record of Western Europe? *Palaeontology*, **50**, 765–774.

STANLEY, S. M. & HARDIE, L. A. 1998. Secular oscillations in the carbonate mineralogy of reef-building and sediment-producing organisms driven by tectonically forced shifts in seawater chemistry. *Palaeogeography Palaeoclimatology Palaeoecology*, **144**, 3–19.

STANLEY, S. M., RIES, J. B. & HARDIE, L. A. 2010. Increased production of calcite and slower growth for the major sediment-producing alga *Halimeda* as the Mg/Ca ratio of seawater is lowered to a 'calcite sea' level. *Journal of Sedimentary Research*, **80**, 6–16.

STEVENS, C. H. 1971. Distribution and diversity of Pennsylvanian marine faunas relative to water depth and distance from shore. *Lethaia*, **4**, 403–412.

TAYLOR, P. D. 2008. Seawater chemistry, biomineralization and the fossil record of calcareous organisms. *In*: OKADA, H., MAWATARI, S. F., SUZUKI, N. & GAUTAM, P. (eds) *Origin and Evolution of Natural Diversity*. Proceedings of International Symposium 'The Origin and Evolution of Natural Diversity', 1–5 October 2007, Sapporo, Hokkaido University, Sapporo, 21–29.

TAYLOR, P. D., JAMES, N. P., BONE, Y., KUKLINSKI, P. & KYSER, T. K. 2009. Evolving mineralogy of cheilostome bryozoans. *Palaios*, **24**, 440–452.

VALENTINE, J. W., JABLONSKI, D., KIDWELL, S. & ROY, K. 2006. Assessing the fidelity of the fossil record by using marine bivalves. *Proceedings of the National Academy of Sciences of the USA*, **103**, 6599–6604.

WALTER, L. M., BISCHOF, S. A., PATTERSON, W. P. & LYONS, T. W. 1993. Dissolution and recrystallization in modern shelf carbonates – evidence from pore-water and solid-phase chemistry. *Philosophical Transactions of the Royal Society of London, Series A*, **344**, 27–36.

WESTPHAL, H., HEAD, M. J. & MUNNECKE, A. 2000. Differential diagenesis of rhythmic limestone alternations supported by palynological evidence. *Journal of Sedimentary Research*, **70**, 715–725.

WRIGHT, V. P. & CHERNS, L. 2004. Are there 'black holes' in carbonate deposystems? *Geologica Acta* **2**, 285–290.

WRIGHT, V. P. & CHERNS, L. 2008. The subtle thief: selective dissolution of aragonite during shallow burial and the implications for carbonate sedimentology. *In*: LUKASIK, J. & SIMO, J. A. (eds) *Controls on Carbonate Platform and Reef Development*. SEPM Special Publication, **54**, SEPM, Tulsa, Oklahoma, 47–54.

WRIGHT, V. P., CHERNS, L. & HODGES, P. 2003. Missing molluscs: field testing taphonomic loss in the Mesozoic through early large-scale aragonite dissolution. *Geology*, **31**, 211–214.

WRIGHT, V. P., AZERÊDO, A. C., CABRAL, M. C. & CHERNS, L. 2009. Taphonomic differences in molluscan preservation in the Upper Jurassic of Portugal: a key to understanding porosity formation in non-marine limestones. *In*: PASCUCCI, V. & ANDREUCCI, S. (eds) *Abstracts, 27th IAS Meeting*, 20–23 September 2009, Alghero, Italy. Edes, Sassari.

YOUNG, J. R., GALE, A. S., KNIGHT, R. I. & SMITH, A. B. 2010. *Fossils of the Gault Clay*. Field Guide to Fossils, **12**, Palaeontological Association, London.

Detecting common-cause relationships with directional information transfer

BJARTE HANNISDAL

*Department of Earth Science, Centre for Geobiology, University of Bergen, Allégaten 41,
N-5007 Bergen, Norway (e-mail: bjarte.hannisdal@geo.uib.no)*

Abstract: Correlations between sedimentary rock and fossil records may involve a combination of rock-record sampling bias and common response to external forcing. Quantifying their relative importance from incomplete and uncertain proxy data is not trivial given the potential complexity of interactions among the underlying processes. This paper shows how a non-parametric method can be used to detect causal interactions directly from incomplete and irregular time series, by quantifying directional information transfer between variables. A numerical experiment illustrates how estimates of the relative strength, scale, and directionality of coupling can correctly distinguish a common-cause variable from a spurious relationship, even in cases where correlations are misleading. With a joint analysis of Phanerozoic rock and fossil records pending, the method is applied to oxygen, carbon, and sulphur stable isotope records from marine carbonates, identifying complex interactions between climatic changes and the cycling of carbon and sulphur over the Phanerozoic.

Palaeontologists currently entertain two explanations for the correlation between the fossil record of marine diversity and estimates of sedimentary rock quantity (Raup 1976; Peters & Foote 2001, 2002; Smith 2001, 2007; Peters 2005, 2006): (1) the diversity of fossil taxa is controlled by the amount of sedimentary rock available at outcrop and the range of habitats sampled by palaeontologists (the bias hypothesis); and (2) the large-scale geological processes responsible for the accumulation and preservation of sedimentary rocks also modulate origination and extinction in the marine realm (the common-cause hypothesis).

The two explanations are not mutually exclusive, and the pertinent question becomes how to assess their relative importance. Although there is no substitute for obtaining higher-quality data and improving sampling protocols, this question ultimately raises the spectre of causal inference and its evil twin, mechanistic modelling. The aim of this paper is to show how causal interactions can be distinguished from correlations by time series analysis of observed records without recourse to modelling. After a brief introduction to the concept of directional information transfer and its implementation, a numerical experiment is used to illustrate its ability to dissect a nontrivial common-cause relationship. Finally, the method is applied to Phanerozoic seawater isotope records, demonstrating how complex and elusive interactions can be characterized quantitatively.

Directional information transfer

A causal interaction can be considered an inherently directional relationship, with a driving variable having an effect on a response variable. In order to detect the directionality of coupling by statistical methods, the relation between the states of two coupled systems needs to be probabilistic, either because the coupling is weak or variable, or because of noise. Statistical inroads to causality therefore commonly rely on a computational, or predictive, definition of causality by which knowledge of a driving variable should improve our ability to predict a response variable (e.g. Granger 1969).

Information theory provides a general framework for implementing a computational approach to causality detection, most commonly applied in the context of time series analysis (see Hlaváčková-Schindler *et al.* 2007 for a review). The starting point is a quantity known as information entropy (Shannon 1948): if X is a vector of real values $x_1, \ldots, x_m$, each with probability $p(x_i)$, $i = 1, \ldots, m$, then the average amount of information gained from measuring a particular value x_i is given by

$$H(X) = -\sum_i^m p(x_i) \log p(x_i). \tag{1}$$

The entropy of X can be interpreted as a measure of our uncertainty regarding the state of X, or how

From: McGowan, A. J. & Smith, A. B. (eds) *Comparing the Geological and Fossil Records: Implications for Biodiversity Studies*. Geological Society, London, Special Publications, **358**, 19–29.
DOI: 10.1144/SP358.3   0305-8719/11/$15.00 © The Geological Society of London 2011.

surprised one should be upon learning the result of a measurement. It bears more than a formal similarity to entropy in statistical thermodynamics, as both measure the spread of the distribution of possible states, whereby a gas has higher entropy than a crystal. Palaeontologists are perhaps most familiar with $H(X)$, or Shannon's H, as one of several indices used by ecologists to measure biodiversity (e.g. Magurran 2004).

The conditional entropy of two variables X and Y is given by

$$H(X|Y) = -\sum_i^{m_X} \sum_j^{m_Y} p(x_i, y_j) \log p(x_i|y_j), \quad (2)$$

where $p(x_i, y_j)$ is the joint probability that X is in state x_i and Y is in state y_j, and $p(x_i|y_i)$ is the conditional probability. $H(X|Y)$ measures the amount of uncertainty that remains in X when Y is known, or equivalently, what Y does not say about X. Given (1) and (2), we can obtain a very general measure of the extent to which one variable improves our knowledge of another variable, known as the mutual information between X and Y (Shannon 1948):

$$I(X; Y) = H(X) - H(X|Y). \quad (3)$$

$I(X; Y)$ thus quantifies how much the uncertainty in X is reduced by knowing Y, and $I(X; Y)$ is zero if X and Y are statistically independent. Because it is sensitive to high-order non-linear relationships, mutual information represents a more general measure of statistical association than standard measures of correlation, which are restricted to linear (e.g. Pearson's R) or monotonic (e.g. Spearman's ρ) relationships. Like correlation, however, mutual information is symmetrical, that is, $I(X; Y) = I(Y; X)$, and cannot detect the directionality of coupling in a drive–response interaction.

Schreiber (2000) introduced transfer entropy, a non-symmetrical measure of information transfer that works on transition probabilities instead of static probabilities. It takes advantage of the fact that the transition probabilities describing the dynamics of two entirely independent processes will have a generalized Markov property (i.e. the state of Y will have no influence on the transition probabilities of X). Schreiber proposed to measure the deviation of an observed system from that expected under total independence by means of another entropy, the Kullback–Leibler divergence:

$$K(P, Q) = \sum_i P(i) \log \frac{P(i)}{Q(i)}, \quad (4)$$

which measures the distance between two probability distributions P and Q. $K(P, Q)$ is not symmetrical under exchange of P and Q, and is commonly applied as an estimate of the information lost when Q is used to approximate P. Palaeontologists will appreciate this as the engine inside the increasingly popular information criteria for model selection used in likelihood-based inference (Burnham & Anderson 2002). Attempts to explain the Kullback–Leibler divergence invariably converge on the phrase 'fundamental quantity'. Fortunately, only two of its properties need to be appreciated here: (1) the distance from P to Q is not equal to the distance from Q to P, and (2) values of $K(P, Q)$ are not based on means and variances, but on the entire distributions P and Q.

Schreiber's (2000) transfer entropy is here used in the modified form proposed by Verdes (2005): Given three time series X, Y, and Z, visiting states x_i, y_i, and z_i with probabilities $p(x_i)$, $p(y_i)$, and $p(z_i)$, the aim is to quantify the relative causal influence Y and Z have on X. The directional information transfer (IT) from Y to X given Z is defined by

$$IT_{Y:X,Z} = \frac{1}{N} \sum_i p(x_i, y_i, z_i) \log \frac{p(x_i|y_i, z_i)}{p(x_i|z_i)}, \quad (5)$$

where N is the length of the time series and the sum extends over all states i visited by the variables. $IT_{Y:X,Z} \neq IT_{X:Y,Z}$, hence the IT measures a directed flow of information. Because $IT_{Y:X,Z}$ is conditioned on the state of Z, it excludes any information transfer from Y to X that could result from shared information (mutual correlation) with Z. This feature might come in handy when trying to disentangle common-cause interactions.

The transition probabilities $p(x_i|y_i, z_i) = p(x_i, y_i, z_i)/p(y_i|z_i)$ are approximated by the frequency

$$p(x_i, y_i, z_i) = \frac{1}{N_p} n(\Delta x_{ij} < k_X, \Delta y_{ij} < k_Y, \Delta z_{ij} < k_Z),$$

$$(6)$$

where $\Delta x_{ij} = |x_i - x_j|$ (similarly for y_{ij} and z_{ij}), N_p is the total number of ij pairs, and $n(\Delta x_{ij} < k_X, \Delta y_{ij} < k_Y, \Delta z_{ij} < k_Z)$ is the number of distances that fall within the neighbourhood region defined by k_X, k_Y, and k_Z. The size of k_X is analogous to the bin size of a histogram and captures the magnitude of transitions in X. The IT will thus vary as a function of the scale of variability k_X, which will be expressed in units of standard deviation. For each value of k_X, the scale of information transfer is sought through a maximization of $p(x_i|y_i)$:

$$k_Y = \arg \max \frac{n(\Delta x_{ij} < k_X, \Delta y_{ij} < k_Y)}{n(\Delta y_{ij} < k_Y)}, \quad (7)$$

and similarly for $p(x_i|z_i)$ (Verdes 2005). Although the IT does not explicitly consider temporal structure, or leads and lags, such effects are implicit in the summation over all possible transitions (Equations 5 & 6). In fact, randomly shuffling the time series prior to analysis will not affect the results.

IT calculated according to Equation 5 quantifies the relative strength of coupling (including nonlinear interactions) between two variables X and Y beyond their common interaction with a third variable Z, and will in the following be referred to as conditional IT. As we shall see, however, testing for significant directionality of coupling between two variables X and Y requires a more stringent significance criterion, and with small data sets the directionality test is limited to a strictly bivariate IT calculation, without conditioning on Z:

$$\text{IT}_{Y \to X} = \frac{1}{N} \sum_i p(x_i, y_i) \log \frac{p(x_i|y_i)}{p(x_i)}. \qquad (8)$$

Causal inference, and the detection of common-cause interactions in particular, will rely on a combination of the two IT calculations. The notation $Y{:}X, Z$ will be used for conditional IT, and $Y \to X$ for directionality.

Correlation does not imply causation

Distinguishing correlation from causation is the centrepiece of the IT approach. To illustrate this, three hypothetical scenarios are compared: a complex causal relationship with no apparent correlation (Fig. 1a), a linear causal relationship with obvious correlation (Fig. 1b), and correlated time series that are not causally related (Fig. 1c). In each case, the IT method correctly detects the presence/absence and directionality of a causal relationship, with only 50 data points. This achievement rests on yet another clever trick.

Even under strictly unidirectional coupling, the statistical association (correlation in the general sense) between two systems goes both ways. An ideal measure of directionality should yield insignificant values in the uncoupled direction (from response to driver), and this requires an appropriate significance criterion (Paluš & Vejmelka 2007; Vejmelka & Paluš 2008). In the conditional IT calculation, significance is evaluated against a large number of randomly shuffled versions of the driver time series (Verdes 2005). Significant directionality, on the other hand, is assessed by comparison with amplitude-adjusted fast Fourier transform (AAFT) surrogate time series (Schreiber & Schmitz 2000; Kantz & Schreiber 2003; Vejmelka & Paluš 2008). AAFT surrogates preserve both the amplitude

distributions and the autocorrelation functions (i.e. frequency power spectra) of the underlying systems, but destroy any dynamic coupling by randomizing the phase of the frequency components. IT is deemed significant if the area under the IT curve exceeds that of the 95th percentile of a large number of IT analyses of surrogates. For convenience, the surrogate mean IT is subtracted from all IT curves to zero-mean normalize the surrogate IT.

In the nonlinear example (Fig. 1a), the relationship between X and Y is hidden from view, because any change in Y can correspond to a large or small increase or decrease in X. Hence, the first differences are scattered in all quadrants (Fig. 1d), and their correlation is near zero, suggesting there is no relationship at all. Directional IT analysis (Fig. 1g) compares the information transfer from Y to X (Equation 8) to that from X to Y, and tests whether they are (1) significant, and (2) significantly different. The IT varies as a function of scale (corresponding to the bin size k_X in Equation 6, expressed in units of standard deviation) and is therefore plotted as a curve. Because of limited data at very small and very large scales, the relevant IT emerges at intermediate scales. Significance is determined by the scale-integrated IT, that is, the total area under the IT curve must exceed that of the 95th percentile of the surrogates (grey lines in Fig. 1g), regardless of the shape of the curve. Finally, if IT in one or both directions is significant, then the difference in the area under the curves is used to test for significant directionality of information flow. In the nonlinear example, the IT correctly identifies the drive–response relationship as a unidirectional information flow from Y to X. Similarly, the IT can enhance correlation analysis of a linearly coupled system by distinguishing the driver from the response (Fig. 1b, e, h).

The ability to detect linear or non-linear drive–response relationship from short and irregularly sampled time series with only relative age information (Fig. 1a, b) suggests that the IT method holds some promise as a tool for geological data analysis (Hannisdal in review). An equally important issue, however, is to avoid false positive results by rendering insignificant any information transfer between systems that are entirely uncoupled.

Pioneering work by the MBL group (Raup et al. 1973; Raup & Gould 1974; Raup 1977) instilled in palaeontologists a deep appreciation, if not fear, of the power of random processes in explaining real-world trends and patterns. In the context of time series, pairs of random walks can be found to show strikingly correlated patterns that imply a strong causal linkage (Fig. 1c, f). A random walk being simply the running sum of a set of normally distributed random numbers, our intuition suggests that there should be no information flow between

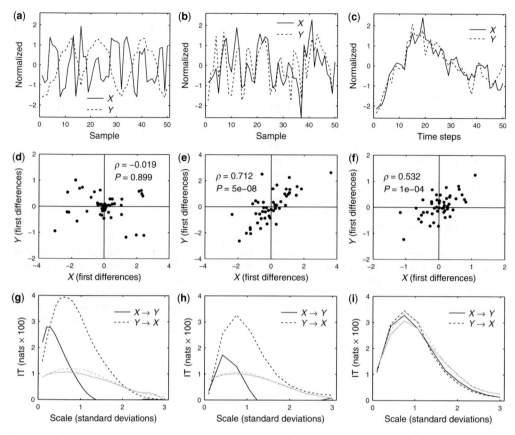

Fig. 1. Distinguishing correlation from causation in short time series. (**a**) A hypothetical example of a nonlinear coupling: $x_i = y_i(18y_i^2 - 27y_i + 10)/2 + z_i(1 - z_i)$, where $y_i = [1 - \cos(2\pi i/315)]/2$, $z_i = [1 + \sin(2\pi i/80)]/2$, and $i = 1, \ldots, 1000$ (from Verdes 2005). The response variable X and its main driver Y are then sampled at 50 randomly spaced time steps and normalized to unit standard deviation. (**b**) A linear coupling: $x_i = (10w_i + 7y_i - 3z_i)/10$, where $w_i = q_i(18q_i^2 - 27q_i + 10)/3$, and $q_i = [1 + \sin(2\pi i/639)]/2$. y_i, z_i, and i are defined as in (a). The X and Y time series are sampled at 50 randomly spaced time steps and normalized. (**c**) An example of two unbiased random walks X and Y that are accidentally correlated. Each time series is the running sum of 50 normally distributed pseudo-random numbers with zero mean and unit standard deviation. (**d–f**) Scatter plots of first differences (the value at one time step minus the preceding value) for the X and Y variables in (a–c). Correlations are given as Spearman's ρ, with P-values representing the probability that the rank-order correlation between X and Y is zero. (**g–i**) Directional information transfer (IT) from Y to X as a function of the scale of variability in X (in units of standard deviation). Grey lines are the 95th percentiles of the IT for 500×500 AAFT surrogates. In (g) and (h), black lines are the mean IT curves of 500 random sampling replicates. Note that the total IT (area under curve) from X to Y is not significantly greater than the surrogates, although the shape of the curve is different. In panel (i), black lines are the 95th percentiles of IT for 500 random walks. All time series are linearly detrended, power transformed (Box–Cox), and normalized prior to analysis.

two random walks if we exclude any correlation induced by coinciding amplitudes and/or trends. This is tested here by applying the AAFT surrogate-based directional IT analysis to a large number of significantly correlated random walks ($P < 0.01$ for both amplitudes and first differences). Indeed, the results show that the area under the 95th percentile of the IT is equal to that of the surrogate distribution in both directions, and there is no directionality (Fig. 1i). Knowing one random walk does not reduce our uncertainty regarding the state of another random walk, even if they look very similar. Because the AAFT surrogates preserve both the static correlation (amplitudes) and the auto-correlation structure (frequency power spectrum) of the original time series, a strong correlation by itself is not sufficient to cause false directionality. Note that all time series analysed in this paper are linearly detrended, power transformed (Box–Cox), and normalized to unit standard deviation before IT analysis to avoid biases associated with non-stationarity (Hannisdal in review).

A common-cause experiment

Consider a situation in which sea-level change is a forcing factor of both diversity and the amount of available sedimentary rock. Even if palaeontological sampling is random with respect to outcrop, the common-cause (confounding) variable might yield a sampled diversity curve that is spuriously correlated with outcrop area. This scenario is here simulated using the process-based numerical sediment dynamics model Sedflux (Syvitski & Hutton 2001; Hutton & Syvitski 2008). Sedflux is run in two-dimensional mode, tracking transport and deposition of river-derived sediment along an onshore–offshore profile in response to evolving boundary conditions (e.g. river discharge, bathymetry, sea-level, shoreline position). A hypothetical sea-level curve (Fig. 2a) provides the dominant control on shelf accommodation, while river sediment flux adds higher-frequency variability along with the gradual filling of the accommodation space.

In addition to the resulting depositional profile, model output comprises time series of water depth and the thickness and grain size of the deposited sediment across the shelf (see Hannisdal 2006;

Hannisdal & Peters 2010 for examples). The water depth history (Fig. 2b) gives the areal extent of available shelf habitat, here defined as the number of horizontal grid cells with a water depth between 10 and 100 m, at any given time (Fig. 2d). Similarly, the chronostratigraphic diagram (Fig. 2c) gives the areal extent of shelf deposition, which, in combination with sedimentation rate, can be used to approximate the amount of sedimentary rock or the probability of a given time step being represented at outcrop (Fig. 2e).

A hypothetical diversity curve is generated by combining the shelf habitat time series with a lower-frequency signal in such a way that the relationship between diversity and habitat is not one-to-one but maintains obvious in-phase variability (and thus a strong linear correlation). Rather than a direct species-area effect, this could be the result of the biota responding to processes that are coupled to changes in sea-level (and thus to changes in shelf habitat) but with some heterogeneity in the relative strength of coupling and possible feedback mechanisms. The diversity curve is then degraded into a fossil diversity curve through a randomly fluctuating preservation probability with a mean value that increases towards the youngest part of the record. Finally, the time series of sea level, outcrop area, and fossil diversity are sampled at 100 randomly spaced time steps (Fig. 3a).

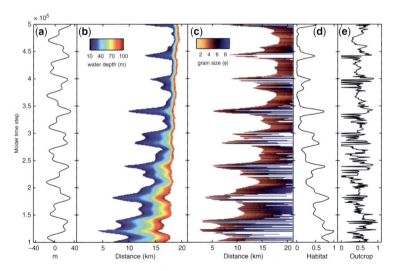

Fig. 2. A numerical common-cause experiment. (**a**) A hypothetical sea-level curve used as input to the Sedflux model. (**b**) Resulting water depth history from Sedflux output, truncated at 10 and 100 m, with the extent of shallow shelf area evolving in response to sea-level change and gradual filling of the accommodation space by river-derived sediment. (**c**) Time series of sediment deposition in the marine domain, with colors representing grain size (ranging from sand in yellow to clay in blue), and white areas representing non-deposition or erosion. (**d**) Time series of available shelf habitat (number of horizontal grid cells between 10 and 100 m water depths), which is used to drive a hypothetical diversity curve. (**e**) Time series of the areal extent of deposition, which is used as a proxy for sedimentary rock potentially available at outcrop.

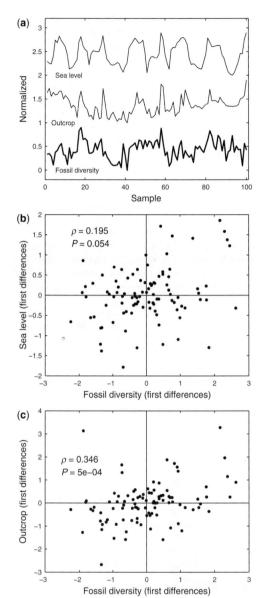

Fig. 3. A non-trivial common-cause interaction. (**a**) Time series of sea-level change (Fig. 2a), available rock outcrop (Fig. 2e), and a hypothetical fossil diversity curve derived by combining the habitat time series (Fig. 2d) with a low-frequency sinusoid and a randomly fluctuating preservation probability whose mean value decreases with age. The three time series are then sub-sampled at 100 randomly spaced time steps and normalized. (**b**) Scatter plot of first differences of the sea-level and fossil diversity records in (a), showing positive and near-significant correlation (Spearman's ρ). (**c**) Scatter plot of first differences of the outcrop and fossil diversity records in (a), showing much stronger correlation, despite preservation/sampling being independent of the outcrop time series.

Given these three records, we wish to evaluate the competing hypotheses of rock-record bias (false) and common-cause interaction (true). Despite sampling being independent of outcrop area, Spearman rank-order correlations and partial correlations suggest that outcrop is more strongly linked with the fossil diversity curve than sea-level, thus favouring rock-record bias over a common-cause explanation. Based on the correlations, one would infer that rock-record bias is the most important factor (Fig. 3b), although the relationship with sea-level is also positive and approaching significance (Fig. 3c). Even worse, partial Spearman correlations would imply a pure rock-record bias, because the partial correlation between outcrop and diversity (given sea-level) is significant ($P = 0.003$), whereas the partial correlation between sea-level and diversity (given outcrop) is not ($P = 0.62$).

In contrast, IT analysis identifies sea-level change as the common-cause variable (Fig. 4). The conditional IT (Equation 5) quantifies information transfer from sea-level (SL) to fossil diversity (Df) while excluding their mutual correlation with outcrop (Oc), and this is significant as indicated by the area under the IT curve (SL:Df, Oc; Fig. 4a). On the other hand, IT from outcrop to fossil diversity is not significant beyond the correlation with sea-level (Oc:Df, SL; Fig. 4a). In other words, outcrop does not provide any information on changes in diversity not already contained in the sea-level curve. A strong interpretation of this result is that the relationship is a pure common-cause interaction, and the correlation between outcrop and diversity is entirely spurious. Although we know this to be true in the present case, a more cautious interpretation would be that common-cause interaction is at least more important than rock-record bias. A directional IT analysis, using AAFT surrogates, lends further support to the inference of a drive–response relationship, showing a significant unidirectional flow of information from sea-level to fossil diversity (Fig. 4b).

Phanerozoic O, C, and S isotope records

In a forthcoming paper, IT analysis is applied to Phanerozoic rock and fossil records to quantify the importance of common-cause interactions relative to rock-record bias. Like other recent attempts to identify possible abiotic forcing of macroevolution (e.g. Mayhew *et al.* 2008; Cárdenas & Harries 2010), this takes advantage of the comprehensive database of isotope ratios from marine carbonates available for the study of Phanerozoic Earth system evolution (Veizer *et al.* 1999; McArthur *et al.* 2001; Shields & Veizer 2002; Kampschulte & Strauss 2004; Prokoph *et al.* 2008). As indirect, proxy variables, these

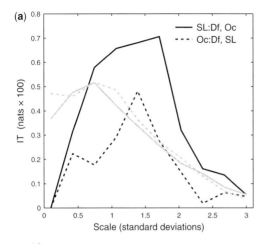

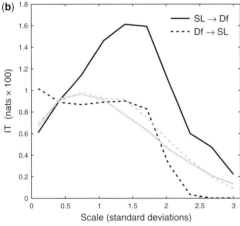

Fig. 4. IT analysis of the common-cause interaction in Figure 3a. (**a**) Conditional IT from sea-level (SL) to fossil diversity (Df), given outcrop (Oc) exceeds the 95th percentile of 1000 IT calculations with randomly shuffled SL time series (grey). Total conditional IT from Oc to Df given SL is not significant. (**b**) Directional IT from sea-level to fossil diversity (SL → Df) is significantly greater than that of 1000 AAFT surrogates (grey), while IT in the opposite direction is not.

records capture a mixture of signals from underlying processes (e.g. climate, tectonism, biotic evolution) that are inextricably linked and may act as confounding (common-cause) factors. Characterizing the interrelationships among such isotope records is not a trivial exercise, and may ultimately call upon biogeochemical modelling (e.g. Berner 2006). IT analysis could nonetheless shed some light on these interactions without having to specify models for the causal relationships.

A comparison of Phanerozoic seawater oxygen ($\delta^{18}O$), carbon ($\delta^{13}C$), and sulphur ($\delta^{34}S$) stable isotope records is facilitated by the recent compilation by Prokoph *et al.* (2008), using their 'low-latitude' subset of the $\delta^{18}O$ and $\delta^{13}C$ database and the global data set of $\delta^{34}S$ (Fig. 5a–c). Although both oxygen and carbon isotopic fractionation is sensitive to environmental and biological heterogeneity (e.g. water depth, salinity, pH, physiology), changes in $\delta^{18}O$ are generally attributed to temperature variations (and ice volume during glacial times), and $\delta^{13}C$ variability reflects changes in biological productivity and/or the efficiency of carbon burial in marine sediments. The $\delta^{34}S$ of carbonate-associated sulphate, generally considered a proper global proxy for seawater sulphate isotopic composition (Kampschulte & Strauss 2004), responds to mid-ocean ridge activity, precipitation/dissolution of evaporites, and burial/weathering of sedimentary pyrite formed through bacterial sulphate reduction. The latter has by far the strongest effect on fractionation, hence sulphur cycling in marine sediments and pyrite burial efficiency play a key role in Phanerozoic $\delta^{34}S$ evolution.

Veizer *et al.* (1999), Prokoph & Veizer (1999) and later Prokoph *et al.* (2008) found covariation and matching periodicities in the C and S records, and interpreted these as evidence for the biologically mediated redox coupling of the global carbon and sulphur cycles. In contrast, they obtained only weak and intermittent relationships between O and C, providing little support for a linkage between climate and carbon cycling on this time-scale. Indeed, the correlations among the three time series in terms of first differences are underwhelming (Fig. 5d–f). IT analysis, on the other hand, gives a very different picture (Fig. 6).

Conditional IT from C to S is clearly significant beyond any mutual association with the O record, and extends to large-scale changes in S (Fig. 6a). This may reflect the importance of enhanced carbon burial on sedimentary pyrite formation and the long-term impact of carbon burial efficiency and its environmental partitioning on the sulphur isotope chemistry of the ocean (Berner & Raiswell 1983; Gill *et al.* 2007). In contrast, the O record has no bearing on changes in S beyond any shared information with the C record. If the direction is reversed, conditional IT from S to C is also significant regardless of any common interaction with O (Fig. 6b). This suggests that the coupling between the global carbon and sulphur cycles is a two-way interaction, or that the S record is sensitive to processes relevant for understanding changes in the C record. For example, the formation of large evaporite deposits and associated drawdown of seawater sulphate concentrations may alter rates of organic matter remineralization by sulphate-reducing bacteria (Wortmann & Chernyavsky 2007). Interestingly, the conditional IT from O to C given S is

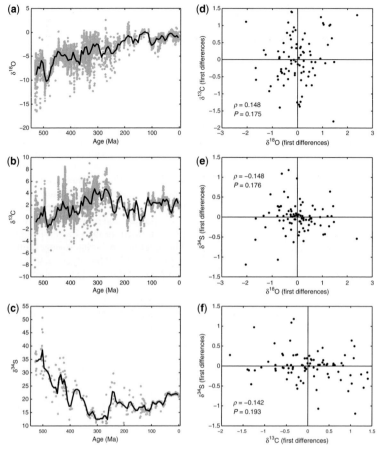

Fig. 5. Phanerozoic seawater isotope records. Raw data (grey) are from Prokoph *et al.* (2008), solid lines are averages for *c.* 6 Ma time bins. $\delta^{18}O$ (**a**) and $\delta^{13}C$ (**b**) data are 'low-latitude' subsets of the database, while $\delta^{34}S$ (**c**) represents a global data set. (**d**–**f**) Scatter plots of first differences for the bin-averaged records in (a–c), with correlations given as Spearman's ρ.

also significant and of a comparable magnitude, implying that O and C are indeed coupled on this time-scale, and that O and S explain different aspects of the C record (Fig. 6b).

Directional IT analysis supports and refines the conditional IT results, indicating a two-way inter-action between C and S with opposite directionality of coupling occurring on different scales (Fig. 6c). Changes in S seem to affect small-scale changes in C, whereas C drives larger-scale variability in S. It is tempting to interpret this result as an indi-cation of scale-dependent feedbacks in the global cycling of carbon and sulphur over the Phanerozoic. The relationship between O and C is also a two-way interaction, with C influencing O across scales, while O is the dominant driver of C on intermediate scales (Fig. 6d). This suggests that quantitative evi-dence for long-term feedbacks between climate

and carbon cycling can be wrested from the Phanerozoic records.

Summary and conclusions

The relative strength of causal dependencies among observed time series will under ideal conditions be reflected in the relative magnitude of partial corre-lations and/or signal frequency power spectra. However, when faced with the added complexity of causal antecedents, scale-dependent coupling, nonlinearities, and indirect proxy measurements, standard correlation is less powerful and in the worst case misleading. Moreover, generalized regression techniques commonly assume a (linear or nonlinear) model of the drive–response relationship. This paper has shown how an information-theoretic approach to time series analysis can distinguish correlation

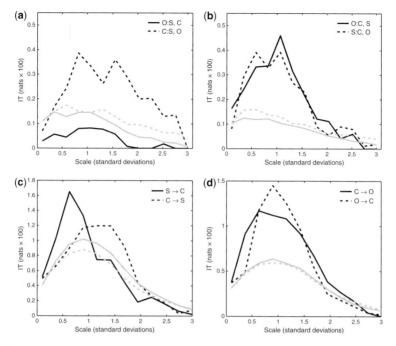

Fig. 6. IT analysis of the Phanerozoic O, C, and S isotope records. (**a**) Conditional IT from C to S is clearly significant beyond any mutual correlations with the O record, while IT from O to S given C is not. Solid and dashed grey lines are 95th percentiles of 1000 randomly shuffled O and C records, respectively. (**b**) In the opposite direction, both O and S yield significant conditional IT to the C record. (**c**) Directional IT between S and C is significant in both directions, but on different scales. Grey lines are 95th percentiles of 1000 AAFT surrogates. (**d**) Directional IT between C and O is also significant in both directions, with C → O spread across scales, whereas O → C dominates on intermediate scales.

from causation in irregular and noisy geological time series, without modelling.

Specifically, the quantification of directional information transfer can provide a more powerful test of a common-cause hypothesis than the analysis of correlations per se, an ability that stems from a combination of advantageous properties: (1) because it is non-symmetric, and an explicit function of scale, the IT can measure scale-dependent strength and directionality of coupling; (2) the IT is non-parametric, sensitive to non-linear relationships, and does not require absolute time; (3) significance is established by means of AAFT surrogates that preserve both correlation and autocorrelation structures of the original time series. The numerical common-cause experiment, albeit contrived, showed how these properties enable the IT to peer through correlations and quantitatively characterize underlying causal relationships. Directional information transfer analysis of Phanerozoic stable isotope records quantifies relationships among coupled components of the Earth system that have eluded previous statistical analyses, such as a long-term coupling between climate and carbon cycling.

Still, the IT method is a statistical tool, not a truth serum. Surrogates replicate noise levels, and even autocorrelated noise will not generate false directionality where none exists (Hannisdal 2011). If a causal relationship is present, however, different signal-to-noise ratios may inflate conditional IT in favor of the least noisy variable or possibly make a two-way interaction appear unidirectional. Two variables responding to a common forcing factor may show significantly directional IT if one variable captures the underlying common cause with greater fidelity than the other. This would represent a challenge for all data-driven approaches to causal inference, and can only be met with additional information on data quality. In principle, such information can be incorporated in a conditional IT analysis.

In conclusion, the IT provides an alternative, model-free test of common-cause interactions in geological time series that shows promise for dissecting the relationship between rock and fossil records.

The author would like to thank two anonymous reviewers for helpful comments.

References

BERNER, R. A. 2006. GEOCARBSULF: a combined model for Phanerozoic atmospheric O_2 and CO_2. *Geochimica et Cosmochimica Acta*, **70**, 5653–5664, doi: 10.1016/j.gca.2005.11.032.

BERNER, R. A. & RAISWELL, R. 1983. Burial of organic-carbon and pyrite sulfur in sediments over Phanerozoic time – a new theory. *Geochimica et Cosmochimica Acta*, **47**, 855–862.

BURNHAM, K. P. & ANDERSON, D. R. 2002. *Model Selection and Multi-Model Inference: A Practical Information-Theoretic Approach*. Springer-Verlag, New York.

CÁRDENAS, A. L. & HARRIES, P. J. 2010. Effect of nutrient availability on marine origination rates throughout the Phanerozoic eon. *Nature Geoscience*, **3**, 430–434, doi: 10.1038/ngeo869.

GILL, B. C., LYONS, T. W. & SALTZMAN, M. R. 2007. Parallel, high-resolution carbon and sulfur isotope records of the evolving Paleozoic marine sulfur reservoir. *Palaeogeography Palaeoclimatology Palaeoecology*, **256**, 156–173, doi: 10.1016/j.palaeo.2007.02.030.

GRANGER, C. W. J. 1969. Investigating causal relations by econometric models and cross-spectral methods. *Econometrica*, **37**, 414–438.

HANNISDAL, B. 2006. Phenotypic evolution in the fossil record: numerical experiments. *Journal of Geology*, **114**, 133–153.

HANNISDAL, B. 2011. Non-parametric inference of causal interactions from geological records. *American Journal of Science*, **311**, 315–334, doi: 10.2475/04.2011.02.

HANNISDAL, B. & PETERS, S. E. 2010. On the relationship between macrostratigraphy and geological processes: quantitative information capture and sampling robustness. *Journal of Geology*, **118**, 111–130, doi: 10.1086/650180.

HLAVÁČKOVÁ-SCHINDLER, K., PALUŠ, M., VEJMELKA, M. & BHATTACHARYA, J. 2007. Causality detection based on information-theoretic approaches in time series analysis. *Physics Reports-Review Section of Physics Letters*, **441**, 1–46, doi: 10.1016/j.physrep.2006.12.004.

HUTTON, E. W. H. & SYVITSKI, J. P. M. 2008. SedFlux 2.0: an advanced process-response model that generates three-dimensional stratigraphy. *Computers & Geosciences*, **34**, 1319–1337.

KAMPSCHULTE, A. & STRAUSS, H. 2004. The sulfur isotopic evolution of Phanerozoic seawater based on the analysis of structurally substituted sulfate in carbonates. *Chemical Geology*, **204**, 255–286, doi: 10.1016/j.chemgeo.2003.11.013.

KANTZ, H. & SCHREIBER, T. 2003. *Nonlinear Time Series Analysis*. 2nd edn. Cambridge University Press, Cambridge.

MAGURRAN, A. E. 2004. *Measuring Biological Diversity*. Blackwell, Oxford.

MAYHEW, P. J., JENKINS, G. B. & BENTON, T. G. 2008. A long-term association between global temperature and biodiversity, origination and extinction in the fossil record. *Proceedings of the Royal Society, B*, **275**, 47–53, doi: 10.1098/rspb.2007.1302.

MCARTHUR, J. M., HOWARTH, R. J. & BAILEY, T. R. 2001. Strontium isotope stratigraphy: LOWESS version 3: Best fit to the marine Sr-isotope curve for 0–509 Ma and accompanying look-up table for deriving numerical age. *Journal of Geology*, **109**, 155–170.

PALUŠ, M. & VEJMELKA, M. 2007. Directionality of coupling from bivariate time series: how to avoid false causalities and missed connections. *Physical Review E*, **75**, 1–14.

PETERS, S. E. 2005. Geologic constraints on the macroevolutionary history of marine animals. *Proceedings of the National Academy of Sciences of the United States of America*, **102**, 12 326–12 331.

PETERS, S. E. 2006. Genus extinction, origination, and the durations of sedimentary hiatuses. *Paleobiology*, **32**, 387–407.

PETERS, S. E. & FOOTE, M. 2001. Biodiversity in the Phanerozoic: a reinterpretation. *Paleobiology*, **27**, 583–601.

PETERS, S. E. & FOOTE, M. 2002. Determinants of extinction in the fossil record. *Nature*, **416**, 420–424.

PROKOPH, A., SHIELDS, G. A. & VEIZER, J. 2008. Compilation and time-series analysis of a marine carbonate $\delta^{18}O$, $\delta^{13}C$, $^{87}Sr/^{86}Sr$ and $\delta^{34}S$ database through Earth history. *Earth-Science Reviews*, **87**, 113–133, doi: 10.1016/j.earscirev.2007.12.003.

PROKOPH, A. & VEIZER, J. 1999. Trends, cycles and non-stationarities in isotope signals of Phanerozoic seawater. *Chemical Geology*, **161**, 225–240.

RAUP, D. M. 1976. Species diversity in the Phanerozoic: an interpretation. *Paleobiology*, **2**, 289–297.

RAUP, D. M. 1977. Stochastic models in evolutionary paleontology. *In*: HALLAM, A. (ed.) *Patterns of Evolution as Illustrated by the Fossil Record*. Elsevier, Amsterdam, 59–78.

RAUP, D. M. & GOULD, S. J. 1974. Stochastic simulation and evolution of morphology: towards a nomothetic paleontology. *Systematic Zoology*, **23**, 305–322.

RAUP, D. M., GOULD, S. J., SCHOPF, T. J. M. & SIMBERLOFF, D. S. 1973. Stochastic models of phylogeny and the evolution of diversity. *Journal of Geology*, **81**, 525–542.

SCHREIBER, T. 2000. Measuring information transfer. *Physical Review Letters*, **85**, 461–464.

SCHREIBER, T. & SCHMITZ, A. 2000. Surrogate time series. *Physica D-Nonlinear Phenomena*, **142**, 346–382.

SHANNON, C. E. 1948. A mathematical theory of communication. *Bell Systems Technical Journal*, **27**, 379–423 and 623–656.

SHIELDS, G. & VEIZER, J. 2002. Precambrian marine carbonate isotope database: Version 1.1. *Geochemistry Geophysics Geosystems*, **3**, 1–12, doi: 103110.1029/2001gc000266.

SMITH, A. B. 2001. Large-scale heterogeneity of the fossil record: implications for Phanerozoic biodiversity studies. *Philosophical Transactions of the Royal Society of London Series B-Biological Sciences*, **356**, 351–367.

SMITH, A. B. 2007. Marine diversity through the Phanerozoic: problems and prospects. *Journal of the Geological Society, London*, **164**, 731–745.

SYVITSKI, J. P. M. & HUTTON, E. W. H. 2001. 2D SEDFLUX 1.0C: an advanced process-response numerical model for the fill of marine sedimentary basins. *Computers & Geosciences*, **27**, 731–753.

VEIZER, J., ALA, D. *ET AL.* 1999. $^{87}Sr/^{86}Sr$, $\delta^{13}C$ and $\delta^{18}O$ evolution of Phanerozoic seawater. *Chemical Geology,* **161**, 59–88.

VEJMELKA, M. & PALUŠ, M. 2008. Inferring the directionality of coupling with conditional mutual information. *Physical Review E,* **77**, 1–12.

VERDES, P. F. 2005. Assessing causality from multivariate time series. *Physical Review E,* **72**, 1–9.

WORTMANN, U. G. & CHERNYAVSKY, B. M. 2007. Effect of evaporite deposition on Early Cretaceous carbon and sulphur cycling. *Nature,* **446**, 654–656, doi: 10.1038/nature05693.

Variation in stratigraphic congruence (GER) through the Phanerozoic and across higher taxa is partially determined by sources of bias

ANNE O'CONNOR[1], CLIVE MONCRIEFF[2] & MATTHEW A. WILLS[1]*

[1]*Department of Biology and Biochemistry, The University of Bath, The Avenue, Claverton Down, Bath BA2 7AY, UK*

[2]*Biometry Research Group, The Natural History Museum, Cromwell Road, London SW7 5BD, UK*

*Corresponding author (e-mail: m.a.wills@bath.ac.uk)

Abstract: Many published cladograms report measures of stratigraphic congruence. Strong congruence between cladistic branching order and the order of first fossil occurrences is taken to support both the accuracy of cladograms and the fidelity of the record. Poor congruence may reflect inaccurate trees, a misleading fossil record, or both. However, it has been demonstrated that most congruence indices are logically or empirically biased by parameters that are not uniformly distributed across taxa or through time. These include tree size and balance, mean ghost range duration (gap size) and the range and distribution of origination dates. This study used 650 published cladograms to investigate the influence of these variables on the Gap Excess Ratio (GER). In a range of multivariate models, factors other than congruence *per se* explained up to 74.5% of the observed variance in GER amongst trees. Arthropods typically have poorer GER values than other groups, but the residual differences from our models are much less striking. The models also show no clear residual trend in GER through the Phanerozoic. Because the GER is strongly influenced by parameters related to cladogram size, balance and duration, comparisons across trees should be made with caution.

Supplementary material: Data legends are available at http://www.geolsoc.org.uk/SUP18484

It is estimated that only 2% of the species that have existed on Earth are alive today. Fossils provide the only direct evidence of the remaining 98%, but their particular utility may be in documenting transitional forms and sequences of character acquisition within the deepest branches of the tree (Wills & Fortey 2000). Morphological data from fossils may therefore be vital for accurate cladistics (Donoghue *et al.* 1989). In addition to preserving extinct combinations of character states (Donoghue *et al.* 1989; Wagner 1999; Grantham 2004; Cobbett *et al.* 2007) fossils also occur in rocks that can be dated in relative and absolute time (Springer *et al.* 2001; Crane *et al.* 2004; Donoghue & Purnell 2009). Classically, temporal data do not contribute to phylogenetic inferences (but see Wagner 1995 and Fisher 2008), and hence the order in which taxa branch within a cladogram can be compared legitimately with the sequence in which they first appear through the fossil record. Both should reflect the same underlying evolutionary history, but neither is logically contingent on the other. Significant congruence is consistent with an accurate phylogeny that is mapped onto a sequence of first fossil occurrences that document

reliably the order in which groups evolved. Poor congruence is amenable to a variety of explanations, either singly or in concert. It might result from a spurious tree, the misinterpretation of particular fossils, or from probabilities of preservation that are too low or variable between lineages or through time to record the origination of groups in the correct temporal sequence.

Stratigraphic congruence indices are now routinely reported for published trees that include fossils. They are utilized in two ways: for refining/testing particular phylogenies, and for the statistical treatment of large samples of cladograms in order to find trends. The first application includes the use of stratigraphy as an ancillary criterion for choosing between or filtering large numbers of otherwise equally parsimonious trees. Stratigraphic data are not included in the original optimizations, but merely utilized post-hoc. However, the use of such a 'stratigraphic yardstick' is only defensible where there is reason to trust the fidelity of the fossil record on other grounds (Wills 1999; Benton *et al.* 2000). There is an inherent need for a measure of congruence between the fossil record

From: McGOWAN, A. J. & SMITH, A. B. (eds) *Comparing the Geological and Fossil Records: Implications for Biodiversity Studies.* Geological Society, London, Special Publications, **358**, 31–52.
DOI: 10.1144/SP358.4   0305-8719/11/$15.00 © The Geological Society of London 2011.

and phylogeny: not least for studies that date events and calibrate evolutionary rates by superimposing fossil records onto trees (Norell & Novacek 1992). These usually employ molecular data (e.g. Smith *et al.* 2006), but there is now also a burgeoning literature on morphology (Ruta *et al.* 2006; Brusatte *et al.* 2008; Friedman 2010). These applications do not utilize a 'yardstick' approach, but rather use cladistic inferences and stratigraphic dates for cross-validation: accurate knowledge of the sequence of appearance of phylogenetic terminals as implied by significant congruence is a minimum requirement.

The second use for measures of congruence is in statistical studies of hundreds of trees that look for widespread trends (Wills 2002). Here, there has been a tendency to use *cladograms* as the 'yardstick', against which to measure something about the fidelity of the fossil record (*precisely* the reverse logic to that employed above). Most investigations have focused on differences in congruence for higher taxa, or for taxa from different habitats (e.g. terrestrial v. marine or freshwater). Benton & Hitchin (1996) and Benton & Simms (1995) both argued that the quality of the continental vertebrate fossil record is comparable to that of echinoderms. However, Benton *et al.* (1999) demonstrated that the congruence values for echinoderm cladograms were significantly better than those for fish and tetrapods, while Wills (1999) suggested that the fossil record of arthropods was less complete than tetrapods and fish. More recently, there has been an acknowledgment that while we cannot assume trees to be of uniform quality across taxa (there are, in fact, good reasons to suppose otherwise), taxonomic differences in congruence (necessarily conflating aspects of fossil record fidelity and phylogenetic accuracy) might nonetheless be informative. The contrast between the situation in dinosaurs (Wills *et al.* 2008, where most trees approach maximal congruence) and that in malacostracan crustaceans (Wills *et al.* 2009, where congruence is actually worse than for most random permutations of stratigraphic range data across many trees) is particularly striking. A much smaller number of studies have used stratigraphic congruence as an index of fossil record 'quality' through time. The first of these (Benton *et al.* 2000) used three indices calculated for 1000 cladograms of animals and plants, concluding that the fossil record is of uniform quality (principally, the probability of preservation in this context) in Palaeozoic, Mesozoic and Cenozoic time bins, and at taxonomic levels of (predominantly) families and above. This contrasts with the expectation of some authors (e.g. Benton 1994; Benton *et al.* 2000) that quality should deteriorate with increasing geological age, since the probability of preservation is thought to decrease while the

chances of deformation and subduction increase (but see Smith & McGowan 2007). Moreover, the probability of gaps in the fossil record can be inversely related to the intensity of study (Donoghue *et al.* 1989). Wills (2007) reworked these data, assessing the pattern through time using 77 series and stages, and a refined measure of congruence (the Gap Index: GI). At this improved resolution, congruence was seen to be at its highest in the Mesozoic, deteriorating back into the Palaeozoic but also, paradoxically towards the Recent (Fig. 1a, b). How can this be?

Unfortunately, there is no universal measure of stratigraphic congruence that can be compared across trees and data sets. This is precisely analogous to the problems encountered with measures of the quality of phylogenetic data such as the CI (consistency index) (Kluge & Farris 1969) and RI (retention index) (Farris 1989), which are biased by numbers of characters and numbers of taxa (Sanderson & Donoghue 1989; Archie & Felsenstein 1993). Hence, in the same way that the CI cannot be compared directly for data sets of different dimensions, so existing measures of stratigraphic congruence are logically or empirically found to be biased by a range of factors relating not only to the size and balance of the tree, but also to the distribution of observed stratigraphic ranges.

The gap excess ratio (GER) was proposed by Wills (1999) and is now widely used as a measure of congruence derived from the inferred extent of ghost ranges relative to a theoretical maximum and minimum. For the most recent applications, see Brusatte & Serreno (2008), Lelièvre *et al.* (2008), Tetlie & Poschmann (2008), Wills *et al.* (2008, 2009), Dyke *et al.* (2009), Frobisch & Schoch (2009), Tsyganov-Bodounov *et al.* (2009), Brusatte *et al.* (2010), Campione & Reisz (2010), Kroh & Smith (2010) and Lamsdell *et al.* (2010).

Where sister terminals or nodes are first represented in the fossil record at different times, a ghost range is subtended between them. These ghost ranges can be summed throughout a tree to yield a total minimum implied gap (denoted $\sum$MIG (minimum implied gap) in Benton & Storrs (1994), or simply MIG in Wills (1999)). The GER scales the observed MIG between the sum of ghost ranges for the worst (G_{max}) and best (G_{min}) possible fit of a given set of stratigraphic data onto *any* tree topology (Fig. 2).

Because the GER is calculated over an entire tree, studies that use it to investigate patterns of congruence through time must assign each tree to a particular time bin (e.g. Benton *et al.* 1999, 2000). This approach is problematic where cladograms span more than one time bin, which disproportionately affects larger and more deeply rooted trees. As a means to obviate this, Wills (2007) devised the

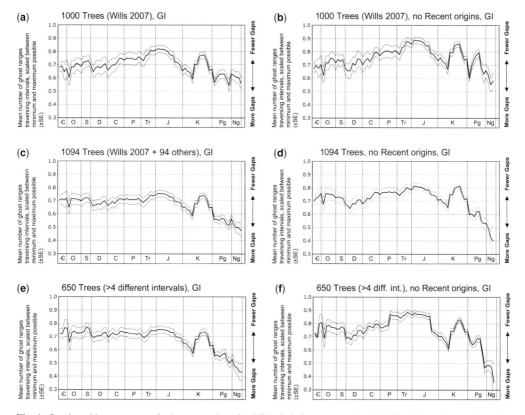

Fig. 1. Stratigraphic congruence for large samples of published cladograms, resolved to 77 series and stages through the Phanerozoic. Congruence is measured using the gap index (GI: see Fig. 2), Bold lines indicate means, ±SE. All plots omit the contributions of given intervals for given trees where there are fewer than six possible GI values. Most trajectories have a convex pattern, with congruence greatest in the Triassic and Jurassic, and declining both towards the Cambrian, and towards the Recent. Plots a, c and e are for all ghost ranges in all cladograms, while plots b, d and f omit ghost ranges subtended between terminals originating in the Recent (i.e. those with no fossil record and others). (**a**) 1000 Trees from Benton *et al.* (2000) and Wills (2007). (**b**) 1000 Trees from Benton *et al.* (2000) and Wills (2007), but omitting taxa with no fossil record. This has little impact on the decline in GI from the Jurassic to the Recent, but rather raises mean GI throughout the profile. (**c**) 1094 Trees, including 1000 trees as above, plus 94 trees of molluscs and birds compiled by AOC (see materials and methods section). The addition of a modest number of additional trees removes much of the apparent decline in GI from the Triassic back to the Cambrian seen in a. (**d**) 1094 trees (as in c), but omitting taxa with no fossil record. (**e**) 650 Trees (as in d), but omitting those cladograms with taxa originating in fewer than five different stratigraphic intervals. This filters out very small trees, as well as those with originations concentrated in a limited number of strata. This was the sample of trees used in the main set of analyses presented in the rest of this paper. (**f**) 650 Trees (as in e), but omitting taxa with no fossil record.

gap index (GI): a measure precisely analogous to the GER, but calculated for each stratigraphic interval through which a tree passes (Figs 1 & 2). Mean gap index values have been used to investigate the distributions of ghost ranges through time, without needing to place a given cladogram into a single time bin.

This paper has three purposes:

(1) To quantify the magnitude and direction of several sources of bias on the GER (Wills 1999)

in the largest empirical data set published to date (Benton *et al.* 2000; Wills 2007). The calculation of GER statistics (and that of many of its putative sources of bias) has been automated for an unlimited number of stratigraphic ranges in the program 'Ghosts' (Wills 1999, and subsequent revisions). This is also able to handle large batches of trees and output tables of summary statistics.

(2) To determine the extent to which differences in median GER values previously reported

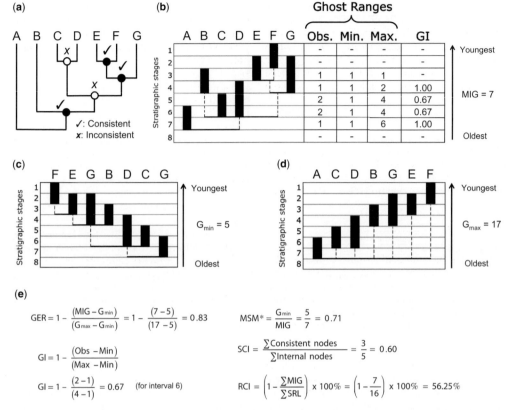

Fig. 2. Calculation of the stratigraphic consistency index (SCI), gap index (GI), gap excess ratio (GER), modified Manhattan stratigraphic measure (MSM*) and the relative completeness index (RCI) for a simple, hypothetical case. (**a**) Hypothetical phylogeny for terminals A to G, showing stratigraphically consistent (✓) and inconsistent (✗) nodes. (**b**) Cladogram in 'a' plotted onto hypothetical 'observed' stratigraphic ranges (thick, vertical black bars) and showing ghost ranges (vertical broken lines). The table shows observed (Obs.), minimum (Min.) and maximum (Max.) possible numbers of ghost ranges traversing each stratigraphic interval for *any* topology. These bounds are used to calculate the gap index (GI: see panel e), scaling the number of ghost ranges in each stratigraphic interval between these maxima and minima. The sum of the observed ghost ranges across the tree is the minimum implied gap (MIG) (**c**) Minimum possible ghost ranges summed over all intervals (G_{min}) on a pectinate tree. (**d**) Maximum possible ghost ranges summed over all intervals on a pectinate tree (G_{max}). (**e**) Formulae for all indices, with worked examples for all of the above. The GER is precisely analogous to the GI, except that ghost ranges are summed over the entire tree, rather than being calculated for single intervals. GI calculation is for interval 6. The MSM* was originally described in terms of the consistency index of an irreversible and ordered stratigraphic character. However, it is precisely equivalent to the formulation given here. The SCI is simply the number of internal cladogram nodes with a sister group or terminal of equivalent age or older, expressed as a fraction of the total number of internal nodes. The RCI differs from all of the other indices in expressing a ratio between the sum of observed ranges (standard range length or SRL, represented by the thick black bars) and the sum of ghost ranges (ΣMIG, equivalent to the MIG in other formulae). It is also the only index not scaled between zero and one.

between major taxonomic groups can be explained by the distribution of the sources of bias identified above. Do excellent values for many groups of vertebrates and poor values for arthropods reflect differences in congruence alone?

(3) To test whether the pattern of congruence throughout the Phanerozoic is closer to that reported by Benton *et al.* (2000) (a uniform distribution, albeit at a very coarse temporal

scale) or that found by Wills (2007) (complex, but higher in the Mesozoic than the Palaeozoic and Cenozoic).

Parameters that influence indices of stratigraphic congruence

Since the present paper concentrates on the GER, a lengthy discussion of the behaviour of other metrics is beyond its scope. However, a number of tree parameters are repeatedly cited in the literature as confounding comparisons of indices between cladograms. These parameters concern both the nature and structure of cladograms, and the stratigraphic ranges of their constituent taxa. They include biases deduced logically and, less frequently, those observed empirically.

Tree balance

The balance of a tree refers to the degree of symmetry in its topological branching structure. Heard's index (I_m) is one of the most widely used indices of *imbalance*, and is given by (Heard 1992):

$$I_m = \frac{\sum\limits_{\text{all interior nodes}} |T_R - T_L|}{(n-1)(n-2)/2}$$

where T_R is the number of terminal taxa subtended to the right of an internal node, T_L is the number of terminal taxa subtended to the left of that node, and n is the total number of terminals in the tree. A perfectly balanced tree is one that bifurcates symmetrically at each internal node, always having equal numbers of taxa on either side ($I_m = 0.0$). A completely imbalanced or pectinate (comb-like) tree is one in which every bifurcation leads to at least one terminal taxon ($I_m = 1.0$). A comprehensive simulation and review of the effects of tree balance was provided by Siddall (1996), later partially evaluated empirically by Benton & Hitchin (1997). In summary, all congruence indices are theoretically biased by tree balance/imbalance (or can only be calculated for special cases), but this has not been reported from empirical studies.

The first described and perhaps most intuitive index of congruence is the Spearman rank correlation (by convention, SRC in this context) between age rank (the order in which taxa first appear in the fossil record) and clade rank (the order of branching in a pectinate or completely unbalanced tree) (Gauthier *et al.* 1988; Norell & Novacek 1992). This metric is only applicable to completely pectinate trees (or those that can be simplified or subdivided into a pectinate structure) (Benton *et al.* 1999; Pol *et al.* 2004) thereby excluding the majority with a more complex topology *a priori* (Pearson 1999; Wills *et al.* 2008).

Siddall (1996) and Hitchin & Benton (1997) focused on the stratigraphic consistency index (SCI) of Huelsenbeck (1994). This expresses the number of consistent internal nodes as a fraction of the total number of resolved internal nodes in the tree, where a consistent node is one with a sister node or terminal of the same age or older (Fig. 2). Values of the SCI range between 0.0 (completely incongruent) and 1.0 (completely

congruent), but these are only achievable in fully pectinate trees, or those in which groups of terminals originate penecontemporeneously (Siddall 1996; Wills 1999). If no taxa originate at the same time, then pairs of sister nodes cannot both be consistent. Therefore, on a fully balanced tree, the SCI will always be 0.5 (i.e. 50% of nodes will always be consistent and 50% inconsistent), but on a pectinate tree it can vary between zero and one. Similarly, in cases where all taxa originate at the same time (such that congruence becomes meaningless), all nodes will be considered consistent, yielding an SCI of 1.0 (Wills 1999; Pol *et al.* 2004). Benton & Hitchin (1996) did not find a significant relationship between the SCI and balance for their empirical sample of 376 cladograms of echinoderms, fishes and tetrapods.

The GER is also subject to logical biases relating to tree balance, and for reasons similar to those afflicting the SCI (Wills 1999). Because G_{max} and G_{min} are defined by minimally and maximally congruent pectinate trees respectively, it may be impossible to achieve them on more balanced topologies. In a series of simulations, Pol *et al.* (2004) demonstrated that there is a marked difference in the distributions of possible stratigraphic debt values for fully pectinate and balanced trees. Simple simulations for the GER illustrate a similar point. A completely balanced (Fig. 3a) and a completely imbalanced (Fig. 3b) topology, both of 32 terminals, were randomly assigned the same set of regularly spaced stratigraphic dates 5000 times. Initially, there were four evenly spaced dates (0, 5, 10 and 15 units), with eight terminals given each date. On the balanced topology (Fig. 3a), GER values of 0.00 and 1.00 are never achieved, and the distribution has a mode of 0.356 and a median of 0.378. A moderately low GER is therefore easier to obtain (by chance) than a high one. On the pectinate topology (Fig. 3b), the mode is 0.000 and the median 0.022: a very low GER is now very easy to obtain, and a high one *very* unlikely (although values of 1.00 are now *theoretically* possible).

Tree size

Benton & Storrs (1994) and Benton & Simms (1995) reported unsurprising biases in the SRC related to the number of terminals: larger trees have higher values of Spearman's rho. Consequently, they excluded trees with less than four taxa from their analyses of 74 vertebrate cladograms, with more equivocal results than those previously reported by Norell & Novacek (1992).

We distinguish between topologies and trees in this context. A topology is simply a branching structure, or a tree without the terminals labelled. Many distinct trees can therefore share the same

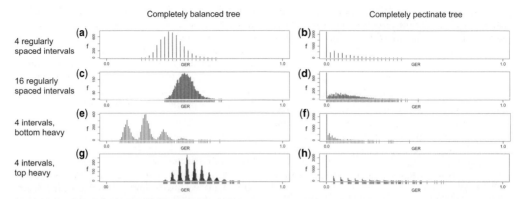

Fig. 3. Simulation of the effects of four variables on the distribution of possible GER values for a tree of 32 taxa. Panels a, c, e and g are for a perfectly balanced tree, while panels b, d, f and h are for a completely pectinate tree. Distributions of GER values are from 5000 random reassignments of the range data across each tree. (**a, b**) Four regularly spaced first occurrence dates (ages 0, 5, 10 and 15 units), with eight terminals given each date. (**c, d**) Sixteen regularly spaced first occurrence dates (ages 0 to 15 units in increments of 1), with two terminals given each date. Increasing the number of different first occurrence dates (without changing the range of first occurrences) causes the median GER to increase. (**e, f**) Four first occurrence dates, but with ages 0, 13, 14 and 15 units (eight terminals given each date). These bottom-heavy origination dates yield a multimodal distribution of GER values with a lower median than the regularly spaced dates. (**g, h**) Four first occurrence dates, with ages 0, 1, 2 and 15 units (eight terminals given each date). Top-heavy origination dates yield a multimodal distribution of GER values with a higher median than the regularly spaced dates.

topology, differing only in the locations of the terminals. Siddall (1996) demonstrated that as the *number* of terminals increases, the mean level of imbalance for randomly resolved topologies decreases. More precisely, Heard's index of imbalance (I_m) decreased logarithmically as the number of taxa (*n*) increased. For example, when $n = 3$, there is only one possible resolved topology, and this is completely unbalanced. However, when $n = 4$, there are two possible topologies: one completely imbalanced or pectinate, and one completely balanced. At $n = 6$ there are six possible topologies, with only one of these completely unbalanced (Hitchin & Benton 1997). Consequently, because tree size can 'drive' tree imbalance, and tree imbalance in turn influences congruence indices (see above), there is also a theoretical relationship between tree size and congruence mediated by imbalance (Wills 1999, 2001; Wagner & Sidor 2000; Pol *et al.* 2004; Lelièvre *et al.* 2008; Wills *et al.* 2008). However, this relationship has not been universally reported in empirical studies. Hitchin & Benton (1997) found that the SCI is biased by tree size but not tree shape, while Benton & Storrs (1994) found that shape *was* important. The later, much larger empirical study by Benton *et al.* (1999) did not find the anticipated effects of either tree shape or tree size. There appears to be no consensus on the sensitivity of the SCI to either parameter (Pol *et al.* 2004).

The Manhattan Stratigraphic Measure (MSM) of Siddall (1998) is simply the consistency index

(ci) (Kluge & Farris 1969) of an ordered, stratigraphic character, while its modification (MSM*: Pol & Norell 2001) is equivalent to the ci of an irreversible and ordered stratigraphic character. It can be given by:

$$MSM^* = \frac{L_m}{L_o}$$

where L_m is the minimum length for the age character on *any* tree (i.e. the number of states, minus one) and L_o is the length obtained by optimizing the age character onto the *actual* phylogenetic tree. Siddall (1998) used randomization tests to demonstrate that raw MSM values were not biased by tree shape, but rather by tree size. Pol *et al.* (2004) carried out extensive testing of the MSM*, in addition to the GER and the SCI, to examine their sensitivity to tree shape, tree size and the number of possible ages of first appearance among terminal taxa. They employed the same randomization procedure used by Siddall (1998) but found that the indices seemed to be affected by all three parameters: the GER was found to be influenced by a marginally greater degree than the MSM*, but not as much as the SCI. Lelièvre *et al.* (2008) noted that because the MSM* is equivalent to the consistency index of an irreversible character, it must theoretically be subject to the same biases as the ci and CI, including the number of taxa in the tree (Farris 1989).

We also note that the GER can be formulated in a manner analogous to the retention index of the age character for comparison with the MSM* (Pol & Norell 2006):

$$GER = \frac{L_M - L_o}{L_M - L_m}$$

where L_M and L_m are the maximum and minimum lengths for the age character on *any* topology and L_o is the length obtained by optimizing the age character on the *actual* phylogenetic tree. As such, it is also subject to the same biases as the retention index of an irreversible age character (Finarelli & Clyde 2002).

Number of different origination dates

Pol *et al.* (2004) also demonstrated that the number of different first occurrence dates correlates positively with the median number of steps expected in a reversible stratigraphic character. When stratigraphic data were permuted across a given topology, increasing the number of coded stratigraphic intervals (while holding the total stratigraphic range of the tree constant) caused the randomized distribution of step lengths to shift to the right. A similar phenomenon can also be demonstrated for the GER, which is closer in concept to the use of an irreversible stratigraphic character. Figure 3c, d show the effects of increasing the number of different origination dates to sixteen (relative to the four used in panels 3a, b), but with no change in the range or interquartile range of those dates. For the balanced tree (Fig. 3c), the distribution of GER values moves significantly to the right (mode of 0.409 and median of 0.449) (medians significantly different: Kruskal–Wallis chi-squared = 3898.92, $P < 2.2e\text{-}16$). For the pectinate tree (Fig. 3d), the change is much less apparent (mode of 0.000 and median of 0.089), but there is still a highly significant increase in the median (Kruskal–Wallis chi-squared = 1184.31, $P < 2.2e\text{-}16$).

Range of origination dates

We are aware of no purely computational reason, why the number of stratigraphic intervals spanned by a tree *per se* (the number of divisions between the oldest and youngest origin) should influence the distribution of possible congruence metrics. In practice, however, the number of intervals spanned correlates positively with the size of the tree and the number of different intervals coded; parameters that *do* have an effect. Benton & Storrs (1994) and Hitchin & Benton (1997) observed an empirical relationship for the SRC: trees with longer temporal spans tended to have higher correlation than those in

which the range of origins was more constrained. If all other parameters are invariant, and first occurrences are dated within some constant margin of error, then a wider span of origins may ensure that first occurrences resolve in the correct order (Benton 1995; Mannion *et al.* 2010). This relates more directly to the mean gap or ghost range size, which we also consider in our model.

Distribution of origination dates

Irrespective of the range of origins and the number of coded intervals, the distribution of first occurrence dates can also be shown to influence the distribution of possible GER values. We considered two extreme cases: one 'bottom-heavy' (in which most first occurrences were close to the oldest origin) (Fig. 3e, f) and one 'top-heavy' (a mirror image of the first) (Fig. 3g, h). Instead of four groups of eight origin dates at 0, 5, 10, and 15 (as in 3a, b), we used 0, 13, 14 and 15 (bottom-heavy) and 0, 1, 2 and 15 (top-heavy). All four simulations show a harmonic pattern in the distribution of GER values. Differences are most visible in the balanced trees: the bottom-heavy balanced tree (Fig. 3e) has a much lower median (0.217) than the even-spaced balanced tree (0.378) (Fig. 3a), and the range of values observed is greater. By contrast, the top-heavy balanced tree (Fig. 3g) has a significantly higher median (0.467) than the even-spaced balanced tree. The pectinate trees follow a similar general pattern: the bottom heavy pectinate tree (Fig. 3f) has a lower median (0.008) than the even-spaced pectinate tree (0.022) (Fig. 3b), while the top-heavy pectinate tree (Fig. 3h) has the highest median of all (0.041).

Materials and methods

The data set

Our initial data set consisted of 1094 published cladograms. One thousand were taken from the analyses of Benton *et al.* (2000) and Wills (2007), and we supplemented these with 94 cladograms of the hitherto underrepresented molluscs and birds (supplementary publication, table 1 and references therein). Most of the results presented here are for a large subset of 650 of these cladograms: removing trees where the number of different first occurrence dates was fewer than five. This filtered out trivially small cladograms, as well as those where originations were concentrated in a very small number of intervals. Time slice analyses using the GI show patterns of 'gappiness' for 650 trees that differ little from those from all 1094. Indeed, the addition of 94 mollusc and bird trees appears to have had more affect on the results than the subsequent filtering. By removing very small trees, however, we obtained a distribution of overall GER values (and ultimately, residuals) much more amenable to linear modelling.

The resulting set of 650 trees encompassed a wide range of organisms, at various taxonomic levels, from both animal and plant phyla. The *Paleobiology database* (http://paleodb.org) served as the primary source of stratigraphic range data for genera and higher taxa. Additional sources of stratigraphic information were *The Fossil Record 2* (Benton 1993) and *Sepkoski's online genus database* (Sepkoski 2002). Data for lower taxonomic levels were taken directly from the literature where possible.

Stratigraphic ranges were coded to the nearest series and stage after Benton *et al.* (2000), the Phanerozoic being divided into 77 intervals with an average duration of 7.04 million years. Where regional or archaic stage names were used these were reconciled with international stratigraphic standards using the *International stratigraphic chart* (Remane & Ogg 2009), *The Geologic Timescale 2004* (Gradstein *et al.* 2004) and the *GeoWhen database* (http://www.stratigraphy.org/bak/geowhen/index.html). These data are summarized in the supplementary publication, table 2. Perl

scripts were used to automate much of the data setup and file formatting. GER values and all other primary statistics (e.g. tree balance, mean gap size, gap variance, range centre of gravity, etc.) were calculated using a modified version of Ghosts 2.3 (Wills 1999, 2007).

Independent variables

A summary of all twelve 'predictor' (potentially influential) variables used in this study is given in Table 1. Two principal aims of this paper are to determine whether there are differences in median GER values for trees of different higher taxa, and for trees from different geological periods. These comparisons are not straightforward because other parameters are not distributed homogeneously across these categories. For example, the finding that trees of arthropods have poorer GER values than those of tetrapods (Wills 2001) cannot be interpreted as a straightforward taxonomic effect if we also know that the arthropod trees in our sample are smaller than those of tetrapods on average, and also that arthropods trees tend to investigate the relationships between taxa of a higher rank than those of tetrapods. A number of such parameters are therefore included in our models if only so that we can discount their importance.

Table 1. *Predictor variables*

Predictor variable	Description	Type	Transform
Number taxa	The number of taxa in the cladogram	Continuous	log
Heard's index	Measure of the tree balance using Heard's Index (I_m): varies from 0.0 (un-balanced) to 1.0 (perfectly balanced)	Continuous	N/A
Mean origins	Mean age of originations of taxa in the cladogram	Continuous	N/A
Range origins	Range of originations of taxa in the cladogram	Continuous	log
Mean gap size	Mean gap size (ghost range) of taxa in the cladogram	Continuous	log
Gap standard deviation	Standard deviation of the gap size (ghost range)	Continuous	log
Range centre gravity	Centre of gravity of the observed stratigraphic ranges of taxa in the cladogram)	Continuous	N/A
% no fossils	Percentage of taxa in the cladogram with no fossil record	Continuous	N/A
% extend recent	Percentage of taxa in the cladogram that extend to the recent	Continuous	log
Taxon rank	Median taxonomic rank of the cladogram. 6 levels: species, genus, family, order, class, phylum and above	Categorical	N/A
Taxon group	Taxonomic group of the cladogram. 11 levels: plants, cnidaria, mollusca, arthropoda, echinodermata, bryozoa, graphtolites, fishes, tetrapoda, all life, brachiopoda	Categorical	N/A
Number strat intervals	Number of different stratigraphic intervals taxa in the cladogram are contained within continuous	Continuous	N/A

The number of taxa simply records the size of each tree; the number of terminals irrespective of whether these are species, higher taxa, or a mixture. The mean age of originations offers a proxy for the age of the tree, and has been removed from the model where we have investigated patterns in the GER through time.

The range of originations records the temporal span of that portion of the tree capable of subtending ghost ranges. It has been suggested that cladograms with a narrow range of originations will tend to be more stratigraphically incongruent than cladograms with a more widely spaced range of originations (Benton & Storrs 1994; Hitchin & Benton 1997; Benton et al. 1999; Wills 1999).

Taxonomic rank has been approximated by scoring each terminal in each tree in one of six categories (species $= 1$, genus $= 2$, family $= 3$, order $= 4$, class $= 5$, phylum or above $= 6$). The median of these values has been scored as the level of analysis for the entire tree. This variable has subsequently been treated as categorical and unordered in our models.

The mean gap size is the mean number of stratigraphic intervals traversed by each ghost range in the tree (mean ghost range length): its inclusion may require some explanation. It is true that *for a given tree*, longer ghost ranges imply poorer congruence and a lower GER. However, when looking across a sample of different trees, it is not the case that trees with longer ghost ranges must necessarily have lower GER values. Indeed, it is straightforward to simulate examples where different gap sizes yield *identical* distributions of GER values: precisely because the scaling in simulations is entirely arbitrary. Rather, the mean gap size attempts to offer a proxy for the actual time (number of stages) between taxonomic sampling events on the tree. Simply dividing the range of origination dates by the number of terminals (minus one) would be a weaker proxy for this (not least because of differences in tree balance), although we note that these two parameters are strongly correlated across our 1094 trees ($r = 0.751$, $P < 2.2$e-16). The gap standard deviation is simply the standard deviation of gap values calculated above, and will be zero if all gaps are of the same length. Thus, a high gap standard deviation indicates a wide disparity of taxonomic sampling frequency through time. Since we acknowledge that mean gap size can only be inferred as part of the process by which GER values are calculated (plotting range data onto a particular tree and inferring ghost ranges) we also present a subset of analyses omitting mean gap size and gap standard deviation.

The range centre of gravity is a proxy for the overall shape of the clade (based on observed ranges) through time. It is calculated using the formula of Gould et al. (1987):

$$CG = \frac{\sum_{i=1}^{n} N_i t_i}{\sum_{i=1}^{n} N_i}$$

where N_i is the number of observed ranges in the ith interval, and t_i is the age of the ith interval. We then express this relative to the range of intervals spanned by the entire clade, scaling between 0.0 and 1.0. Hence, clades with a CG (centre of gravity) of 0.5 are balanced at the midpoint between their origin and extinction. Those with values greater than 0.5 are top heavy, while those below 0.5 are bottom heavy. Clade shape is itself a proxy for the mode of radiation: bottom heavy clades are those with more rapid initial diversification. We note that extant clades will have ranges truncated by the present for some fraction of their constituent taxa. This will tend to raise the centre of gravity artificially.

Taxa with no fossil record may have legitimately originated in the Recent, although this becomes increasingly unlikely as their taxonomic rank increases. Often, however, complete absence from the record reflects other factors such as the size of individuals, population size, habitat, and the volatility of tissues. Such taxa could potentially subtend ghost ranges deep into trees, thereby lowering GER values. By including the percentage of taxa with no fossil record as a variable, we will test if this effect is detectable. The percentage of taxa that extend to the Recent is also included because the effects of this parameter were investigated by Wills (2007).

Exploratory, non-linear models (CART and GAM)

All analyses and tests were conducted in R version 2.10.1 (R Development Core Team 2009). Several of our independent variables had markedly non-normal distributions: these were initially transformed as indicated in Table 1, which has greatly improved the distribution of GER residuals in the resulting models. As outlined by Crawley (2007), classification and regression trees (CART) (Breiman et al., 1984) (created using the packages 'tree' and 'rpart') were used to highlight 'important' variables (i.e. those that have the greatest influence on the response) and complex variable interactions, while generalized additive models (GAM) (Hastie & Tibshirani 1986) (using 'mgcv') were used to identify non-linearities.

A classification and regression tree (CART) is a binary decision tree, commonly used in data mining to create a model that will predict the value of a dependent (or response) variable based on a number of independent (or predictor) variables. It is constructed by splitting a node into two daughter nodes repeatedly, beginning at the root. At the first node a threshold level is selected for one of the predictor variables and the mean value of the response is calculated above and below this threshold value, both of which are then used to calculate the deviance. This is repeated for all possible values of the threshold for this predictor. The 'best split' for the predictor variable is the threshold value that has the highest deviance, that is, explains the greatest amount of deviance in the response (Crawley 2007). This is calculated for each of the predictors and the one with the highest overall deviance is chosen as the split for that node. The procedure is repeated at each node. The tree is initially overfit and then pruned using cross-validation, starting from the terminal nodes and moving up the tree (Breiman *et al.* 1984). In the final topology, the values at the terminals represent the mean value of the response variable, given the values of the predictors represented by the path from the root of the tree to the terminal node. We also implemented a random forests approach using the package 'randomForest'. This is a powerful statistical tool for quantifying the importance of variables, and which has increased predictive power compared with individual regression trees (Breiman 2001; Strobl *et al.* 2007). A random forest of 500 bootstrapped trees was generated and the predictions for each tree used in a 'voting' process to establish the relative importance of each predictor variable on the response (GER). Two different measures of importance were used: the mean decrease in accuracy (%IncMean) and the mean decrease Gini (IncNodeImpurity). Both of these measures were calculated separately for each tree in the forest and then averaged. There is no consensus on which measure is preferable.

Generalized additive models (GAMs) are an extension of generalized linear models (GLMs), but implement additive rather than linear regression (Wood 2006). The linear function of a predictor is replaced with smoothed functions, which can reveal non-linearities in the predictors (O'Brien & Rago 1996). One of the main justifications for using non-parametric methods is that they model complex relationships between variables without strong model assumptions (Wood 2006). Unlike other non-parametric models, their output is relatively straightforward to interpret. Results from these analyses were used to test whether the use of linear regression was appropriate, while the CART output provided a graphical representation of the interactions between predictors. Due to the large number of variables, this enabled us to reduce the interactions in the initial linear model to include only the variables that were shown to interact in the CART output (Crawley 2007).

Linear models

The data were modelled six times, always with the GER as the dependent variable:

(1)   All twelve of the independent variables in Table 1. This model was used to determine the effects of all variables on the GER.

(2)   All of the independent variables in Table 1, *except* the mean gap size and the gap standard deviation. These two variables were removed experimentally for the reasons discussed above.

(3)   As 1, but omitting the categorical variable for taxonomic group. This model was used to investigate GER residuals in different taxonomic groups.

(4)   As 2, but also omitting the categorical variable for taxonomic group. This model was used to investigate GER residuals in different taxonomic groups, but without modelling out the effects of mean gap size and gap standard deviation.

(5)   As 1, but omitting the mean age of origin. This variable was removed so that we could investigate GER residuals for the other variables in each geological period.

(6)   As 2, but also omitting the mean age of origin. This model was also used to investigate GER residuals in each geological period, but without modelling out the effects of mean gap size and gap standard deviation.

For each analysis, we used reverse stepwise regression, starting with the most complex model (i.e. including all of the variables specified above) and removing non-significant terms one by one, highest *P*-value first. Within this constraint, interaction terms were removed first, followed by main effects, unless the non-significant main effect was part of a significant interaction. Factor (categorical) terms were not removed if at least one level was significant. This process was repeated until all remaining terms were significant. By comparison, models were also selected automatically using the 'step' function in R. This uses the standard Akaike information criterion (AIC) as its optimality criterion (Akaike 1974). Both of these mechanisms yielded adequate models, omitting some of the independent variables. Model diagnostics, including the residuals against the fitted values and the normal quantile–quantile (Q–Q) plots, were examined to ensure the resulting fit was acceptably linear.

Trees were assigned to one of the twelve geological periods according to their mean age of origin. Kruskal–Wallis and subsequent post-hoc tests were carried out to determine whether any of the differences observed through time or by taxonomic group were significantly different from any others.

Results

Regression trees and linear models

The regression trees reveal many complex inter-actions between parameters, with several terms appearing more than once. In the analysis including mean gap size and gap standard deviation (Fig. 4), mean gap size itself emerges as the most significant variable. Partitions of trees with greater mean ghost range lengths consistently have lower mean GER values. The range of origins of taxa in the cladogram is also highly influential, along with the range centre of gravity. Other important variables, to a lesser degree, are Heard's index of tree balance and the mean origination date.

In the regression tree omitting mean gap size and gap standard deviation (Fig. 5), taxonomic group becomes by far the most important variable. This is because different taxonomic groups typically contain trees with different mean gap sizes, taxonomic group becoming a proxy for gap size when it is not explicitly present in the model. Other, much less important variables are the range of origins, the percentage of taxa extending to the

Recent, the percentage of taxa with no fossil record, and the mean origination date.

Random forest analyses including all variables (Fig. 6a, b) yields results similar to those for the simple regression tree. Overall, mean gap size, range of origins and gap standard deviation are found to be the most important predictors, with the range centre of gravity and the mean of origins important to a lesser degree. Random forest analyses for all variables except the mean gap size and the standard deviation of gaps (Fig. 6c, d), show that taxonomic group, the range centre of gravity, the mean of origination dates and the range of origination dates become the most important predictors.

With all variables included, GAM plots (supplementary publication, fig. 1) of each continuous linear predictor against component smooth functions of the fitted GAM object show little evidence of curvature, with narrow 95% confidence intervals. This meant that we were able to use linear models for the rest of our analyses. The GER model diagnostic plots (supplementary publication, fig. 2) of the residuals v. the fitted values are scattered above and below the zero-line indicating no heterosce-dasticity (problematic trends in the data caused by variables with markedly different variances: often resulting in a funnel-shaped residual plot). Moreover, the normal Q–Q plots show a linear

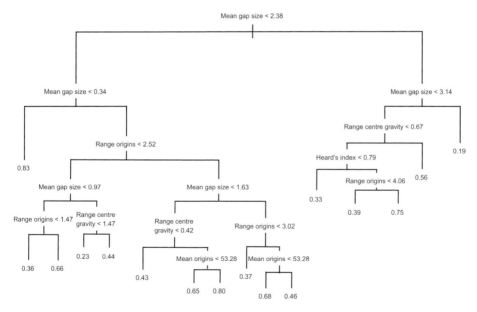

Fig. 4. Regression tree for the GER. The split value is shown beside the predictor at each node. The longer the branch, the greater the deviance explained. If the value of the predictor is less than the split value at the node, then the next step is taken down the left branch of the tree. If it is greater, the next step is along the right branch. For example, if the mean gap size is <2.38 then refer to the left side of the tree. At the next node if the mean gap size is <0.34, then the mean GER value is 0.83: hence for a tree with very small gaps the mean GER value is quite high.

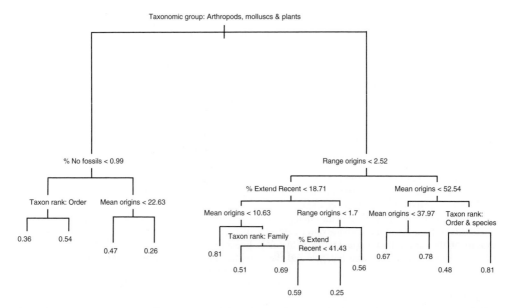

Fig. 5. Regression tree for the GER excluding mean gap size and gap standard deviation (see Fig. 4 for explanation).

pattern, as the matched quantiles generally lie along a straight line (with a modest number of outliers).

The final linear model including all twelve variables (Model 1) contains many complex and significant interaction terms (Table 2). The adjusted r^2 is 0.7445, meaning that 74.5% of the total variance in the GER is accounted for by variation in the independent variables. The mean gap size, the gap standard deviation, the range of origins, and the range centre of gravity are highly significant, and the mean of origins is significant: exactly as would be predicted from the regression tree and random forest analyses. However, the number of taxa and Heard's index are also highly significant: variables not highlighted above. Results from the automated linear modelling are given in the supplementary publication, table 3. These models were broadly similar to those obtained manually, albeit containing more terms, which is often the case (Crawley 2007).

Differences in GER between major taxa

The variation in median GER values across different taxonomic groups is illustrated in Figure 7a. Arthropods have the lowest median GER (see also Wills 2001), followed by plants and molluscs Kruskal–Wallis and post-hoc tests revealed that there are significant differences (Kruskal–Wallis chi-squared $= 86.916$, $P < 2.2E-16$) between arthropods and all other taxonomic groups except plants (Mann–Whitney tests with Holm's sequential Bonferroni corrections, $P < 3.029E-03$; Tukey HSD $P < 0.043$), between plants and both echinoderms

(Mann–Whitney tests with Holm's sequential Bonferroni corrections, $P < 1.456E-03$; Tukey HSD, $P = 0.005$) and tetrapods (Mann–Whitney tests with Holm's sequential Bonferroni corrections, $P < 1.006E-04$; Tukey HSD, $P = 0.0004$) and between molluscs and both echinoderms (Mann–Whitney tests with Holm's sequential Bonferroni corrections, $P < 4.273E-03$; Tukey HSD, $P = 0.017$) and tetrapods (Mann–Whitney tests with Holm's sequential Bonferroni corrections, $P < 1.092E-04$; Tukey HSD, $P = 0.001$). Figure 7b shows the variation in residual GER values by taxonomic group, with all other variables modelled out (Model 3). Differences are much less marked, and significant (Kruskal–Wallis chi-squared $= 16.404$, P-value $= 0.006$) only between arthropods and fishes (Mann–Whitney tests with Holm's sequential Bonferroni corrections, $P = 0.004$; Tukey HSD, $P = 0.029$) and between arthropods and tetrapods (Mann–Whitney tests with Holm's sequential Bonferroni corrections, $P = 0.002$; Tukey HSD, $P = 0.005$). Finally, with mean gap and gap standard deviation removed (Model 4), much of the residual variance in GER values is restored (Fig. 7c).

Differences in GER between geological periods

Figure 8a illustrates the variation in GER over the eleven periods of the Phanerozoic, with trees dated according to the mean date of origin of their constituent taxa. The convex pattern (with greatest

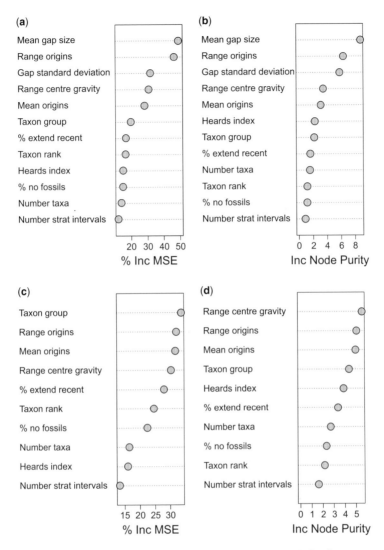

Fig. 6. Random forest analyses of the importance of variables in predicting the GER. Predictors are sorted in decreasing order of importance from top to bottom. All plots are derived from 500 bootstrap replicates. (**a, b**) Analyses including all 12 variables (i.e. model 1). (**c, d**) Analyses including only 10 variables (omitting mean gap size and gap standard deviation) (i.e. model 2) (a/c) Mean decrease in accuracy (% Inc MSE) and (b/d) Mean decrease Gini (Inc Node Purity: based on the Gini gain criterion algorithm employed by the 'randomForest' package in R).

congruence in the late Palaeozoic and Mesozoic closely resembles that observed by Wills (2007) for the GI, and investigated in more detail in Figure 1. Figure 8b, c show that there is little variation in the model residuals through time for the GER.

Discussion

Factors influencing the GER

Analyses using regression trees, random forests and linear modelling all found that the gap size and gap standard deviation have a marked and significant effect on the GER. Trees with shorter ghost ranges tend to have better GER values than those with longer ones. At first sight, this may seem hardly surprising: the GER is supposed to record something about the extent of ghost ranges, after all. However, the GER is scaled relative to a theoretical maximum and minimum extent of ghost ranges and so is a ratio, whereas the average gap length is measured in units of stratigraphic stages. This means that trees with identical topologies and identical *relative* distributions of origination dates can

Table 2. *Linear model for the GER*

Coefficients	Estimate	Std. Error	t value	Pr (>|t|)	
(Intercept)	−1.08	0.31	−3.49	0.000527	***
Number taxa	−0.19	0.02	−12.27	<2.00E-16	***
Heards index	0.89	0.24	3.70	0.000232	***
Mean origins	0.02	0.01	3.11	0.001937	**
Range origins	0.90	0.13	6.70	4.78E-11	***
Taxon rank genus	−0.04	0.03	−1.43	0.154262	
Taxon rank family	−0.04	0.03	−1.28	0.202237	
Taxon rank order	−0.06	0.03	−2.10	0.036496	*
Taxon rank class	−0.04	0.04	−1.04	0.298979	
Taxon rank phylum	−0.15	0.05	−2.98	0.003014	**
Taxon group Echinodermata	0.06	0.03	2.28	0.022804	*
Taxon group Fishes	0.09	0.02	3.98	0.000078	***
Taxon group Mollusca	0.09	0.03	3.34	0.000883	***
Taxon group Plants	0.10	0.03	2.89	0.004055	**
Taxon group Tetrapoda	0.11	0.02	4.90	0.000001	***
Range centre gravity	2.57	0.48	5.36	1.2E-07	***
Mean gap size	−0.84	0.09	−9.21	<2.00E-16	***
Gap standard deviation	−0.13	0.02	−6.62	8.00E-11	***
Ranger centre gravity: Mean gap size	0.41	0.15	2.79	0.005473	**
Heards index: Range centre gravity	−2.50	0.51	−4.93	0.000001	***
Range origins: Mean gap size	0.10	0.02	4.75	0.000003	***
Range origins: Range centre gravity	−0.79	0.19	−4.12	0.000043	***
Mean origins: Range centre gravity	−0.03	0.01	−2.72	0.006716	**
Mean origins: Range origins	−0.01	0.00	−2.74	0.006239	**
Heards index: Range origins: Mean gap size	−0.08	0.02	−3.91	0.000102	***
Heards index: Range origins: Range centre gravity	0.55	0.13	4.19	0.000033	***
Mean origins: Range origins: Mean gap size	0.00	0.00	−2.67	0.007890	**
Mean origins: Range origins: Range centre gravity	0.01	0.00	3.24	0.001242	**
Heards index: Mean origins: Range origins	−0.01	0.00	−3.76	0.000189	***
Heards index: Mean origins: Range centre gravity: Mean gap size	0.01	0.00	3.36	0.000822	***
Heards index: Mean origins: Range origins: Mean gap size	0.01	0.00	6.12	0.000000	***
Heards index: Mean origins: Range origins: Range centre gravity	0.02	0.01	2.94	0.003406	**
Heards index: Mean origins: Range origins: Range centre gravity: Mean gap size	−0.01	0.00	−5.90	0.000000	***

Significant codes: 0 '***' 0.001 '**' 0.01 '*' 0.05 '.' 0.1 ' ' 1
Residual standard error: 0.1245 on 616 degrees of freedom.
Multiple R-squared: 0.7571, Adjusted R-squared: 0.7445.
F-statistic: 60 on 32 and 616 DF, P-value: >2.2E-16.
GER = Number taxa + Heards index + Mean origins + Range origins + Taxon rank + Taxon group + Range centre gravity + Mean gap size + Gap standard deviation + Range centre gravity: Mean gap size + Heards index: Range centre gravity + Range origins: Mean gap size + Range origins: Range centre gravity + Mean origins: Range centre gravity + Mean origins: Range origins + Heards index: Range origins: Mean gap size + Heards index: Range origins: Range centre gravity + Mean origins: Range origins: Mean gap size + Mean origins: Range origins: Range centre gravity + Heards index: Mean origins: Range origins + Heards index: Mean origins: Range centre gravity: Mean gap size + Heards index: Mean origins: Range origins: Mean gap size + Heards index: Mean origins: Range origins: Range centre gravity + Heards index: Mean origins: Range origins: Range centre gravity: Mean gap size.
Top panel shows each of the terms in the model along with their statistics. Bottom panel shows the model equation and the overall statistics for the model indicating that 74.5% of the total variance in the GER is accounted for by variation in the independent variables.

have very *different* mean ghost range lengths, yet ultimately identical GER values. In other words, there is no theoretical reason why mean ghost range length should correlate with the GER when looking across a sample of *different* trees (as we are doing here). This correlation is only expected for permutations of range data across a *given* tree. Our empirical result is informative, therefore: it tells us that as the actual (not relative) durations of gaps between sister taxa increase in real trees, so the sums of those ghost ranges tend towards their theoretical minimum (a change in their relative extent).

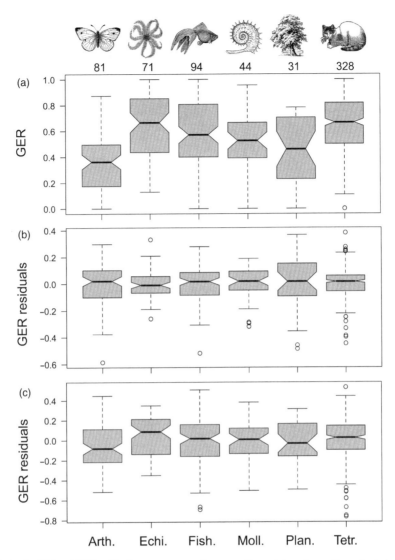

Fig. 7. Variation in GER and GER residuals across taxonomic groups. For each box, the median value is indicated by a black horizontal bar, the shaded area represents upper and lower quartiles and the dashed lines connect to the most eccentric points within 1.5 interquartile ranges of the median. Outliers are shown as circles. Numbers below icons at the top indicate the number of trees in each taxonomic group. (**a**) Raw GER values. (**b**) Residual GER values for the model initially incorporating all 11 independent variables (except taxonomic group). (**c**) Residual GER values for the model initially incorporating 9 independent variables (except taxonomic group, mean gap size and gap standard deviation). Abbreviations: Arth., arthropods; Echi., echinoderms; Fish., fishes; Moll., molluscs; Plan., plants; Tetr., tetrapods. Clipart courtesy of http://etc.usf.edu

While this finding is not wholly unexpected, it is not a necessary one. We also found that ghost ranges of a more uniform length tend to yield higher GERs than highly variable ones. Hence, the highest GER values are found when the gaps between sister taxa are relatively small and of a similar size.

When rerunning the linear models, but omitting mean gap size and gap standard deviation from the

outset, the amount of variance explained declined from 74.5 to 26.6%, highlighting the predictive value of these two variables. In their absence, taxonomic group becomes by far the most important predictor: arthropods, molluscs and plants have lower GER values than other groups on average. For these latter taxa, the range of origins and the mean age of origins subsequently become important

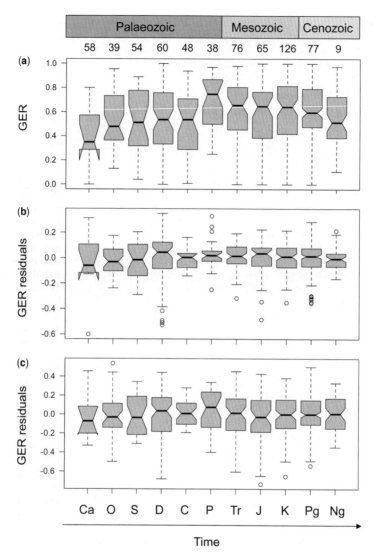

Fig. 8. Variation in GER and GER residuals through time. Trees are dated according to the mean age of origin of their constituent taxa. For each box the median value is indicated by a black horizontal bar, the shaded area represents upper and lower quartiles and the dashed lines connect to the most eccentric points within 1.5 interquartile ranges of the median. Outliers are shown as circles. Numbers above periods indicate the number of trees in each geological time period. (**a**) Raw GER values. (**b**) Residual GER values for the model initially incorporating all 11 independent variables (except mean of origins). (**c**) Residual GER values for the model initially incorporating 9 independent variables (except mean of origins mean gap size and gap standard deviation). Abbreviations: Ca, Cambrian; O, Ordovician; S, Silurian; D, Devonian; C, Carboniferous; P, Permian; Tr, Triassic; J, Jurassic; K, Cretaceous; Pg, Palaeogene; Ng, Neogene.

predictors of the GER, while the percentage of extant terminals is important for all groups. Many authors have observed differences in stratigraphic congruence between higher taxa (Benton & Storrs 1994; Benton & Simms 1995; Hitchin & Benton 1997; Wills 1999, 2001; Wagner & Sidor 2000; Pol *et al.* 2004; Lelièvre *et al.* 2008; Wills *et al.* 2008, 2009). However, taxonomic group may also be a

proxy for differences in the gap size (and other variables) highlighted above, so that when gap size is omitted from the model as an explicit variable, its effects emerge through the taxonomic group.

Our simple simulations and previous studies all suggested that tree balance should be an important predictor of the GER. It emerges as a moderately important variable in random forest analyses based on node purity, but is only decisive at a low level in regression trees. However, tree balance is significant in our linear models, both on its own and in interaction with other parameters: mean gap size, the mean age of origins, the range of origins and the range centre of gravity. The bivariate relationship between tree balance and the range centre of gravity is a significantly negative one: balanced trees have a slightly lower range CG on average.

The total stratigraphic range of origins emerges as the second most important variable in the random forest analysis based on all variables. This is much higher than its rank in the simple regression tree: a difference explicable by the greater stability of the former. Random forests grow a large number of bootstrapped trees (typically 500 or more). Rather than using a single regression tree, the values for variable importance are aggregated by averaging across all trees. Random forests have been shown to provide better models for prediction than simple regression trees, which can be very sensitive to minor perturbations in the data (Prasad et al. 2006). The range of origins is also highly significant in our final linear models, appearing in several complex interactions, as well as making its own unique contribution. The relationship to the GER is slightly positive when all other variables and interactions are factored out: cladograms spanning a wider range of origins have higher GER values on average (see Benton et al. (1999) for a similar observation in their bivariate analyses). The *bivariate* relationship for our sample of 650 cladograms is negative, however ($r = -0.248$, $P = 1.611E-10$). One reason for the apparent discrepancy is the very strong positive correlation between the range of origins and both gap size ($r = 0.801$, $P < 2.2-10$) and gap standard deviation ($r = 0.933$, $P < 2.2E-16$) in our sample: hence the total span of the tree becomes a proxy for these and other variables.

The range centre of gravity emerges as the fourth most important variable in the random forest analyses of all twelve predictors, and first or fourth in the analyses omitting mean gap size and gap standard deviation. In the linear model of all variables and the model omitting mean gap size and gap standard deviation, it is highly significant. As predicted by our simulations, the overall relationship is a positive one: as the CG of clades increases (moving from more bottom-heavy shapes to more top-heavy ones) the mean GER increases also. Moreover, bottom-heaviness may be indicative of a rapid radiation, which may itself result in a narrow range of origins (known to correlate with a poorer GER). However, we note that our CG calculations were performed on all clades, irrespective of whether they are truncated by the Recent or not.

Extant clades will inevitably be 'flat topped': this will raise the CG of those with an approximately symmetrical diversity profile through time, and have less predictable effects on others.

Differences in GER between major taxa

As in previous studies, there are significant differences in median GER scores between major taxonomic groups. Figure 7a shows the characteristically poor values for arthropods (Wills 2001) (median 0.36) and plants (median 0.46), and much higher values for tetrapods (0.67) and echinoderms (0.67) (Benton et al. 1999). Many of these differences are significant, most notably between arthropods and all other taxa, except plants.

Most arthropod taxa are known to have a poor fossil record. The exoskeleton is rarely heavily mineralized (the heavily calcified trilobites and decapods are the exceptions) and their preservation potential is therefore low. Many arthropod groups are also very small (Wills 2001; Wills et al. 2009). In addition, Wills (2007) reported that levels of homoplasy are particularly high in arthropods, and certainly much higher than in vertebrates. Although homoplasy and other measures of data quality do not necessarily relate directly to the accuracy of cladograms (which is unknowable in all of our examples), some correlation is probable. If the cladograms for some taxonomic groups are less accurate than others, then stratigraphic congruence is also likely to be lower.

Many plant groups (particularly those from the Silurian and Devonian) appear to have been subject to rapid tissue degradation and therefore also have a relatively poor fossil record (Niklas 1988; Friedman & Cook 2000). Some mollusc groups, such as the bivalves, appear to have an excellent fossil record (Jablonski et al. 2003) while other groups are relatively sparse (Reid et al. 1996; Brayard et al. 2010). The low GER values obtained for our sample of trees are nonetheless somewhat surprising.

Although major taxa have a characteristic signature in terms of their stratigraphic congruence, it also appears that much of the difference is explicable in terms of the distributions of other variables. Figure 7b shows the residual GER scores by taxonomic group from the model (3) initially including all predictors except taxonomic group: most of the variation between groups is removed and the only significant differences are between fishes and arthropods and between tetrapods and arthropods. A similar model, but removing mean gap size and gap standard deviation (4), reveals a pattern of GER residuals more closely resembling that in Figure 7a. Hence, much of the variation in GER values between major taxa is attributable to variation in gap size and standard deviation. Molluscs, arthropods and plants have much larger median gap sizes and gap standard deviation than echinoderms, fishes and tetrapods, with all of these contrasts being significant. Finally, plants, molluscs and arthropods also have a much greater median range of origins than echinoderms, fishes and tetrapods, although this variable has a slightly positive effect on the GER once other variables are controlled.

These results highlight the fact that the while the GER is a ratio, and does not index the *absolute* extent of ghost ranges (still less the completeness of the fossil record), it is nonetheless influenced by measures that *are* related to absolute time in our empirical sample. For example, trees with longer ghost ranges (numbers of intervals) tend to have a poorer GER, but not because this is a necessary consequence of the way the index is calculated. Rather, in our empirical sample of real trees, longer mean ghost ranges (in absolute terms) predict poorer stratigraphic congruence. The inclusion of variables measured in absolute time reflected a desire to factor out heterogeneity across major taxa in our empirical sample. However, this is a highly conservative approach. Large gaps in absolute terms may imply a patchy fossil record and/or sparse sampling of taxa through time and across the tree: factors that may legitimately depress congruence and the GER. Hence, while Figure 7b shows there is relatively little difference in the *residual* GER values between taxa, *absolute* differences are still marked (it is simply that many of these differences can be explained by variation in parameters such as the range of origination dates, tree balance, etc.).

Differences in GER between geological periods

The overall pattern of GER values binned into periods (according to the mean date of origin of each tree) is similar to that obtained for the gap index (GI) (Fig. 1): higher in the late Palaeozoic and Mesozoic than the early Palaeozoic and Cenozoic. We note that the precise pattern obtained for the GI is sensitive to the inclusion of our additional mollusc and bird datasets, as well as the filtering out of very small trees. Variation in the GER through time is at least partially a response to changes in the proportions of higher taxa through time (Wills 2007). For example, arthropods, molluscs and echinoderms, radiated during the Cambrian, while tetrapods diversified during the Carboniferous (Benton & King 1989; Briggs & Fortey 1989). Since we know that different taxa have significantly different GER values, we would expect the values in each period to track this turnover in diversity. The

irregular increase in median GER from the Cambrian to the Carboniferous (Fig. 8) may partly reflect declining proportions of arthropod, mollusc and plant trees, and increasing numbers of fishes and tetrapods. The decline in congruence from the Triassic to the Recent is more difficult to explain, because tetrapods (whose congruence is generally very good) constitute an increasing fraction of the sample (Fig. 9). Wills (2007) experimented with the removal of extant taxa and those with no fossil record: the latter often subtend extensive ghost ranges deep into trees. These 'pull of the Recent' type effects (Jablonski *et al.* 2003) appear to make relatively little contribution to the late Mesozoic and Cenozoic decline, although removing them (Fig. 1b, d, e) increases mean GER values systemically over the entire Phanerozoic. From simulations, Wills (2007) concluded that the decline to the Recent was more probably a result of the manner in which the GI is calculated: tending to slice across the bottom of trees nearer the Cambrian, and preferentially through their tops towards the Recent. Our binning approach calculates congruence values for whole trees, and so is not subject to this effect. If data for the Palaeozoic, Mesozoic and Cenozoic are pooled, there is a significant difference in median GER value between the Mesozoic and both the Palaeozoic and Cenozoic (Mann–Whitney tests with Holm's sequential Bonferroni corrections, $P = 0.004$ and 0.010 respectively). The number of comparisons entailed, however, mean that the Kruskal–Wallis does not detect differences between individual periods.

Our models of residual GER by period, omitting just mean date of origin (model 5) or this in addition to mean gap size and gap standard deviation (model 6) show much less variation through time. With all significant conflating variables factored out (model 5, Fig. 8b) the highest residuals occur in the Silurian and Permian, with the lowest values in the Cambrian and Palaeogene. Omitting the mean gap size and the gap standard deviation from the outset (model 6, Fig. 8c) increases the spread of values in each period, but restores little of the pattern seen for the raw GER values.

Conclusions

(1)   *The GER cannot be used straightforwardly to compare congruence for different trees.* The GER is not a measure of fossil record completeness, but rather a ratio of the total extent of ghost ranges relative to their theoretical maximum and minimum on any tree of the same size. Simulations demonstrate that the distribution of possible GER values is influenced by a number of variables, including

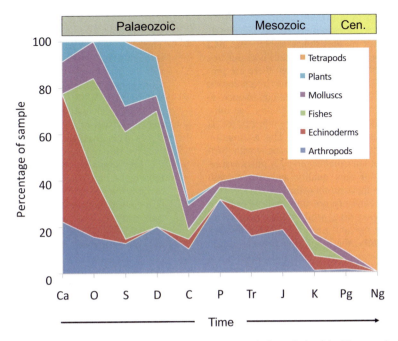

Fig. 9. Taxonomic composition of our sample of 650 cladograms through the periods of the Phanerozoic. Trees are assigned to periods according to the mean date of origin of their constituent taxa. A large fraction of our sample is tetrapods, and their proportional contribution increases in successive eras.

the number of terminals and the distribution of origination dates. As documented elsewhere, this means it is not straightforward to compare GER values derived from different trees and different data sets. This problem is precisely analogous to the effects of character matrix dimensions on measures of cladogram quality such as the ensemble consistency index (CI) and ensemble retention index (RI) (Archie 1989; Naylor & Kraus 1995). However, because the distribution of possible GER values is also influenced by tree balance, the use of the GER as an ancillary criterion for selecting among otherwise equally optimal trees is also problematic (Wills 1999). These difficulties do not uniquely afflict the GER, but are reported to various degrees for other congruence metrics such as the SCI (Benton & Storrs 1995; Siddall 1996; Wagner & Sidor 2000; Wills *et al.* 2008) and MSM* (Pol *et al.* 2004; Wills *et al.* 2008).

(2) *The GER is influenced by many factors besides congruence per se.* Values of twelve variables reasoned or observed to influence the GER were calculated for our sample of 650 animal and plant cladograms. Their empirical relationship with the GER was investigated using a variety of non-parametric (regression trees, random forests, and generalized additive

models) and parametric (linear models) approaches. There are many significant interactions between these variables, which, taken together, account for approximately three quarters of the variation in the GER. Mean ghost range size, the standard deviation of ghost range sizes and the range of origination dates (temporal range of the tree) emerge as the most important variables, followed by the overall shape of the clade (whether bottom or top heavy). Models omitting ghost range size and standard deviation explained only 26.6% of the variance in the GER.

(3) *The GER is not a straightforward measure of the quality of the fossil record.* Empirical studies of congruence indices for large samples of cladograms have been used to address two questions: is congruence different across major taxonomic or environmental subdivisions (Benton 1995; Benton & Simms 1995; Benton & Hitchin 1996, 1997; Benton *et al.* 1999), and does congruence change through geological time (Benton *et al.* 2000; Wills 2007)? In most of these studies, congruence is assumed to be a proxy for the quality of the fossil record. This is an oversimplification for three reasons. Firstly, metrics such as the SCI, GER and MSM* variously measure the agreement between two sequences: all can

return high or maximal values when fossil occurrences are extremely sparse, so long as their orders of occurrence agree. The RCI (Benton & Storrs 1994) (Fig. 2) is closer to a measure of record quality, but there are problems with its scaling and interpretation (Wills 1999). Secondly, for congruence to provide a proxy for record quality, we must assume that cladograms are correct, or at least of uniform accuracy. The assumption of a 'cladistic yardstick' is fraught with difficulties, not least that accuracy is unknowable. Measures of data quality and tree support (insofar as these might offer a proxy for accuracy) are certainly not uniformly distributed across higher taxa, while higher taxa themselves are not uniformly distributed through time. Thirdly, the other sources of bias discussed above are not uniformly distributed, either across higher taxa or through time.

(4) *The GER varies very significantly between higher taxa, but some of this variation can be explained by other independent variables.* Previous studies have reported significant differences in GER values across higher taxa, with arthropods and tetrapods being characteristically poor (Wills 2001; Wills *et al.* 2009), and vertebrates very good (Benton & Hitchin 1996; Wills *et al.* 2008). Our study replicates these findings, additionally reporting poor values for molluscs. However, when our linear models were used to remove the effects of eleven of the variables above (omitting taxonomic group), very few residual group differences remained significant (arthropods compared with fishes, and arthropods compared with tetrapods). This is because several important sources of bias are not distributed uniformly across higher taxa. For example, arthropod cladograms tend to have a longer absolute duration and longer ghost ranges than those of vertebrates: both factors associated empirically with lower GER values. There are few differences in *congruence* between higher taxa that cannot be explained by variation in other parameters, therefore. However, this does not imply that ghost ranges are equally extensive in different taxa, still less that the fossil record is of uniform quality across taxa.

(5) *Although the GER varies considerably between periods (with highest values in the Carboniferous, Triassic and Jurassic), some of this variation can be explained by other independent variables.* The results of empirical studies of congruence through time depend largely upon the methods used. Benton *et al.* (2001) binned cladograms into

eras, but at such a coarse temporal scale, were unable to detect any significant trend. They concluded that congruence is uniform through time, which was also taken to imply that the fossil record does not deteriorate (at a predominantly familial scale) with increasing antiquity. Wills (2007) used an alternative approach, calculating the congruence (gap index) in each of 77 stratigraphic series and stages. This revealed a convex pattern of congruence through the Phanerozoic: highest in the Mesozoic, but declining into the early Palaeozoic and paradoxically also declining towards the Recent. We note that the addition of a modest number of trees is able to perturb this pattern. The analysis presented here returned to a binning approach, and replicated the convex pattern observed by Wills (2007) (albeit using a rather different method). However, when the effects of eleven of our twelve independent variables were factored out (omitting a proxy for the age of the tree), the residual GER values were much less variable.

We are very grateful to A. Smith and A. J. McGowan for their invitation to contribute to the Lyell Meeting 2010, and to this special volume. We are also grateful to M. Benton for making the data from Benton *et al.* (2000) available. Thanks to R. Benson and A. Smith for thorough and constructive reviews that enabled us to significantly improve the clarity of our manuscript. Thanks also to S. Wood, J. Faraway and V. Olson for their help in implementing GAMs. Our research was funded by the Leverhulme Trust (Grant F/00 351/Z).

References

AKAIKE, H. 1974. A new look at the statistical model identification. *IEEE Transactions on Automatic Control*, **19**, 716–723.

ARCHIE, J. W. 1989. Homoplasy excess ratios: new indices for measuring levels of homoplasy in phylogenetic systematics and a critique of the consistency index. *Systematic Zoology*, **38**, 253–269.

ARCHIE, J. W. & FELSENSTEIN, J. 1993. The number of evolutionary steps on random and minimum length trees for random evolutionary data. *Theoretical Population Biology*, **43**, 52–79.

BENTON, M. J. 1993. *The Fossil Record 2.* Chapman & Hall, London, UK.

BENTON, M. J. 1994. Palaeontological data and identifying mass extinctions. *Trends in Ecology and Evolution*, **9**, 181–185.

BENTON, M. J. 1995. Testing the time axis of phylogenies. *Philosophical Transactions: Biological Sciences*, **349**, 5–10.

BENTON, M. J. & HITCHIN, R. 1996. Testing the quality of the fossil record by groups and by major habitats. *Historical Biology*, **12**, 111–157.

BENTON, M. J. & HITCHIN, R. 1997. Congruence between phylogenetic and stratigraphic data on the history of life. *Proceedings of the Royal Society, B*, **264**, 885–890.

BENTON, M. J. & KING, P. W. 1989. Mass extinctions among tetrapods and the quality of the fossil record [and discussion]. *Philosophical Transactions of the Royal Society of London, Series B*, **325**, 369–386.

BENTON, M. J. & SIMMS, M. J. 1995. Testing the marine and continental fossil records. *Geology*, **23**, 601–604.

BENTON, M. J. & STORRS, G. W. 1994. Testing the quality of the fossil record: paleontological knowledge is improving. *Geology*, **22**, 111–114.

BENTON, M. J., HITCHIN, R. & WILLS, M. A. 1999. Assessing congruence between cladistic and stratigraphic data. *Systematic Biology*, **48**, 581–596.

BENTON, M. J., WILLS, M. A. & HITCHIN, R. 2000. Quality of the fossil record through time. *Nature*, **403**, 534–537.

BRAYARD, A., NUTZEL, A., STEPHEN, D. A., BYLUND, K. G., JENKS, J. & BUCHER, H. 2010. Gastropod evidence against Early Triassic Lilliput effect. *Geology*, **38**, 147–150.

BREIMAN, L. 2001. Random forests. *Machine Learning*, **45**, 5–32.

BREIMAN, L., FRIEDMAN, R. A., OLSHEN, R. A. & STONE, C. G. 1984. *Classification and Regression Trees*. Wadsworth, Pacific Grove, USA.

BRIGGS, D. E. G. & FORTEY, R. A. 1989. The early radiation and relationships of the major arthropod groups. *Science*, **246**, 241–243.

BRUSATTE, S. L., BENTON, M. J., DESOIO, J. B. & LANGER, M. C. 2010. The higher-level phylogeny of Archosauria (Tetrapoda: Diapsida). *Journal of Systematic Palaeontology*, **8**, 3–47.

BRUSATTE, S. L. & SERRENO, P. C. 2008. Phylogeny of Allosauroidea (Dinosauria: Theropoda): comparative analysis and resolution. *Journal of Systematic Palaeontology*, **6**, 155–182.

BRUSATTE, S. L., BENTON, M. J., RUTA, M. & LLOYD, G. T. 2008. Superiority, competition, and opportunism in the evolutionary radiation of dinosaurs. *Science*, **321**, 1485–1488.

CAMPIONE, N. E. & REISZ, R. R. 2010. *Varanops brevirostris* (Eupelycosauria: Varanopidae) from the Lower Permian of Texas, with discussion of varanopid morphology and interrelationships. *Journal of Vertebrate Paleontology*, **30**, 724–746.

COBBETT, A., WILKINSON, M. & WILLS, M. A. 2007. Fossils impact as hard as living taxa in parsimony analyses of morphology. *Systematic Biology*, **56**, 753–766.

CRANE, P. R., HERENDEEN, P. & FRIIS, E. M. 2004. Fossils and plant phylogeny. *American Journal of Botany*, **91**, 1683–1699.

CRAWLEY, M. J. 2007. *The R book*. Wiley, Chichester, UK.

DONOGHUE, P. C. J. & PURNELL, M. A. 2009. Distinguishing heat from light in debate over controversial fossils. *BioEssays*, **31**, 178–189.

DONOGHUE, M. J., DOYLE, J. A., GAUTHIER, J., KLUGE, A. J. & ROWE, T. 1989. The importance of fossils in phylogeny reconstruction. *Annual Review of Ecology and Systematics*, **20**, 431–460.

DYKE, G. J., MCGOWAN, A. J., NUDDS, R. L. & SMITH, D. 2009. The shape of pterosaur: evolution evidence from the fossil record. *Journal of Evolutionary Biology*, **22**, 890–898.

FARRIS, J. S. 1989. The retention index and the rescaled consistency index. *Cladistics*, **5**, 417–419.

FINARELLI, J. A. & CLYDE, W. C. 2002. Comparing the gap excess ratio and the retention index of the stratigraphic character. *Systematic Biology*, **51**, 166–176.

FISHER, D. C. 2008. Stratocladistics: integrating temporal data and character data in phylogenetic inference. *Annual Review of Ecology, Evolution and Systematics*, **39**, 365–385.

FRIEDMAN, M. 2010. Explosive morphological diversification of spiny-finned teleost fishes in the aftermath of the end-Cretaceous extinction. *Proceedings of the Royal Society, B*, **277**, 1675–1683.

FRIEDMAN, W. E. & COOK, M. E. 2000. The origin and early evolution of tracheids in vascular plants: integration of palaeobotanical and neobotanical data. *Proceedings of the Royal Society, B*, **355**, 857–868.

FROBISCH, N. B. & SCHOCH, R. R. 2009. Testing the impact of miniaturization on phylogeny: Paleozoic dissorophoid amphibians. *Systematic Biology*, **58**, 312–327.

GAUTHIER, J., KLUGE, A. & ROWE, T. 1988. Amniote phylogeny and the importance of fossils. *Cladistics*, **4**, 105–209.

GOULD, S. J., GILINSKY, N. L. & GERMAN, R. Z. 1987. Asymmetry of lineages and the direction of evolutionary time. *Science*, **236**, 1437–1441.

GRADSTEIN, F. M., OGG, J. G. & SMITH, A. G. (eds) 2004. *A Geologic Timescale 2004*. Cambridge University Press, Cambridge, UK.

GRANTHAM, T. 2004. The role of fossils in phylogeny reconstruction: Why is it so difficult to integrate paleobiological and neontological evolutionary biology? *Biology and Philosophy*, **19**, 687–720.

HASTIE, T. & TIBSHIRANI, R. 1986. Generalised additive models. *Statistical Science*, **1**, 297–310.

HEARD, S. B. 1992. Patterns in tree balance among cladistic, phenetic, and randomly generated phylogenetic trees. *Evolution*, **46**, 1818–1826.

HITCHIN, R. & BENTON, M. J. 1997. Stratigraphic indices and tree balance. *Systematic Biology*, **46**, 563–569.

HUELSENBECK, J. P. 1994. Comparing the stratigraphic record to estimates of phylogeny. *Paleobiology*, **20**, 470–483.

JABLONSKI, D., ROY, K., VALENTINE, J. W., PRICE, R. M. & ANDERSON, P. S. 2003. The impact of the pull of the Recent on the history of marine diversity. *Science*, **300**, 1133–1135.

KLUGE, A. & FARRIS, J. S. 1969. Quantitative phyletics and the evolution of anurans. *Systematic Zoology*, **18**, 1–32.

KROH, A. & SMITH, A. B. 2010. The phylogeny and classification of post-Palaeozoic echinoids. *Journal of Systematic Palaeontology*, **8**, 147–212.

LAMSDELL, J. C., BRADDY, S. J. & TETLIE, O. E. 2010. The systematics and phylogeny of the Stylonurina (Arthropoda: Chelicerata: Eurypterida). *Journal of Systematic Palaeontology*, **8**, 49–61.

LELIÈVRE, H., ZARAGÜETA BAGILS, R. & ROUGET, I. 2008. Temporal information, fossil record and phylogeny. *Comptes Rendus Palevol*, **7**, 27–36.

MANNION, P. D., UPCHURCH, P., CARRANO, M. T. & BARRETT, P. M. 2010. Testing the effect of the rock record on diversity: a multidisciplinary approach to elucidating the generic richness of sauropodomorph dinosaurs through time. *Biological Reviews*, **86**, 157–181.

NAYLOR, G. & KRAUS, F. 1995. The relationship between *s* and *m* and the retention index. *Systematic Biology*, **44**, 559–562.

NIKLAS, K. J. 1988. Patterns of vascular plant diversification in the fossil record: proof and conjecture. *Annals of the Missouri Botanical Garden*, **75**, 35–54.

NORELL, M. A. & NOVACEK, M. J. 1992. Congruence between superpositional and phylogenetic patterns: comparing cladistic patterns with fossil records. *Cladistics*, **8**, 319–337.

O'BRIEN, L. & RAGO, P. 1996. An application of the Generalized Additive Model to groundfish survey data with Atlantic cod off the northeast coast of the United States as an example. *NAFO Scientific Council Studies*, **28**, 79–95.

PEARSON, P. N. 1999. Apomorphy distribution is an important aspect of cladogram symmetry. *Systematic Biology*, **48**, 399–406.

POL, D. & NORELL, M. A. 2001. Comments on the Manhattan stratigraphic measure. *Cladistics*, **17**, 285–289.

POL, D. & NORELL, M. A. 2006. Uncertainty in the age of fossils and the stratigraphic fit to phylogenies. *Systematic Biology*, **55**, 512–521.

POL, D., NORELL, M. A. & SIDDALL, M. E. 2004. Measures of stratigraphic fit to phylogeny and their sensitivity to tree size, tree shape, and scale. *Cladistics*, **20**, 64–75.

PRASAD, A. M., IVERSON, L. R. & LIAW, A. 2006. Newer classification and regression tree techniques: bagging and random forests for ecological prediction. *Ecosystems*, **9**, 181–199.

R DEVELOPMENT CORE TEAM. 2009. *R: A language and Environment for Statistical Computing*. R Foundation for Statistical Computing, Vienna, Austria. http://www.R-project.org.

REID, D. G., RUMBAK, E. & THOMAS, R. H. 1996. DNA, morphology and fossils: phylogeny and evolutionary rates of the gastropod genus *Littorina*. *Philosophical Transactions of the Royal Society B*, **351**, 877–895.

REMANE, J. & OGG, J. (eds) 2009. *International Stratigraphic Chart*. International Union of Geological Sciences, International Commission on Stratigraphy. Paris, France. http://stratigraphy.org/upload/ISChart2009.pdf

RUTA, M., WAGNER, P. J. & COATES, M. I. 2006. Evolutionary patterns in early tetrapods. I. Rapid initial diversification followed by decrease in rates of character change. *Proceedings of the Royal Society, B*, **273**, 2107–2111.

SANDERSON, M. J. & DONOGHUE, M. J. 1989. Patterns of variation in levels of homoplasy. *Evolution*, **43**, 1781–1795.

SEPKOSKI, J. J. 2002. A compendium of fossil marine animal genera. *Bulletins of American Paleontology*, **363**, 1–560. http://strata.geology.wisc.edu/jack/

SIDDALL, M. E. 1996. Stratigraphic consistency and the shape of things. *Systematic Biology*, **45**, 111–115.

SIDDALL, M. E. 1998. Stratigraphic fit to phylogenies: a proposed solution. *Cladistics*, **14**, 201–208.

SMITH, A. B. & MCGOWAN, A. J. 2007. The shape of the Phanerozoic marine palaeodiversity curve: how much can be predicted from the sedimentary rock record of western Europe? *Palaeontology*, **50**, 765–774.

SMITH, A. B., PISANI, D., MACKENZIE-DODDS, J. A., STOCKLEY, B., WEBSTER, B. L. & LITTLEWOOD, D. T. J. 2006. Testing the molecular clock: molecular and paleontological estimates of divergence times in the Echinoidea (Echinodermata). *Molecular Biology and Evolution*, **23**, 1832–1851.

SPRINGER, M. S., TEELING, E. C., MADSEN, O., STANHOPE, M. J. & DE JONG, W. W. 2001. Integrated fossil and molecular data reconstruct bat echolocation. *Proceedings of the National Academy of Sciences*, **98**, 6241–6246.

STROBL, C., BOULESTEIX, A.-L., ZEILEIS, A. & HOTHORN, T. 2007. Bias in random forest variable importance measures: Illustrations, sources and a solution. *BMC Bioinformatics*, **8**, 25.

TETLIE, O. E. & POSCHMANN, M. 2008. Phylogeny and palaeoecology of the Adelophthalmoidea (Arthropoda: Chelicerata: Eurypterida). *Journal of Systematic Palaeontology*, **6**, 237–249.

TSYGANOV-BODOUNOV, A., HAYWARD, P. J., PORTER, J. S. & SKIBINSKI, D. O. F. 2009. Bayesian phylogenetics of Bryozoa. *Molecular Phylogenetics and Evolution*, **52**, 904–910.

WAGNER, P. J. 1995. Stratigraphic tests of cladistic hypotheses. *Paleobiology*, **21**, 153–178.

WAGNER, P. J. 1999. The utility of fossil data in phylogenetic analyses. *American Malacological Bulletin*, **15**, 1–31.

WAGNER, P. J. & SIDOR, C. A. 2000. Age rank/clade rank metrics – sampling, taxonomy, and the meaning of 'stratigraphic consistency'. *Systematic Biology*, **49**, 463–479.

WILLS, M. A. 1999. Congruence between phylogeny and stratigraphy: randomization tests and the gap excess ratio. *Systematic Biology*, **48**, 559–580.

WILLS, M. A. 2001. How good is the fossil record of arthropods? An assessment using the stratigraphic congruence of cladograms. *Geological Journal*, **36**, 187–210.

WILLS, M. A. 2002. The tree of life and the rock of ages: Are we getting better at estimating phylogeny? *BioEssays*, **24**, 203–207.

WILLS, M. A. 2007. Fossil ghost ranges are most common in some of the oldest and some of the youngest strata. *Proceedings of the Royal Society, B*, **274**, 2421–2427.

WILLS, M. A. & FORTEY, R. A. 2000. The shape of life: how much is written in stone? *BioEssays*, **22**, 1142–1152.

WILLS, M. A., BARRETT, P. M. & HEATHCOTE, J. F. 2008. The modified gap excess ratio (GER*) and the stratigraphic congruence of dinosaur phylogenies. *Systematic Biology*, **57**, 891–904.

WILLS, M. A., JENNER, R. A. & NI DHUBHGHAILL, C. 2009. Eumalacostracan evolution: conflict between three sources of data. *Arthropod Systematics and Phylogeny*, **67**, 71–90.

WOOD, S. N. 2006. *Generalised additive models: An introduction with R*. Chapman & Hall/CRC, Boca Raton, USA.

Impact of outcrop area on estimates of Phanerozoic terrestrial biodiversity trends

P. D. WALL[1,2]*, L. C. IVANY[1] & B. H. WILKINSON[1]

[1]*Department of Earth Sciences, Syracuse University, Syracuse, NY 13244, USA*

[2]*ExxonMobil, Houston, Texas, USA*

*Corresponding author (e-mail: pdwall@syr.edu)

Abstract: Studies of biodiversity through time have primarily focused on the marine realm which is generally considered to have a more complete record than terrestrial environments. Recently, this assumption has been challenged, and it has been argued that the record of life on land is comparable or even more robust than that of the shallow oceans. Moreover, it has been claimed that terrestrial successions preserve an exponential rise in diversity, even when corrected for sampling biases such as changes in continental rock volume through time. We evaluate relations between terrestrial diversity and exposed areas of terrestrial sediments using our compiled data on areas of global continental outcrops and generic diversity from the Paleobiology Database. Terrestrial global generic diversity and terrestrial outcrop area are highly correlated following a linear relation. No significant correlation is observed between habitat area and either outcrop area or biodiversity, suggesting that the observed relation between diversity and outcrop area is not driven by a common cause, such as eustasy. We do find evidence for a small residual increase in diversity through time after removing the effect of outcrop area, but caution that this may be driven by an increased proportion of terrestrial fauna with high preservation potential.

Palaeontologists have traditionally assumed that the fossil record of marine organisms, especially that of skeletonized invertebrates, is superior to the terrestrial record (e.g. Padian & Clemens 1985; Vermeij & Leighton 2003). This assumption, along with the invertebrate palaeontology backgrounds of important early researchers in the field, has focused studies of biodiversity through time on the marine fossil record (e.g. Valentine 1969, 1970; Raup 1972, 1976a, b; Sepkoski 1976, 1981, 1982; Bambach 1977). A major exception to this trend was the familial compendium of tetrapod biodiversity compiled by Benton (1993), which remains a standard in the field. Following an initial period of debate, the apparent large-scale increases in diversity through time observed in both the marine and terrestrial realms were taken at face value for much of the two decades following initial publication. A resurgence of interest in sampling biases in the record in the late 1990s led to a fresh flowering of work on the subject, again focused primarily on marine biodiversity (e.g. Miller & Foote 1996; Miller 1997; Alroy et al. 2001; Peters & Foote 2001; Crampton et al. 2003, 2006a, b). Recent terrestrial studies have largely focused on small taxonomic groups, local regions, or select time periods (e.g. Alroy [1999] for Cenozoic North American mammals; Barrett et al. 2009 for dinosaurs, Butler et al. 2009 for pterosaurs, 2010 for dinosaurs; Fara 2004 for lissamphibians,

Mannion et al. 2011 for sauropods, but see Kalmar & Currie 2010 for a more global outlook). Here we attempt to expand upon these studies by assessing how variations in the amount of exposed terrestrial rock affect estimates of global terrestrial biodiversity through the Phanerozoic.

The quantity of rock available for sampling has long been postulated as one of the primary biases of estimates of fossil biodiversity. While attempts to militate its influence date back to the earliest days of palaeontology with the work of Phillips in 1860, modern interest can largely be traced to Raup's (1972, 1976a, b) concerns about initial attempts at tabulating global trends in marine biodiversity through time (e.g. Valentine 1969). Raup cautioned that a great deal of the observed increase in diversity over the course of the Phanerozoic could be attributed to the concurrent increase in exposed marine sediments. The publication of the so-called consensus, or 'kiss and make up', paper (Sepkoski et al. 1981), which argued for a real if somewhat more limited increase in biodiversity through time, assuaged concerns about biases and attention instead turned to biological explanations for the observed trend.

In the late 1990s a new wave of research appeared (e.g. Miller & Foote 1996; Alroy 1999; Alroy et al. 2001) demonstrating that variations in sampling in fact do have the potential to profoundly alter our view of biodiversity through

From: McGowan, A. J. & Smith, A. B. (eds) *Comparing the Geological and Fossil Records: Implications for Biodiversity Studies*. Geological Society, London, Special Publications, **358**, 53–62.
DOI: 10.1144/SP358.5   0305-8719/11/$15.00 © The Geological Society of London 2011.

time. Arguments have continued, however, as to the importance of these biases and how to mitigate them. The Paleobiology Database (PaleoDB) project has emphasized sample standardization methodologies (Alroy *et al.* 2001, 2008). Other authors (e.g. Peters & Foote 2001, 2002; Smith 2001; Smith *et al.* 2001; Crampton *et al.* 2003, 2006b; Smith & McGowan 2007; Wall *et al.* 2009) have focused more specifically on measures of outcrop area and, like David Raup thirty years before, demonstrated a strong correlation between outcrop area and biodiversity. These recent papers have also attempted to disentangle the issue of 'common cause' (Peters & Foote 2001) – do biodiversity and outcrop area co-vary because outcrop area is a biasing factor, or do they co-vary because both factors are driven by a third factor, such as eustasy? Peters (2005) demonstrated that observed patterns of origination and extinction are consistent with changes in sea level, which in turn drive changes in both biodiversity and outcrop area. Wall *et al.* (2009) also provide evidence that fluctuations in habitable area affect diversification. They note, however, that at the scale of global biodiversity over the course of the Phanerozoic, variance in sampling potential driven by the subsequent burial and erosion of sediments overwhelms the biological signal.

Work on terrestrial diversity has also noted sampling biases. Wang & Dodson (2006) determined that the perceived decline in dinosaur diversity during the end of the Cretaceous is likely a sampling artefact, and Lloyd *et al.* (2008) perceived the apparent diversification of dinosaurs during the mid-Cretaceous to be largely driven by sampling bias. Barrett *et al.* (2009) also found that dinosaur diversity was affected by sampling artefacts, although they thought the end-Cretaceous diversity decline to be robust. The results of both Mannion *et al.* (2011) and Butler *et al.* (2011) are similar for sauropods and all non-avian dinosaurs, respectively. The last two studies also assessed evidence for eustatic controls on biodiversity observed in the marine realm but found equivocal support. At a global scale, Kalmar & Currie (2010), using Alexander Ronov's estimates of volumes of sedimentary rocks (Ronov *et al.* 1980) and Benton's (1993) estimates of familial diversity through time, found that the sedimentary record, although correlated, did not explain the exponential rise in terrestrial diversity through time. In this study, we also compare the terrestrial rock and diversity records, but rely on two different estimates – generic occurrence data from the Paleobiology Database and outcrop data derived from the UNESCO world geological maps (Choubert & Faure-Muret 1976). The results of this study suggest that, contrary to findings by Kalmar &

Currie (2010), outcrop area acts as a large-scale bias, largely obscuring the terrestrial diversity signal. At a global scale there is little evidence for common cause as observed in the marine record, and evidence for a long-term increase in terrestrial biodiversity through time is weak.

Data

Diversity data

The compilation by Benton (1993) has long been, deservingly, the gold standard for terrestrial biodiversity studies. Much like the Sepkoski compendia (Sepkoski 1982, 2002), it consists of extensively vetted first and last occurrences, in this case at a familial level. For methodological reasons, however, we have chosen to rely on the generic sampled occurrences compiled by the PaleoDB. While the PaleoDB lacks the single authoritative voice and extensive standardization that accompanies the Benton compendium, its structure allows us to more accurately estimate the number of taxa recovered from the sediments of any time interval. The methods of counting fossil occurrences from a time interval have proven to be a surprisingly durable and contentious subject (e.g. Alroy *et al.* 2001, 2008). Different methods are robust to different statistical and sampling biases and are thus useful under different circumstances. First and last occurrence data necessitate the use of range-through methodologies. One advantage of these methodologies is that they capture taxa not actually sampled within a time period. While this can mitigate random fluctuations in sampling, the counting methodologies are still affected by biases. Sampled-in-bin metrics reflect the actual data from which the first-and-last occurrence metrics are derived and thus give us a better picture of the density of sampling within a range and therefore how biases are affecting the data.

For this study we focused on both plants and animals found in non-marine environments. Occurrence data were downloaded from the PaleoDB on May 30, 2010. Data were restricted to terrestrial environments. However, for plants and tetrapods, where environmental preferences are well known, shallow-water marine deposits that incorporate allochthonous terrestrial material were included. This is a conservative approach, as the inclusion of terrestrial organisms from marine deposits should serve to weaken any correlation between terrestrial outcrop area and richness while at the same time strengthens any correlation between terrestrial diversity and habitat area if one exists. In these cases the occurrence data were further parsed to exclude marine taxa (e.g. Cetacea, Mosasauroidea, etc.). For plants, palynotaxa were not included. The final list consists of over 47 000 generic occurrences of 3710 genera, for approximately 4500 unique time-interval/genus combinations. The resulting diversity curve is largely consistent with that of Benton (1993), showing low terrestrial diversity in the Palaeozoic and steep increase in diversity towards the Recent. The greater degree of variability in the generic curve in comparison to Benton's familial curve is expected due to the finer taxonomic resolution, and can also be attributed to the smoothing effect of the range-through assumption implicit in using first and last occurrences to tabulate diversity rather than counting taxa sampled within bins. The resulting curve gives a good approximation of the content of the terrestrial fossil record as it exists today. Whether or how this curve is influenced by sampling biases of some sort cannot be determined without standardizing for intensity and/or geographical scope of sampling. Below, we compile data on outcrop areas of terrestrial rocks through time to assess the effect of changes in the area available for sampling on sampled diversity.

Outcrop area

There is an on-going debate as to the validity of outcrop area as an estimate of the amount of rock available for fossil collection, or rock exposure area. Recent work (Dunhill 2011; Benton *et al.* 2011) has demonstrated a poor to non-existent relation between the two variables, at least at the scale of outcrop areas of several square kilometres. It is unclear, however, how well this extrapolates to the scale of global sedimentary outcrops. In areas with little vegetation and overburden outcrop area and rock exposure area should exhibit a very high correlation. It is also likely that at larger spatial scales (i.e. 100s or 1000s of km^2) and coarser temporal scales (10^6 ma) correlations should improve. As this study was conducted at these larger scales we feel that the study is still valid. Additionally, the spatial distribution of taxa and samples also affect our estimates of biodiversity (Bush *et al.* 2004; Wall *et al.* 2009), and while outcrop maps may not perfectly reflect the areal extent of exposure, they certainly capture information on its spatial distribution.

The outcrop area data used here are described in detail in Wall *et al.* (2009). Total sedimentary outcrop areas were calculated from the Geologic Atlas of the World (Choubert & Faure-Muret 1976), which was produced by the United Nations Education, Scientific and Cultural Organization (UNESCO) at a scale of 1:10 000 000. As the atlas lacks detailed lithological information, the Ronov dataset, also utilized by Kalmar & Currie (2010), was employed. These data consist of estimates by

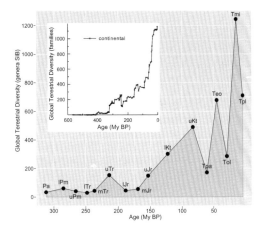

Fig. 1. Global terrestrial diversity (genera sampled in bin; SIB) through time (Ma BP). Note the general increase in diversity through time. Inset figure shows data from Benton (1993) showing long-term increase in diversity through time. Inset modified from Kalmar & Currie (2010). Abbreviations in this and subsequent figures are: Pa, Pennsylvanian; lPm, Lower Permian; uPm, Upper Permian; lTr, Lower Triassic; mTr, Middle Triassic; uTr, Upper Triassic; lJr, Lower Jurassic; mJr, Middle Jurassic; uJr, Upper Jurassic; lKt, Lower Cretaceous; uKt, Upper Cretaceous; Tpa, Palaeocene; Teo, Eocene; Tol, Oligocene; Tmi, Miocene; Tpl, Pliocene.

both Ronov (1964, 1978, 1980, 1994) and co-workers (Ronov & Khain 1954, 1955, 1956, 1961, 1962; Ronov *et al.* 1974a, b, 1976; Khain *et al.* 1975, Khain & Seslavinskiy 1977; Khain & Balukhovskiy 1979) of total volumes and areas of rock both exposed and in the subsurface. We parse the dataset into terrestrial and marine lithologies and assume that the relative volumetric proportion of terrestrial and marine lithologies in the subsurface is the same as the relative areal proportion on the surface. This assumption allows us to combine the two datasets to arrive at estimates of the areal extents of terrestrial outcrop by time interval (Fig. 2). Terrestrial outcrop is relatively constant through the Palaeozoic and early Mesozoic. Cretaceous outcrop is abundant, but this is partially driven by the longer durations of the upper and lower Cretaceous relative to other time intervals. Outcrop area decreases in the Palaeocene and gradually increases over the remainder of the Cenozoic, peaking in the Miocene.

Habitat area

The correlation between outcrop area and diversity is generally accepted for shallow-water marine environments, but the discussion is on-going as to

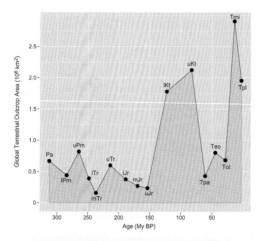

Fig. 2. Global terrestrial outcrop area through time. Note that the Miocene, Pliocene, and Cretaceous make up the majority of all outcrop area. Abbreviations as in Figure 1.

whether this is due to a common driving mechanism or a bias. The simplest hypothesis for a common-cause mechanism in the marine realm postulates that, as the seas flood the continental shelf, the enlarged area allows for a diversification of marine life as organisms move into new regions and also allows for a greater area available for the deposition of marine sediments. Hence, outcrop area and diversity are correlated, but the correlation reflects more than a simple bias but instead reflects underlying Earth-systems processes.

Potential mechanisms behind common cause in terrestrial environments are less clear. It is possible that the relationship is simply the inverse of that postulated for the marine realm – receding oceans open up new land area to habitation and the deposition of terrestrial sediments. As terrestrial organisms expand into previously subaqueous portions of the shelf they diversify, following the oft-noted diversity-area relationship (e.g. Rosenzweig 1995). At the same time, the expanded land area increases the potential area for sediments to be deposited in terrestrial depositional environments such as fluvial plains. These larger expanses of terrestrial sediments should be captured in the present-day rock record as larger areas of outcrop, and so in this scenario the correlation between outcrop area and diversity is real and a bias is not required to explain the observations. This hypothesis predicts a positive correlation between all three factors (outcrop area, habitat area, and biodiversity) and it is the hypothesis we test in this work.

One assumption of the previous logic is that the mean habitat quality (e.g. the proportion of desert vs rainforest) of the two time intervals is roughly

equivalent, or at least does not vary in a systematic way (i.e. times of eustatic fall are not systematically drier). This may not be a valid assumption – Pleistocene glacials were characterized by drier climates (e.g. Van der Hammen 1974; Porter 2001) and it appears that this is also true for the Pennsylvanian (Rankey 1997), another time of high amplitude and frequency glacially controlled eustatic change. During non-glacially driven eustatic changes this rule appears less robust – Antarctica appears to have experienced increased rainfall during the Palaeocene relative to the Late Cretaceous (Robert & Kennett 1992). If there is a systematic degradation of terrestrial habitat quality during marine regressions it would subdue any common-cause connections between habitat area and sampled diversity. This is to say that the common-cause mechanism (diversity increasing because of increasing habitat area) would still be a valid biological driver, but it is counteracted by the larger habitat area being of poorer quality resulting in a less-than-expected increase in diversity. Although the mechanics postulated by the common-cause hypothesis would be valid in this scenario, the net result would be no common-cause linkage between outcrop and diversity and any correlation would likely remain the effect of sampling bias.

There are alternative hypotheses connecting eustasy, biodiversity, and habitat area. For example expanding oceans fragment subaerial continental exposure, allowing increased diversification. Under the assumption that increased subaerial exposure of the continents leads to increased terrestrial deposition, a negative correlation between diversity and outcrop area is expected. This is inconsistent with the noted positive correlation between generic richness and outcrop area. The disagreement may be due to our simplified assumptions about terrestrial deposition or it may be due to sampling biases obscuring the true relation. This manuscript primarily focuses on whether the observed increase in terrestrial diversity is real or simply a sampling artefact and less on true eustatic effects on diversity. Because of this, we focus on the first hypothesis in which falling sea-levels spur a rise in both terrestrial diversity and outcrop. For this study we rely on the palaeogeographical maps of Scotese & Golonka (1992) for first-order estimates of habitat area. Digitization of these maps, as described in Walker *et al.* (2002), yields estimates of lowland and mountain terrestrial habitat area through time (Fig. 3).

Results

We first address the quantity and temporal distribution of terrestrial sediments. Terrestrial rocks from the time interval studied – the Pennsylvanian through Pliocene – total just over 14.6 million km^2,

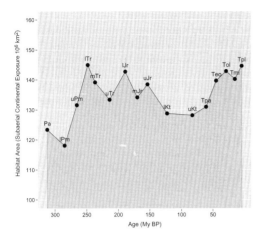

Fig. 3. Available habitat area (sub-aerial continental exposure, 10^6 km^2) through time. Exposed land area exhibits two peaks in the Early Mesozoic and Tertiary separated by a low in the Cretaceous. Note the y-axis begins at 100×10^6 km^2. Abbreviations as in Figure 1.

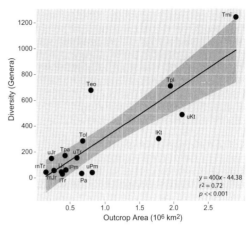

Fig. 4. Outcrop area (ordinate) v. diversity (abscissa). Outcrop area and diversity show a strong positive correlation. Black line represents correlation; grey area indicates 95% confidence interval for correlation. Abbreviations as in Figure 1.

approximately 10% of the Earth's land area and approximately the same size as Antarctica (13.7 million km^2). This sum excludes the early Palaeozoic, of which there is very little terrestrial rock preserved, as well as the extensive Quaternary deposits draping most of the continents. This total area is distributed unevenly between the periods and epochs. The Miocene, at approximately 2.9 million km^2, represents almost 20% of all exposed rocks, an area roughly equivalent to that of India. The Early and Late Cretaceous, 1.8 and 2.1 million km^2 respectively, make up an additional 27%. Thus almost half of all terrestrial rocks come from these three epochs. The middle Triassic is the most poorly represented epoch, having only 156 000 km^2, an area barely larger than Bangladesh and comprising only 1% of all exposed terrestrial rocks.

The proportion of any time interval's original continental area now exposed on Earth's surface as area of outcrop is similarly variable. The proportion of continental area represented as exposed outcrops follows a similar trend to that of outcrop area because the extent of sub-aerial continental landmass is relatively unchanged through time when compared to outcrop area. Exposed rock areas represent about 0.46% of the original continental land mass on average (median). For comparison with the modern areas of continents, this degree of preservation is roughly equivalent to modern continental area being represented by a sample the same size as Egypt. Miocene outcrops represent approximately 2.1% original areal extent. At the other extreme, the Middle Triassic is represented by only 0.1% of its original area. With these statistics

in mind, we turn to the observed relations between terrestrial outcrop area and diversity.

Outcrop area (Oa, 10^6 km^2) and diversity (Dv, genera) are strongly correlated (Fig. 4, Table 1). A linear relation has the best fit (Dv = 400 Oa -44.38, $r^2 = 0.72$, $P \ll 0.001$) followed by the power-law relation (Dv = 9.05×10^{-5} Oa$^{1.06}$, $r^2 = 0.53$, $P = 0.001$). The correlation is robust. Non-parametric correlation ($rho = 0.66$, $P = 0.007$, Table 1) is significant, as is a correlation of the first differences ($r^2 = 0.65$, $P < 0.001$, Table 1). First differences remove net trends in time series – outcrop area and sampled diversity change in step – increasing and decreasing in tandem. Neither outcrop area nor sampled terrestrial diversity are significantly correlated with interval duration ($r^2 = 0.08$, $P = 0.23$ and $r^2 \ll 0.01$, $P = 0.93$, respectively), indicating that variable interval durations are not driving the observed correlation between outcrop area and terrestrial diversity. Relations between diversity and habitat area (Fig. 5) and between outcrop area and habitat area (Fig. 6) are much weaker and insignificant in all models and tests at $\alpha = 0.05$ (Table 1).

Discussion

Quality of the terrestrial record

In terms of preservation of original areal extent, the continental record is far worse than that of the shallow marine realm. Based on the values discussed in Wall *et al.* (2009), the median percentage of the marine shelf preserved and exposed on the

Table 1. *Correlations between terrestrial biodiversity, outcrop area, and habitat area*

Model	Corr. type	r^2/rho	Adjusted r^2	P
Diversity–Outcrop	Partial linear	0.72	0.70	≪0.001
	Partial power	0.53	0.50	0.001
	Spearman	0.66	n/a	0.007
	1st differences	0.65	0.62	<0.001
Diversity–Habitat	Partial linear	0.09	0.02	0.26
	Spearman	0.11	n/a	0.67
	1st differences	0.04	−0.04	0.50
Outcrop–Habitat	Parametric	0.001	−0.07	0.89
	Spearman	−0.09	n/a	0.75
	1st differences	0.10	0.03	0.26
Diversity–Outcrop + Habitat	Parametric	0.79	0.76	≪0.001
			Outcrop area	≪0.001
			Habitat area	0.06
	1st differences		n/a	
			Outcrop area	<0.001
			Habitat area	0.70

surface today is barely less than two percent – a low value to be sure, but over four times higher than that for the terrestrial median. For all time intervals prior to the Miocene, shelfal marine sediments make up a larger proportion of total outcrop than do terrestrial sediments, despite the fact that flooded continental shelf area was far less than that exposed subaerially. The propensity for marine sediments to be preserved and exposed has resulted in a far more globally complete picture of biodiversity and sedimentation in the oceans than on dry land, which is unsurprising given the propensity for subaerial environments to experience erosion rather than deposition.

There are other estimates of the completeness of the fossil record that indicate only minor differences between marine and terrestrial records. Terrestrial and marine cladograms exhibit similar degrees of stratigraphic congruence (e.g. Benton *et al.* 1999; Benton 2001). While this observation is interesting, it is important to differentiate between stratigraphic and spatial completeness. Several authors (Sadler 1981; Schindel 1982) have suggested stratigraphic completeness is similar for both marine and continental rocks. If the descendant taxa generally inhabit the same overall region as their parents, then the observation of similar stratigraphic congruence is expected. The relative completeness of the record at a single point is a different question than the relative completeness of the record at a global scale. The marine and terrestrial records may have

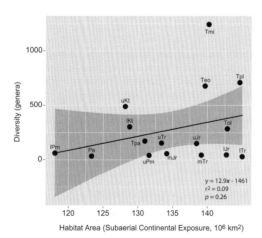

Fig. 5. Habitat Area (ordinate) v. diversity (abscissa). Terrestrial biodiversity does not, at first glance, exhibit a strong relation with original habitat area. Abbreviations as in Figure 1.

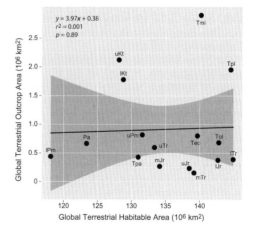

Fig. 6. Terrestrial Habitat Area (sub-aerial continental exposure, 10^6 km^2, ordinate), v. terrestrial outcrop area (10^6 km^2 abscissa). There is no evidence that habitable area influences modern outcrop areas. Abbreviations as in Figure 1.

Table 2. *Comparison of modern and sampled Pleistocene generic diversities for major tetrapod groups*

Taxonomic group	Modern		Pleistocene	
	Genera	(%)	Genera	(%)
Mammals	1231	23.3	626	72.5
Birds	2064	39.1	147	17.0
Squamates	1880	35.7	57	6.6
Turtles	90	1.7	28	3.2
Crocodilians	8	0.2	5	0.6

Number of genera and relative proportions of major tetrapod groups in the Recent and the Pleistocene. See text for sources.

similar degrees of completeness temporally but quite different degrees of completeness spatially. In simple terms of sample size and relative completeness of record, the marine realm presents a far more comprehensive picture than dry land.

Kalmar & Currie (2010) have argued that the terrestrial record is relatively representative because taxa abundant in the modern are also abundant in the fossil record. In both the fossil and modern record, Tetrapoda, Chelicerata, and Hexapoda make up the majority of records. However, this congruence fails when extended to the generic level. Data on major Pleistocene terrestrial tetrapods groups were downloaded from the PaleoDB (Table 2). Mammals are the most diverse group, (626 genera, 72% of all genera) followed by birds (147 genera, 17%), lizards and snakes (28 genera, 7%), turtles (28 genera, 3%), and crocodilians (5 genera, 0.6%). Modern estimates of generic diversity were derived from a number of sources. Monroe & Sibley's checklist (1997) contains 2064 genera of birds. Modern mammal diversity is based on Wilson & Reeder (2005), which contains 1231 extant genera. Generic numbers for squamates (1440 genera of lizards and 440 genera of snakes), turtles (90 genera) and crocodilians (8 genera) were derived from Grzimek *et al.* (2004). Compared to the modern, Pleistocene birds are badly underrepresented. Today birds make up approximately half of all vertebrate genera but they make up less than a fifth of all Pleistocene genera. Mammals are comparatively over-represented, making up almost three quarters of all known Pleistocene genera but only less than a third of modern genera. Assuming that there has not been a tremendous differential between avian and mammalian diversification and extinction rates following the retreat of the ice sheets, it seems reasonable to conclude that mammals, with their diagnostic and enamelled teeth, are over-represented in the fossil record relative to the lightly-boned birds. Squamates (snakes) are also under-represented while crocodilians and turtles are slightly over-represented. While Kalmar

& Currie's (2010) observation is true, the fossil record is not as representative as it might first appear, and it may behove future studies to attempt to account for differences in preservation potential between terrestrial taxa.

Terrestrial biodiversity and outcrop area

Consistent with previous studies, terrestrial outcrop area serves as a good predictor of sampled biodiversity. In the marine realm, the best-fit observed relation was a power-law (Wall *et al.* 2009); for terrestrial fossils, a simple linear relation yields a higher r^2 and lower P-value. The driving factors behind this difference are unclear. Power-law or logarithmic relations are more frequently encountered in analyses of diversity-area relations (e.g. Rosenzweig 1995). The asymptotic shapes of these relations are in part driven by the sampling of a single, defined pool – as sample sizes increase, the number of new taxa encountered with each additional step must decrease. Assuming the calculated linear relation adequately describes the relationship between outcrop area and terrestrial diversity, diversity instead apparently continues to rise consistently with increasing sample size. This implies that, at a global scale, the nested, interrelated nature of samples observed in the marine realm disintegrates. The implication for the realworld is that in the marine realm when palaeontologists explore a new set of marine rocks they can reasonably expect to find a rather large number of previously identified fossils. A new set of terrestrial rocks, however, yields predominantly new taxa, at a mean rate of four new genera for every 1000 km^2 of new terrestrial outcrop. This linear relation is consistent with a relatively poor terrestrial fossil record; in a more complete record we might expect that larger sample sizes should capture previously identified taxa and have an asymptotic relation between diversity and outcrop area.

The outcrop-diversity relationship does not seem to be driven, at least at the global scale, by common cause. Terrestrial habitat area does not appear to influence terrestrial outcrop area. We thus reject the idea that a common cause mechanism is driving both sampled diversity and outcrop area and instead view outcrop area as solely a biasing agent. This is consistent for marine rocks as well and is likely because the imprint of burial and erosion plays such a large role in determining outcrop area (Wall *et al.* 2009). Palaeozoic sediments are not lacking on the Earth's surface because there were fewer Palaeozoic sediments deposited – rather their paucity relative to Mesozoic and Cainozoic sediments is a result of a longer history of destruction via erosion and metamorphosis combined with an extensive cloak of overburden. This is clear in the habitat area and outcrop area curves through time. The epochs of the Cretaceous exhibit greater terrestrial outcrop than almost any other time interval, despite the fact that epicontinental seas restricted the deposition of terrestrial sediments to a relatively small portion of the continents. Instead, the ubiquity of Cretaceous terrestrial rocks is a function of a history of limited burial and that the Cretaceous sediments were draped over earlier sediments. Eustatic sea-level change (which controls habitable area) appears to influence both turnover (Peters 2005) and standing diversity (Wall *et al.* 2009) of organisms in the marine realm. In contrast, while this manuscript does not focus on turnover, standing terrestrial diversity does not appear to show any relation with original habitable area at first glance. Nevertheless, the correlation between terrestrial diversity and habitat area is positive, though insignificant, and a close examination of the data reveals that the time intervals displaying higher than anticipated diversity are also time intervals with large outcrop areas (e.g. Miocene, Pliocene, Upper Cretaceous, and Lower Cretaceous). A multivariate correlation between terrestrial diversity, outcrop area, and original habitat area (Table 1) is significant as a whole ($P \ll 0.001$) and displays an improved correlation v. a single variable model ($r^2 = 0.79$ v. 0.72). The significance of the individual coefficients vary; that of outcrop area is highly significant as before ($P \ll 0.001$); that of habitat area is not as strong ($P = 0.06$) but it a substantial improvement over the univariate model's *P*-value (0.26). The habitat-area coefficient is non-significant for a correlation of first differences. In the marine realm habitat area appears to influence total global diversity (Wall *et al.* 2009). In the terrestrial realm, this statement has less support, especially for relatively small-scale (stage level) changes (Butler *et al.* 2011; Mannion *et al.* 2011). The above observations indicate that outcrop area has a strong influence on recorded diversity and hint that on large time-scales, habitat area may drive terrestrial biodiversification.

If the observed relation between terrestrial diversity and outcrop area is simply a matter of bias the residuals of the correlation between the two variables should provide us insight into the trend of terrestrial biodiversity through time (Fig. 7). Residual diversity (Rd) shows a significant increase through time (*t*; Ma) following the linear equation Rd $= -1.48\,t + 306$ ($r^2 = 0.51$, $P = 0.001$). The rate of increase is small, approximately 1.5 genera per million years and we do not find evidence for the massive exponential increase observed in the uncorrected familial data of Benton (1993). There is also potential that at least some of the increase in generic richness is more the product of a preservational bias caused by the diversification of mammals and the decline of archosaurs rather than

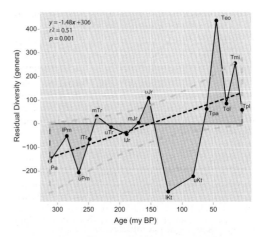

Fig. 7. Residual diversity through time. Residual diversity from the correlation of diversity and outcrop area (Fig. 4), which should remove the effect of variable sampling due to outcrop area, does show a long-term increase in diversity through time (Diversity $= -1.48$ Age $+306$, $r^2 = 0.51$, $P = 0.001$). The thin black line shows the path of diversity through time, the dashed black line indicates the linear relation and the dashed grey lines indicate the 95% confidence intervals for the correlation. Abbreviations as in Figure 1.

a true increase in the generic diversity of life on land. As mentioned above, it appears that mammals are over-represented in the fossil record of the Pleistocene, most likely due to their highly durable and diagnostic teeth. In the modern there are roughly 1.3 mammals for every turtle, crocodilian, snake, or lizard. If we assume that this ratio was the same in the Pleistocene, we would expect to find roughly 117 mammals in the PaleoDB for the 90 reptiles attributed to the time period. Instead the database contains 626 mammalian genera, more than five times the expected value. While there are potential non-preservational explanations for the discrepancy, such as unequal collector interest in the two groups, it is difficult to explain the entire difference by such means. This observation further decreases the chances that diversity has increased through time.

Conclusions

The observed correlation between terrestrial outcrop area and biodiversity appears to be driven solely through sampling bias and not common cause. The areal extents of exposed continental rocks are not representative of the original areas of sub-aerial continents, indicating that the observed correlation between diversity and outcrop area is more likely due to outcrop area acting as a sampling bias. This is further supported by the poor quality of the terrestrial fossil record, which on average represents less than 0.5% of the original continental area. After the bias of variable outcrop area is removed, there remains some evidence that terrestrial biodiversity has increased through time. The resulting picture of biodiversity may not be fully accurate, however, as it fails to take into account the variable preservation potentials of the organisms involved, which may inflate the Cenozoic due to the presence and diversification of mammals.

Many thanks to S. Peters, J. Alroy, J. Handley, J. Brower, and especially H.L.B. Wall for thoughts and assistance. This work was supported by a University Fellowship from the Graduate School at Syracuse University. Generation of geological map data was supported by National Science Foundation grants to B.H.W. This is Paleobiology Database Publication 141.

References

ALROY, J. 1999. The fossil record of North American mammals: evidence for a Palaeocene evolutionary radiation. *Systematic Biology*, **48**, 107–118.

ALROY, J., MARSHALL, C. R. *ET AL.* 2001. Effects of sampling standardization on estimates of Phanerozoic marine diversification. *Proceedings of the National Academy of Sciences USA*, **98**, 6261–6266.

ALROY, J., ABERHAN, M. *ET AL.* 2008. Phanerozoic trends in the global diversity of marine invertebrates. *Science*, **321**, 97–100.

BAMBACH, R. K. 1977. Species richness in marine benthic habitats through the Phanerozoic. *Paleobiology*, **3**, 152–167.

BARRETT, P. M., MCGOWAN, A. J. & PAGE, V. 2009. Dinosaur diversity and the rock record. *Proceedings of the Royal Society, B*, **276**, 2667–2674.

BENTON, M. J. 1993. *The Fossil Record 2*. Chapman and Hall, London.

BENTON, M. J. 2001. Biodiversity on land and in the sea. *Geological Journal*, **36**, 211–230.

BENTON, M. J., HITCHIN, R. & WILLS, M. A. 1999. Assessing congruence between cladistics and stratigraphic data. *Systematic Biology*, **48**, 581–596.

BENTON, M. J., DUNHILL, A. M., LLOYD, G. T. & MARX, F. G. 2011. Assesing the quality of the fossil record: insights from vertebrates. *In*: MCGOWAN, A. J. & SMITH, A. B. (eds) *Comparing the Geological and Fossil Records: Implications for Biodiversity Studies*. Geological Society, London, Special Publications, **358**, 63–94.

BUSH, A. M., MARKEY, M. J. & MARSHALL, C. R. 2004. Removing bias from diversity curves: the effects of spatially organized biodiversity on sampling standardization. *Paleobiology*, **30**, 666–686.

BUTLER, R. J., BARRETT, P. M., NOWBATH, S. & UPCHURCH, P. 2009. Estimating the effects of sampling biases on pterosaur diversity patters: implications for hypotheses of bird/pterosaur competitive replacement. *Paleobiology*, **35**, 432–446.

BUTLER, R. J., BENSON, R. B. J., CARRANO, M. T., MANNION, P. D. & UPCHURCH, P. 2011. Sea level,

dinosaur diversity and sampling biases: investigating the 'common cause' hypothesis in the terrestrial realm. *Proceedings of the Royal Society, B*, **278**, 1165–1170.

CHOUBERT, G. & FAURE-MURET, A. 1976. *Geological Atlas of the World. 1:10,000,000. 22 Sheets with Explanations*. UNESCO/Commission for the Geologic Map of the World, Paris.

CRAMPTON, J. S., BEU, A. G., COOPER, R. A., JONES, C. M., MARSHALL, B. A. & MAXWELL, P. A. 2003. Estimating the rock volume bias in paleobiodiversity studies. *Science*, **301**, 358–360.

CRAMPTON, J. S., FOOTE, M. *ET AL.* 2006a. Second-order sequence stratigraphic controls on the quality of the fossil record at an active margin: New Zealand Eocene to Recent shelf molluscs. *Palaios*, **21**, 86–105.

CRAMPTON, J. S., FOOTE, M. *ET AL.* 2006b. The ark was full! Constant to declining Cenozoic shallow marine biodiversity on an isolated midlatitude continent. *Paleobiology*, **32**, 509–532.

DUNHILL, A. 2011. Using remote sensing and GIS to quantify rock exposure area in the United Kingdom: implications for paleodiversity studies. *Geology*, **39**, 111–114.

FARA, E. 2004. Estimating minimum global species diversity for groups with a poor fossil record: a case study of Late Jurassic–Eocene lissamphibians. *Palaeogeography, Palaeoclimatology, Palaeoecology*, **207**, 59–82.

GRZIMEK, B., KLEIMAN, D. G., GEIST, V. & MCDADE, M. C. (eds) 2004. *Grzimek's Animal Life Encyclopedia*. Thompson-Gale, Detroit.

KALMAR, A. & CURRIE, D. J. 2010. The completeness of the continental fossil record and its impact on patterns of diversification. *Paleobiology*, **36**, 51–60.

KHAIN, V. YE. & BALUKHOVSKIY, A. N. 1979. Neogene lithologic associations of the world. *Soviet Geology*, **10**, 15–23.

KHAIN, V. YE. & SESLAVINSKIY, K. B. 1977. Silurian lithologic associations of the world. *Soviet Geology*, **5**, 21–42.

KHAIN, V. YE., RONOV, A. B. & BALUKHOVSKIY, A. N. 1975. Cretaceous lithologic associations of the world. *Soviet Geology*, **11**, 10–39.

LLOYD, G. T., DAVIS, K. E. *ET AL.* 2008. Dinosaurs and the Cretaceous terrestrial revolution. *Proceedings of the Royal Society, B*, **275**, 2483–2490.

MANNION, P. D., UPCHURCH, P., CARRANO, M. T. & BARRETT, P. M. 2010. Testing the effect of the rock record on diversity: a multidisciplinary approach to elucidating the generic richness of sauropodomorph dinosaurs through time. *Biological Reviews*, **86**, 157–181.

MILLER, A. I. 1997. Dissecting global diversity patterns: examples from the Ordovician radiation. *Annual Review of Ecology and Systematics*, **28**, 85–104.

MILLER, A. I. & FOOTE, M. 1996. Calibrating the Ordovician radiation of marine life: implications for Phanerozoic diversity trends. *Paleobiology*, **22**, 304–309.

MONROE, B. & SIBLEY, C. G. 1997. *A World Checklist of Birds*. Yale University Press, New Haven CN.

PADIAN, K. & CLEMENS, W. A. 1985. Terrestrial vertebrate diversity: episodes and insights. *In*: VALENTINE, J. W. (ed.) *Phanerozoic Diversity Patterns*. Princeton University Press, Princeton, NJ, 41–96.

PETERS, S. E. 2005. Geologic constraints on the macro-evolutionary history of marine animals. *Proceedings of the National Academy of Science USA*, **102**, 12326–12331.

PETERS, S. E. & FOOTE, M. 2001. Biodiversity in the Phanerozoic: a reinterpretation. *Paleobiology*, **27**, 583–601.

PETERS, S. E. & FOOTE, M. 2002. Determinants of extinction in the fossil record. *Nature*, **416**, 420–424.

PHILLIPS, J. 1860. *Life on the Earth: Its Origin and Succession*. Macmillan, Cambridge, UK.

PORTER, S. C. 2001. Chinese loess record of monsoon climate during the last glacial–interglacial cycle. *Earth-Science Reviews*, **54**, 115–128.

RANKEY, E. C. 1997. Relations between relative changes in sea level and climate shifts: Pennsylvanian-Permian mixed carbonate-siliciclastic strata, western United States. *GSA Bulletin*, **109**, 1089–1100.

RAUP, D. M. 1972. Taxonomic diversity during the Phanerozoic. *Science*, **177**, 1065–1071.

RAUP, D. M. 1976a. Species diversity in the Phanerozoic: a tabulation. *Paleobiology*, **2**, 279–288.

RAUP, D. M. 1976b. Species diversity in the Phanerozoic: an interpretation. *Paleobiology*, **2**, 289–297.

ROBERT, C. & KENNETT, J. P. 1992. Paleocene and Eocene kaolinite distribution in the South Atlantic and Southern Ocean: Antarctic climate and paleoceanographic implications. *Marine Geology*, **103**, 99–110.

RONOV, A. B. 1964. General trends in the composition of the crust, ocean, and atmosphere. *Geokhimiya*, **8**, 714–743.

RONOV, A. B. 1978. The Earth's sedimentary shell. *International Geology Review*, **24**, 1313–1363.

RONOV, A. B. 1980. The Earth's sedimentary shell: quantitative patterns of its structures, compositions and evolution (the 20th V. I. Vernadskiy Lecture). *In*: YAROSHEVSKIY, A. A. (ed.) *The Earth's Sedimentary Shell*. Nauka, Moscow, 1–80. [American Geological Institute Reprint Series 5, 1–73.]

RONOV, A. B. 1994. Phanerozoic transgressions and regressions on the continents: a quantitative approach based on areas flooded by the sea and areas of marine and continental deposition. *American Journal of Science*, **294**, 777–801.

RONOV, A. B. & KHAIN, V. Ye. 1954. Devonian lithologic associations of the world. *Soviet Geology*, **41**, 47–76.

RONOV, A. B. & KHAIN, V. Ye. 1955. Carboniferous lithologic associations of the world. *Soviet Geology*, **48**, 92–117.

RONOV, A. B. & KHAIN, V. Ye. 1956. Permian lithologic associations of the world. *Soviet Geology*, **54**, 20–36.

RONOV, A. B. & KHAIN, V. Ye. 1961. Triassic lithologic associations of the world. *Soviet Geology*, **1**, 27–48.

RONOV, A. B. & KHAIN, V. Ye. 1962. Jurassic lithologic associations of the world. *Soviet Geology*, **1**, 9–34.

RONOV, A. B., KHAIN, V. Ye. & BALUKHOVSKIY, A. N. 1974a. Paleogene lithologic associations of the world. *Soviet Geology*, **3**, 10–42.

RONOV, A. B., SESLAVINSKIY, K. B. & KHAIN, V. Ye. 1974b. Cambrian lithologic associations of the world. *Soviet Geology*, **12**, 10–33.

RONOV, A. B., KHAIN, V. Ye. & SESLAVINSKIY, K. B. 1976. Ordovician lithologic associations of the world. *Soviet Geology*, **1**, 7–27.

RONOV, A. B., KHAIN, V. Ye., BALUKHOVSKY, A. N. & SESLAVINSKY, K. B. 1980. Quantitative analysis of Phanerozoic sedimentation. *Sedimentary Geology*, **25**, 311–325.

ROSENZWEIG, M. L. 1995. *Species Diversity in Space and Time*. Cambridge University Press, Cambridge, UK.

SADLER, P. M. 1981. Sediment accumulation rates and the completeness of stratigraphical sections. *Journal of Geology*, **98**, 569–584.

SCHINDEL, D. E. 1982. The gaps in the fossil record. *Nature*, **297**, 282–284.

SCOTESE, C. R. & GOLONKA, J. 1992. *Paleogeographic atlas. PALEOMAP Progress Report 20;c-0692*. University of Texas, Arlington.

SEPKOSKI, J. J., JR. 1976. Species diversity in the Phanerozoic: species-area effects. *Paleobiology*, **2**, 298–303.

SEPKOSKI, J. J., JR. 1981. A factor analytic description of the Phanerozoic marine fossil record. *Paleobiology*, **7**, 36–53.

SEPKOSKI, J. J., JR. 1982. *A compendium of fossil marine families*. Milwaukee Public Museum Contributions in Biology and Geology, **51**.

SEPKOSKI, J. J., JR. 2002. A compendium of fossil marine animal genera. *Bulletins of American Paleontology*, **363**.

SEPKOSKI, J. J., JR., BAMBACH, R. K., RAUP, D. M. & VALENTINE, J. W. 1981. Phanerozoic marine diversity and the fossil record. *Nature*, **293**, 435–437.

SMITH, A. B. 2001. Large-scale heterogeneity of the fossil record: implications for Phanerozoic biodiversity studies. *Philosophical Transactions of the Royal Society of London B*, **356**, 351–367.

SMITH, A. B. & MCGOWAN, A. J. 2007. The shape of the Phanerozoic marine palaeodiversity curve: how much can be predicted from the sedimentary rock record of western Europe. *Palaeontology*, **50**, 765–774.

SMITH, A. B., GALE, A. S. & MONKS, N. E. A. 2001. Sea-level change and rock record bias in the Cretaceous: a problem for extinction and biodiversity studies. *Paleobiology*, **27**, 241–253.

VALENTINE, J. W. 1969. Niche diversity and niche size patterns in marine fossils. *Journal of Paleontology*, **43**, 905–915.

VALENTINE, J. W. 1970. How many marine invertebrate fossil species? A new approximation. *Journal of Paleontology*, **44**, 410–415.

VAN DER HAMMEN, T. 1974. The Pleistocene changes of vegetation and climate in tropical South America. *Journal of Biogeography*, **1**, 3–26.

VERMEIJ, G. J. & LEIGHTON, L. R. 2003. Does global diversity mean anything? *Paleobiology*, **29**, 3–7.

WALKER, L. J., WILKINSON, B. H. & IVANY, L. C. 2002. Continental drift and Phanerozoic carbonate accumulation in shallow-shelf and deep-marine settings. *Journal of Geology*, **110**, 75–87.

WALL, P. D., IVANY, L. C. & WILKINSON, B. H. 2009. Revisiting Raup: exploring the influence of outcrop area on diversity in light of modern sample-standardization techniques. *Paleobiology*, **35**, 146–167.

WANG, S. C. & DODSON, P. 2006. Estimating the diversity of dinosaurs. *Proceedings of the National Academy of Science USA*, **103**, 13601–13605.

WILSON, D. E. & REEDER, D. M. (eds) 2005. *Mammal Species of the World*. 3rd edn. John Hopkins University Press, Baltimore MD.

Assessing the quality of the fossil record: insights from vertebrates

MICHAEL J. BENTON[1]*, ALEXANDER M. DUNHILL[1], GRAEME T. LLOYD[2] &
FELIX G. MARX[1,3]

[1]*School of Earth Sciences, University of Bristol, Wills Memorial Building,
Queen's Road, Bristol, BS8 1RJ, UK*

[2]*Department of Palaeontology, The Natural History Museum, Cromwell Road,
London, SW7 5BD, UK*

[3]*Department of Geology, University of Otago, PO Box 56, Dunedin 9054, New Zealand*

Corresponding author (e-mail: mike.benton@bristol.ac.uk)

Abstract: Assessing the quality of the fossil record is notoriously hard, and many recent attempts have used sampling proxies that can be questioned. For example, counts of geological formations and estimated outcrop areas might not be defensible as reliable sampling proxies: geological formations are units of enormously variable dimensions that depend on rock heterogeneity and fossil content (and so are not independent of the fossil record), and outcrop areas are not always proportional to rock exposure, probably a closer indicator of rock availability. It is shown that in many cases formation counts will always correlate with fossil counts, whatever the degree of sampling. It is not clear, in any case, that these proxies provide a good estimate of what is missing in the gap between the known fossil record and reality; rather they largely explore the gap between known and potential fossil records. Further, using simple, single numerical metrics to correct global-scale raw data, or to model sampling-driven patterns may be premature. There are perhaps four approaches to exploring the incompleteness of the fossil record, (1) regional-scale studies of geological completeness; (2) regional- or clade-scale studies of sampling completeness using comprehensive measures of sampling, such as numbers of localities or specimens or fossil quality; (3) phylogenetic and gap-counting methods; and (4) model-based approaches that compare sampling as one of several explanatory variables with measures of environmental change, singly and in combination. We suggest that palaeontologists, like other scientists, should accept that their data are patchy and incomplete, and use appropriate methods to deal with this issue in each analysis. All that matters is whether the data are *adequate* for a designated study or not. A single answer to the question of whether the fossil record is driven by macro-evolution or megabias is unlikely ever to emerge because of temporal, geographical, and taxonomic variance in the data.

The fossil record is far from perfect, and palaeontologists must be concerned about inadequacy and bias (Raup 1972; Benton 1998; Smith 2001, 2007*a*). Fundamental issues concerning the quality and completeness of the fossil record were enunciated clearly by Charles Darwin (1859, pp. 287–288), who wrote:

> That our palæontological collections are very imperfect, is admitted by every one. The remark of that admirable palæontologist, the late Edward Forbes, should not be forgotten, namely, that numbers of our fossil species are known and named from single and often broken specimens, or from a few specimens collected on some one spot. Only a small portion of the surface of the earth has been geologically explored, and no part with sufficient care, as the important discoveries made every year in Europe prove. No organism wholly soft can be preserved. Shells and bones will decay and disappear when left on the bottom of the sea, where sediment is not accumulating... With

respect to the terrestrial productions which lived during the Secondary and Palæozoic periods, it is superfluous to state that our evidence from fossil remains is fragmentary in an extreme degree.

Raup (1972) clarified the situation when he compared the 'empirical' model of Valentine (1969), a literal reading of the fossil record, with his 'bias simulation model' that explained the bulk of the apparent low diversity levels of marine invertebrates in the Palaeozoic as a sampling error. Two opposite viewpoints have been argued, either that the fossil record is good enough (e.g. Sepkoski *et al.* 1981; Benton 1995; Benton *et al.* 2000; Stanley 2007) or not good enough (e.g. Raup 1972; Alroy *et al.* 2001, 2008; Peters & Foote 2002; Alroy 2010) to show the main patterns of global diversification through time. A resolution between these opposite viewpoints does not appear close (Benton 2009; Erwin 2009; Marshall 2010).

From: McGowan, A. J. & Smith, A. B. (eds) *Comparing the Geological and Fossil Records: Implications
for Biodiversity Studies*. Geological Society, London, Special Publications, **358**, 63–94.
DOI: 10.1144/SP358.6   0305-8719/11/$15.00 © The Geological Society of London 2011.

Key objective evidence for bias in the fossil record could be the extraordinary and ubiquitous correlation of sampling proxies and diversity curves: why is there such close tracking of measures of rock volume by palaeodiversity? There are three possible explanations: (1) rock volume/sampling drives the diversity signal (Peters & Foote 2001, 2002; Smith 2001, 2007*a*; Butler *et al.* 2011); (2) both signals reflect a third, or 'common', cause such as sea-level fluctuation (Peters 2005; Peters & Heim 2010); or (3) both signals are entirely or partially redundant (= identical) with each other. In reality, the close correlation probably reflects a combination of all three factors in different proportions in any test case, and so it is probably fruitless to prolong the debate about which of the three models is correct, and which incorrect.

Much of the literature on the quality of the rock and fossil records has focused on marine settings. This reflects the interests of palaeontologists who engage with these questions, and the fact that many marine rock records are more complete than most terrestrial (continental) rock records. However, the terrestrial fossil record is worth considering for several reasons: terrestrial life today is much more diverse than marine life, perhaps representing 85% of modern biodiversity (May 1990; Vermeij & Grosberg 2010), terrestrial life includes many major taxa that are sensitive to atmospheric, temperature, and topographic change and so are key indicator species in studies of global change, and for many terrestrial groups (e.g. angiosperms, insects, vertebrates) there are mature morphological and molecular phylogenies that enable cross-comparison between stratigraphic and cladistic data.

In this paper, we explore the use of sampling proxies, and suggest that some commonly used measures, notably formation counts and outcrop areas, may not be useful or accurate measures of sampling. Indeed, we suggest that there is probably no single numerical metric that captures all aspects of sampling (= rock volume, accessibility, effort), and recent attempts to correct the raw data, or to model sampling-driven patterns, may be premature. We then look at some case studies of patchy fossil records in taxa with good phylogenetic data, and suggest that in some cases at least the rock volume and fossil occurrence measures are identical, and so correlate almost perfectly. Finally, we suggest that such global-scale confrontations of sampling proxies and fossil data are not adequate at present, and recommend instead study-scale approaches to detect and correct sampling, involving direct evidence for missing data (e.g. Lazarus taxa; ghost ranges), direct evidence for sampling (e.g. number of localities or samples per time bin; fossil specimen completeness), and an integrated, model-based

approach to incorporating sampling and explanatory models into explaining particular diversity curves.

The fossil record, reality and sampling

The fossil record, collector curves and assessing reality

The *known fossil record*, meaning our present understanding as represented by the literature and museum collections, is a subsample of the *potential fossil record*, all the fossils in the rocks, including undocumented materials (Fig. 1). The potential fossil record is itself a subsample of *all life that ever existed*, or reality, and this includes many soft-bodied and microscopic organisms that have never been fossilized and so can never be known.

The difference between the potential fossil record and reality may be very large (Paul 1988; Forey *et al.* 2004). An estimate of this difference has been made based on the proportion of fossilizable to non-fossilizable modern animals: of 1.2 million living species named at the time, Nichol (1977) estimated that fewer than 0.1 million (8%)

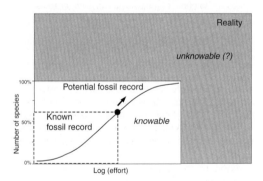

Fig. 1. The known and potential fossil record, and reality. Many sampling methods assess our position on the collector curve, which tracks the accumulation of knowledge, assessed against a measure of effort (e.g. number of specimens, area sampled, number of person-days work), from no knowledge to complete knowledge of the potential fossil record. The moving point shows where current knowledge stands on the collector curve trajectory. Other methods are required to assess the difference between the known fossil record and reality; the proportion of known (or knowable) fossil record (white) to the unknowable portion of reality (grey) is entirely hypothetical, and could range from 100% (where everything is known, or could be found) to infinitesimal (where nothing, or almost nothing, of a particular record at a particular level of focus, taxonomically and stratigraphically, is known); in fact the potential fossil record could represent about 10% of reality (Paul 1988; Forey *et al.* 2004). The collector curve is traced from its original presentation in Preston (1948).

were skeletized, and so potentially preservable as fossils. This kind of figure has also emerged from comparisons of organisms preserved in fossil Lagerstätten, such as the Burgess Shale, of which 10–15% of species are skeletized, which led Paul (1988) to suggest that, in normal conditions of preservation, perhaps 10% of Phanerozoic animal species might be preservable, leaving 90% unknowable, except for hints here and there from soft-bodied organisms glimpsed in sites of exceptional preservation. If the proportions of soft-bodied to hard-bodied organisms have remained constant through geological time, such an estimate may be helpful; if, however, the proportion has varied substantially, then even the best-sampled fossil record may say little about the true global pattern of palaeodiversity through time.

It is assumed that the documented fossil record is improving, as more fossils are found and as more researchers investigate and revise earlier work; doubtless some day all fossils in the rocks could be known scientifically (Maxwell & Benton 1990; Smith 2001; Forey et al. 2004). Certainly, the rate of naming of new species in some fossil groups is approaching saturation, as determined from collector curves (Preston 1948), where the trajectory has apparently reached the asymptote region (Fig. 1; e.g. Maxwell & Benton 1990; Benton & Storrs 1994; Benton 1998, 2008a; Paul 1998; Smith 2007b; Tarver et al. 2007, 2011; Bernard et al. 2010). Further, comparisons of knowledge through research time show that new finds often do not much affect overall macroevolutionary patterns in broad-scale studies (Maxwell & Benton 1990; Sepkoski 1993), although more focused analyses of patchily occurring groups such as dinosaurs may show changing phylogenetic trees as new fossils are found (Tarver et al. 2011). In addition, predicted gaps (e.g. ghost ranges, Lazarus gaps) are often filled by new fossils, and so congruence of cladograms with the fossil record improves (Benton & Storrs 1994) or remains static (Benton 2001) at higher taxonomic levels, although new fossil finds in finer-scale studies can still fill gaps or create new gaps (Weishampel 1996). Finally, comparisons (Foote & Raup 1996; Foote & Sepkoski 1999) of the probability of sampling fossil genera in one, two or three stratigraphic units (the FreqRat method) with proportions of living families with a fossil record show that, as expected, some taxa are apparently well sampled (e.g. ammonoids, conodonts, brachiopods, bryozoans, echinoids, ostracods, bivalves) while others are not (e.g. polychaetes, malacostracans). For well-sampled clades, such as trilobites, bivalves, and mammals, Foote & Raup (1996) estimated that 60% of actual species palaeodiversity had been recorded.

Any metric that measures sampling then is a measure of our trajectory to complete knowledge of the potential fossil record, not of reality, and this is well understood (e.g. Valentine 1969; Paul 1988; Maxwell & Benton 1990; Peters & Foote 2001; Smith 2001, 2007a; Forey et al. 2004). However, it might be easy to forget this, and to assume that comparisons of recorded fossil record data with sampling proxies provide the true pattern of palaeodiversity. This could only be the case if the proportions of soft-bodied to hard-bodied taxa within a clade have remained constant through time, or if the analyst is referring specifically to a single well-skeletonized group, such as dinosaurs or brachiopods. Even for statements about single clades, there are unassessable variables that might make the potential fossil record depart substantially, or unpredictably, from reality, especially if certain key habitats are never, or rarely, sampled. Patchy sampling by geographical region or by stratigraphic unit can be detected by comparative analyses (e.g. Smith 2007b), but the true proportions of different habitats occupied by members of a clade may never be known.

Sampling proxies that measure human effort, such as 'palaeontological interest units' (e.g. Raup 1977; Sheehan 1977) only refer to the unknown part of the potential fossil record. Further, as we argue below, counts of 'number of formations' are so intimately linked with historical aspects of species discovery that they too address only the unknown part of the potential fossil record, and could hardly be used to predict true palaeodiversities. Indeed, this was suggested by the observation (Benton 2008a) that the best way to guarantee to find new genera and species of dinosaurs is to find new basins/formations. Historical studies of the pattern of naming of new species of dinosaurs show that the best determinant of new taxa is the discovery of new formations, not intensified collecting in known sedimentary basins. Whereas the discovery curve for dinosaurs showed evidence of an asymptote ten years ago (Benton 1998), the subsequent, and continuing, exploration of new formations in China and South America has generated a rapid rise in naming of new species in the past decade (Benton 2008a). Nevertheless, the collector curve can never extend beyond the ceiling of the potential fossil record to explore all the dinosaurian species that once lived but were never fossilized (Fig. 1).

A number of attempts have been made to estimate the total species richness of dinosaurs, using a model for estimating future discoveries and generic longevity (Dodson 1990), using assumptions about species-area relationships (Russell 1995), and using relative abundance plots of rare to common species within local faunas (Wang & Dodson 2006). All these methods focus on the potential fossil record, and do not necessarily identify taxa that have never been preserved.

Is the 'unknowable' part of palaeodiversity (Fig. 1, grey) forever unknowable? It might be possible to seek approaches or metrics that at least point to some of the palaeodiversity of soft-bodied organisms or of unpreserved habitats by a variety of means. Some approaches might be (1) *phylogenetic*, using say evidence from a complete molecular phylogeny of living forms to identify likely missing fossils, (2) *stratigraphic*, using information about the known stratigraphic distributions of fossils to predict those that may have been there but can never be sampled, (3) *geographical*, attempting to identify gaps in fossil distributions, or (4) *ecological*, perhaps looking at trophic webs or pyramids to try to pinpoint missing taxa. Some efforts in this area include comparisons of cladograms and stratigraphic order of fossils (e.g. Norell & Novacek 1992; Benton & Storrs 1994; Huelsenbeck 1994; Wills 1999, 2007; Benton *et al.* 2000), comparison of fossil and molecular estimates of clade origins to identify durations and distributions of gaps (e.g. Smith 2007*b*), and calculation of confidence intervals on stratigraphic ranges of fossils using known collecting patterns (e.g. Marshall 1990, 1997). None of these approaches, however, gives an estimate of the difference between the potential fossil record and reality, merely a proportional measure on comparing two time bins or geographical regions, or a measure of whether the difference is likely to be large, or at least likely to bias the data.

Measures of rock volume might provide approaches to bridging the gap between the potential fossil record and reality (Fig. 1), but only if used in models to estimate original areas of habitats and missing taxa predicted to have occupied such habitats, but that have not been, or cannot be, sampled. For example, conservation biologists identify the niche of a modern species, and then plot potential geographical distributions according to the wider distribution of the precise habitats suitable for that species (e.g. Guisan & Zimmermann 2000; Guisan & Thuiller 2005). The assumption is that the species could occupy all of its potential habitat if human and other historical pressures did not prevent it. Such techniques might be used by palaeontologists to identify 'missing species' by combining rock volume data with palaeogeographical maps, and such hypothetical missing species could fall in the potentially knowable or unknowable zones (Fig. 1).

Sampling

Sampling is a set of statistical procedures to explore how subsets of individual observations within a population of individuals may yield wider knowledge about the population of concern, especially for the purpose of making predictions based on statistical inference. In the context of the fossil record, sampling can refer to two research themes, (1) the choice of subsamples of a greater whole as a practical means to determine aspects of the wider sample, and (2) the degree to which the known fossil record is itself a suitable subsample of the greater whole. That 'greater whole' could be simply the potential fossil record (Fig. 1), but it is usually assumed to be reality.

Reasons that the known fossil record falls short of the potential fossil record or reality are a mix of geological and human factors (Raup 1972) that fall into three categories: (1) rock volume, the progressive geological bias against preservation and discovery of ever-older fossils (diagenesis, metamorphism, erosion, covering by younger rocks); (2) accessibility, the currently available rock area or volume; and (3) effort, human factors, such as geographical location, ease of access, and subject interest (by age, location, or fossil group).

Adequacy and applications

An important caveat is that most of the previous discussions of fossil record completeness have concentrated on bias and sampling as they relate to a single use of the fossil record, namely to represent global diversity patterns. Such discussions might, or might not, have an impact on other uses of the fossil record, for example in studies of individual lineages or clades, in local- or regional-scale studies, phylogenetic analyses, ecological studies of communities, or anatomical and functional studies. So, an entirely biased fossil record that provides spurious data on the scaling of a global-scale radiation or mass extinction, might nonetheless provide numerous near-perfect fossil Lagerstätten that represent entire fossil communities and individual fossils that provide remarkable anatomical detail for functional, ecological, and evolutionary studies. As Donovan & Paul (1998) stated, 'the fossil record may be incomplete, but it is entirely adequate for many and most requirements of palaeontology'. Benton *et al.* (2000) made this point in a specific case, when they showed that age v. clade congruence is good at the scale of stratigraphic stages and taxonomic families, and shows no time dependence or bias back through the 550 Ma of the Phanerozoic. At finer temporal scales, Wills (2007) also showed constant levels of completeness through most of the Phanerozoic, but a substantial drop in the Cambrian and, surprisingly, in the Neogene. These may be partly 'edge' effects, but the Cambrian drop likely mixes sampling failure combined with obscure taxa and many soft-bodied forms.

Some observations based on analyses Darwin could not have predicted might be said to suggest that the fossil record, error-ridden and incomplete as it is, is adequate for many purposes, although none of these provides evidence that error in the fossil record is negligible: (1) the order of fossils in the rocks generally matches closely the order of nodes in morphological and molecular trees (Norell & Novacek 1992; Benton & Hitchin 1997); (2) at coarse scales of observation (families and stratigraphic stages), there is no evidence that this matching becomes worse deeper in time (Benton et al. 2000; Wills 2007); (3) macroevolutionary patterns, including posited mass extinctions and diversifications, are largely immune to changes in palaeontological knowledge, even over 100 years of research time (Maxwell & Benton 1990; Sepkoski 1993; Adrain & Westrop 2000); (4) congruence between stratigraphy and phylogeny has also been largely stable through the 20th century, despite an order-of-magnitude increase in the number of fossils (Benton 2001); (5) new fossil finds, even of reputedly poorly sampled groups such as primates and humans, do not always alter perceptions of evolutionary patterns (Tarver et al. 2011); and (6) new post-Cambrian Lagerstätten rarely add new families to existing knowledge, just new species and genera (Benton et al. 2008).

Sampling proxies

Definition

A *sampling proxy* is a metric that represents 'collecting effort' in some way, the x-axis of the species v. effort plot (Fig. 1). Such a sampling proxy should represent some or all of the geological and human factors that can introduce error into interpretations of data from the fossil record, typically time series of diversity. The sampling proxy ought to be a measure of biasing factors such as rock volume, accessibility, or human effort, and it should be *independent of the signal it seeks to correct*, namely the documented fossil record. This might seem evident, but it has rarely been demonstrated, or even discussed, especially in the case of the commonly used universal sampling proxy of formation numbers. Further, sampling proxies generally assess only the difference between the known and potential fossil records, and probably have little to say about the unknowable part of reality (see above).

Sampling proxies can be used in various ways, (1) to assess whether a fossil record is biased, (2) to assess the nature of the bias, and (3) to correct the bias and produce a true evolutionary signal. In the first two cases, the sampling proxy may be partial, in that it documents some aspect of rock volume, accessibility, or human effort. In the third case, however, if the sampling proxy is to be used

as a correction factor or the basis of a sampling-free model, it should be *comprehensive*, and so represent the three key factors of rock volume, accessibility, and effort. In cases where supposedly error-free records have been generated and then used to make statements about evolution (e.g. Peters & Foote 2001, 2002; Smith 2001; Smith & McGowan 2007; Barrett et al. 2009; Wall et al. 2009; Benson et al. 2010; Butler et al. 2011), the sampling proxy, whether formation counts or outcrop areas, was not demonstrated to encapsulate all, or even most, of the error signal – this was an assumption. At best, these papers could be said to have provided examples of how to identify parts of the fossil record that cannot be explained by sampling, and so *might* be real. At worst, they show very little because the universal sampling proxies used may have been partially redundant with the fossil record they sought to correct (formation counts) or were uncertain measures of rock volume/accessibility (outcrop areas). This of course is a criticism of the assumption that the sampling proxies were universal and comprehensive, not a claim that the fossil record is complete and unbiased.

A sampling proxy may be compared for matching with the raw diversity signal in various ways: (1) visual inspection; (2) correlation or rank order correlation (e.g. Fröbisch 2008; Barrett et al. 2009; Butler et al. 2009; Benson et al. 2010); (3) rarefaction to equalize sample sizes in each time bin (e.g. Benton et al. 2004; Lloyd et al. 2008); or (4) modelling a sampling-predicted diversity pattern based on numbers of formations or relative map areas, and comparing this with observed diversity to explore correlations and residuals (e.g. Peters & Foote 2001, 2002; Smith 2001; Smith & McGowan 2007; Barrett et al. 2009; Benson et al. 2010; Butler et al. 2011; Mannion et al. 2011; Lloyd et al. 2011).

There are two commonly used sampling proxies, outcrop area and numbers of formations, and a third might be human effort. These three are discussed here, before assessing the merits of the first two from some recent studies.

Sampling proxies 1: outcrop (map) area

It has been argued (e.g. Smith 2001, 2007a; Crampton et al. 2003; Smith & McGowan 2008; Wall et al. 2009) that map area (= area of outcrop) is a good proxy for sampling. Rocks that cover large areas of the landscape are likely to have been much more sampled than those that do not outcrop widely, so this metric incorporates aspects of rock volume, rock availability at the surface, and human factors.

Practitioners of the map area approach have used a variety of methods. Some (e.g. Wall et al. 2009) have extracted information from global-scale summary geological maps (e.g. Ronov 1978, 1994), although these suffer from enormous generalizations and simplifications both in terms of the ages of the rocks and their areas. Time divisions in such studies have necessarily been broad to accommodate the difficulties of welding together information from disparate and uncoordinated national geological surveys; the OneGeology programme (http://www.onegeology.org/) which aims to produce a single, comprehensive world geological map, may help to rectify this. Others have used maps and memoirs from single countries or groups of countries (e.g. NW Europe; Smith 2007*a*; McGowan & Smith 2008; Smith & McGowan 2007), consulting sheet memoirs to record those stratigraphic units as present that have yielded fossils belonging to particular zones. A problem with the latter approach has been that it was used for comparisons of like with unlike (local or regional map data v. global fossil record). These authors argued that their approach was valid as errors become negligible once the map sample is as large as 1300 – Smith & McGowan (2007) showed extremely high congruence between sampled rock areas through geological time obtained from France and Spain. This presumably indicates a considerable amount of fundamental geology shared between the two countries, relating to bedrock, topography, and soil cover, and does not necessarily say much about accessibility (exposure) or about world geology. Furthermore, we see no evidence that scaling up small-scale data that may contain errors and mismatches will necessarily gloss and over-ride those errors. Nonetheless, when comparing like with like (the fossil record and geological map areas for New Zealand), Crampton *et al.* (2003) found that outcrop area was a good proxy for sampling because it correlated closely with number of collections.

Conclusions from such map area v. palaeodiversity studies have been very different, with some (e.g. Peters 2005, 2008) arguing that the covariation of fossil diversity and outcrop areas indicates a common driver such as sea-level change, others (e.g. Wall *et al.* 2009) arguing for a strong bias on estimates of fossil diversity from outcrop area, and yet others (e.g. Marx 2009) finding only modest evidence for an influence of outcrop area on fossil diversity.

A key issue, perhaps not appreciated fully before, is that outcrop (= map area) is not always correlated with rock exposure (= area of rock exposed at the surface), but is heavily modified by overlying deposits and depends on factors such as topographic elevation and coastal intersection (Dunhill in press). As a substantial proportion of outcropping rock is not exposed at the surface, and

fossil specimens can only be recovered easily from exposed localities, it can be argued that rock exposure area represents a better proxy for the amount of sedimentary rock available for study than outcrop (i.e. map) area. Rock outcrop area can only be regarded as a good sampling proxy if it proves to correlate well with current rock exposure area, and, importantly, to be a good representation of the total historical accessibility of geological units, through a combination of current exposure and past collecting opportunities in old quarries, mines, railway cuttings, and landslips. Previous difficulties in quantifying the effects of exposure failure on sampling (Peters & Heim 2010) have been overcome by using remote sensing, in the form of Google Earth^{TM} imagery, and a Geographic Information System (GIS) to map and quantify rock exposure accurately on a local scale (Dunhill 2011).

Dunhill (2011) showed that rock outcrop and exposure area are not positively correlated across 50 sample localities in England and Wales (Fig. 2a, b), and a further investigation showed these results to be consistent with results obtained, using the same methodology, in New York State (Dunhill in press). However, after data manipulation, rock outcrop and exposure area did display positive correlations in data sets from California and Australia (Dunhill in press), suggesting that variables such as climate and population density may have an influence on the amount of rock exposed at the surface. Further tests showed that proportional rock exposure is dependent on a number of variables that are themselves independent of outcrop area, in particular proximity to the coast, elevation, and bedrock age (Dunhill 2011, in press). Coastal areas exhibit consistently higher proportional exposure than inland localities because of constant high erosion regimes. A common pattern in the data was greater proportional exposure of older bedrock compared to younger bedrock (Fig. 2d), which might be explained by the presence of a greater proportion of harder lithologies of Palaeozoic age and more softer, unlithified sediments of Cenozoic age (Hendy 2009). Areas of higher elevation, both in inland and coastal areas, consistently exhibit higher proportional exposure (Fig. 2c), a result of increased erosion at altitude and greater exposure in areas of high coastal cliffs (Dunhill in press). It is apparent that many of the variables influencing the amount of rock exposed at the surface are not independent of one another, with bedrock age and elevation correlating positively in all four sampled regions (Dunhill in press). This can probably be explained by the simple fact that older rocks have been around for longer and have thus had more time to become uplifted.

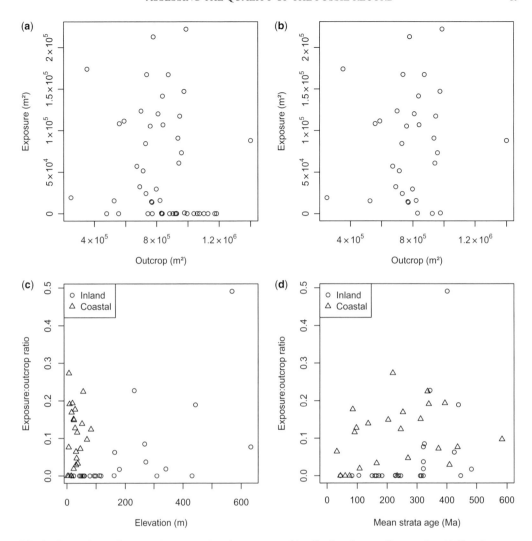

Fig. 2. Comparisons of outcrop (= map area) and exposure, and implications for sampling proxies. (**a**) Negative correlation between outcrop and exposure area for all of the sample localities in the England and Wales data set (Spearman: $r_s = -0.28$, $P = 0.05$). (**b**) Lack of correlation between outcrop and exposure area for the England and Wales data set after areas with zero rock exposure have been removed (Spearman: $r_s = 0.07$, $P = 0.68$). (**c**) Correlation between elevation and exposure:outcrop ratio for all the sample localities in England and Wales (Spearman: $r_s = -0.03$, $P = 0.85$), inland localities only (Spearman: $r_s = 0.72$, $P < 0.001$) and coastal localities only (Spearman: $r_s = 0.15$, $P = 0.47$). (**d**) Correlation between mean strata age and exposure:outcrop ratio for all the sample localities in England and Wales (Spearman: $r_s = 0.37$, $P = 0.009$), inland localities only (Spearman: $r_s = 0.7$, $P < 0.001$) and coastal localities only (Spearman: $r_s = 0.11$, $P = 0.34$) (modified from Dunhill 2011).

The fact that outcrop and exposure area do not consistently correlate across continents, and that rock exposure appears to be controlled by a number of variables that are independent of outcrop area, non-independent of each other, and inconsistent in their effects across different regions of the world, suggests the need for further investigation of the value of outcrop (i.e. map) area as a proxy for sampling. Further, if rock volume is linked with processes that regulate biodiversity, through for example sea-level change on continental shelves (see below, common cause hypothesis), then it might be advisable to exercise caution in interpreting correlations of rock volume sampling metrics and fossil diversity, and especially in using such metrics as the basis for generating a supposedly sampling-free diversity curve (e.g. Smith 2001, 2007*a*; Smith & McGowan 2007; Wall *et al.*

2009). If, as seems likely (see above), these rock volume and area measures (outcrop area, exposure area) assess the potential for future discoveries (i.e. potential minus known fossil record), rather than predict into the unknown part of reality (Fig. 1), then the claim that 'sampling-free' palaeodiversity curves can be generated from them is further put in doubt.

Sampling proxies 2: number of formations

Counts of numbers of geological formations have been used widely as a sampling proxy (e.g. Peters & Foote 2001, 2002; Fröbisch 2008; Barrett *et al.* 2009; Butler *et al.* 2009; Benson *et al.* 2010; Mannion *et al.* 2011). These authors found close or very close covariation between numbers of formations and palaeodiversity, and this has generally been interpreted as evidence for a control by rock volume on apparent biodiversity. For example, Barrett *et al.* (2009), Butler *et al.* (2009), Benson *et al.* (2010), and Mannion *et al.* (2011) used their counts of numbers of formations as a measure of the reality of sampling, and so they modelled 'sampling-corrected' global diversity curves, and based revised interpretations of the history of life on the residuals.

Counts of numbers of formations may reflect a combination of some aspects of exposure area, total thickness, lithological heterogeneity, and intensity of study, and so they might seem to be ideal sampling proxies that encapsulate elements of rock volume, accessibility, and human effort (Peters & Foote 2001). However, formation counts have been widely criticized as sampling proxies because:

(1) Their definitions are arbitrary, being human inventions. The thickness, duration, area covered, and heterogeneity may depend on the customs of geologists, whether by nationality or by main research subject-area, as well as on the recency of revision and definition of formal formation names (Upchurch & Barrett 2005; Peters & Heim 2010).
(2) They do not generally correlate with rock exposure area measurements (Dunhill 2011, in press), which at least offer the simplest guide to accessibility, even though they do not offer a single universal sampling proxy.
(3) They do not necessarily correlate with collection effort (Crampton *et al.* 2003), but see Upchurch *et al.* (2011).
(4) They may largely reflect rock heterogeneity (Crampton *et al.* 2003; Smith 2007a), such that highly varied sediments lead to more and thinner formations, whereas formations are much thicker when the rock types are

unchanging over long time spans, for example, continental red beds or Chalk.
(5) Importantly, they may also depend on fossil abundance and diversity (Wignall & Benton 1999): if fossils are highly abundant and diverse, formations may be subdivided more finely than if fossils are sparse or absent. This means that formation counts are not necessarily independent of the signal they seek to test or correct.
(6) Formations vary enormously in scale: formations vary over several orders of magnitude in thickness and geographical area (e.g. Williams 1901; Peters 2006; Peters & Heim 2010). At the smaller end of the scale are formations within well-studied and highly fossiliferous divisions of the Jurassic or Carboniferous of England that are only metres thick and cover only a few square kilometres. For example, the Beacon Limestone Formation (Lower Jurassic) of the southern UK, reaches a maximum thickness of 5 m (Cox *et al.* 1999), but is typically less than 1 m thick (Simms *et al.* 2004) with an outcrop coverage of around 20 km^2 (BGS digital bedrock geology DiGMapGB-50 of the UK), so representing a maximum volume of 0.073 km^3. At the other end of the scale are units such as the Late Jurassic Morrison Formation of the Midwestern United States that is up to 150 m thick and covers an area of 1.5 million square kilometres (Carpenter 1997), equivalent to a maximum volume of 225 000 km^3. These two examples show eight orders of magnitude difference in rock volume of named formations that might otherwise be treated as equivalent sampling units.

A further issue with the use of formation counts as a measure of sampling is that analysts usually count only *fossiliferous* formations, not all possible formations. Therefore, anything that means organisms are rare and so not fossilized will reduce the fossiliferous formation count, while this figure will rise when fossils are common and abundant. A particular example might be the time bin following a mass extinction, where palaeontologists might recover low numbers of fossils, and indeed those low sample sizes correspond to low numbers of fossiliferous formations. But this does not necessarily indicate sampling bias (Wignall & Benton 1999; *contra* Smith 2001, 2007a, b). Perhaps organisms were truly rare following the mass extinction; therefore fossils are not often found, and so rare fossils mean low numbers of fossiliferous formations. But this is a reversal of the normal sampling assumption. Here fossils determine formation numbers, not the other way round. Smith (2001, p. 355) argued that such a phenomenon could not occur: 'Taxon absences arise because of lack of habitat continuity and/or changing preservational opportunity in the fossil record, not as a result of fluctuating abundance within a uniformly sampled habitat.' Peters & Heim (2010) note, however, that absence of fossils in a stratigraphic unit can reflect sampling failure, human error in reporting, or true absence. Preservation failure is doubtless often the case, but there is no fundamental reason why abundance and diversity of organisms in their communities cannot be reflected in the fossil record.

There are three ways to document the fossiliferous formation count (FFC): a *strict FFC*, consisting of only those formations that have produced named fossils included in the diversity measure; a *wider FFC* consisting of all formations that have ever produced any kind of fossil of the group in question, whether named, unidentified elements, or trace fossils; and a *comprehensive FFC* that includes all formations of the correct facies that have produced, or might produce fossils of the group in question. This allows for future finds, but also documents formations that could contain fossils of the group in question, but do not, and so includes failure of sampling (Wignall & Benton 1999). The need for a wider definition of FFCs was noted by Upchurch & Barrett (2005), when they suggested that a count of dinosaur-bearing formations (DBFs), as used in Barrett *et al.* (2009), and other papers, ought to subsequently sample as widely as possible and include all possible opportunities to observe.

The strict FFC fails as a sampling proxy because it can only assess the difference between the potential and known fossil records (Fig. 1), and nearly always shows strong correlation with palaeodiversity (see below). A wider or comprehensive FFC allows some view into the unknown portion of reality (Fig. 1): for example, in assessing sampling of bird fossils, a comprehensive FFC of all vertebrate-bearing formations will show many that could well have sampled bird fossils, but that entirely lack them, and so can give a proportion of what is missing. However, what is not assessed are the times and places where birds lived in the past, but where rock was not accumulating or has not survived (and so there are no geological formations).

Sampling proxies 3: human effort

Human factors are doubtless hugely important in determining our knowledge of the fossil record, the difference between potential and known (Fig. 1). These human factors were explored especially following Raup's (1972) paper. For example, Sheehan (1977) defined the Paleontologic Interest Unit (PIU) as a measure of the effort devoted to acquiring knowledge, counted in numbers of people, years, or publications on a particular subject, such as fossils of a specified geological interval or taxonomic group. These showed considerable mismatches in terms of effort per million years with, for example, eight times as many palaeontologists (per million years) working on Cenozoic fossils as on Cambrian fossils (Sheehan 1977). Further, it is well understood that the fossils of certain continents have been more intensively collected and studied than those from other continents (e.g. Kiessling 2005; McGowan & Smith 2008) as is confirmed by comparisons of collector curves, where numbers of new taxa identified from Europe and North America have reached an asymptote for many taxa, such as dinosaurs (Benton 1998, 2008a), conodonts (Wickström & Donoghue 2005), echinoids (Smith 2007b), trilobites (Tarver *et al.* 2007), and basal tetrapods (Bernard *et al.* 2010), whereas new taxa from other continents are still on the rising phase of the curve.

The idea of using a measure of human effort as a sampling proxy is reflected in the fossil collections data of the Paleobiology Database (PaleoDB; Alroy *et al.* 2001, 2008; Butler *et al.* 2011), where each collection is an assemblage of fossils collected from a particular formation often by a single palaeontologist or team. Formations with many reported collections then have been very actively sampled. Other measures of effort, as indicated by Sheehan (1977), such as numbers of palaeontologists or numbers of published papers, have proved hard to turn into an acceptable correction algorithm. Further, it would be hard to distinguish whether human factors drive or follow the data: perhaps concentrations of workers reflect abundant and diverse fossils (Raup 1977; Purnell & Donoghue 2005). Until a way can be found to disentangle the direction of causation between human effort and fossil diversity, PIUs may fail as a correction for error in the fossil record because of the probable intimate interconnection of both signals. Further, of course, these measures only assess our current position on the collector curve trajectory to complete knowledge of the potential fossil record (Fig. 1). Other measures of 'effort' though, such as numbers of collections, numbers of sampled localities, numbers of individual fossils (for rarer groups), or specimen quality (e.g. Benton *et al.* 2004; Smith 2007b) may provide useful insights into sampling of reality (see below).

Bias, common cause or redundancy

Models

Many measures of palaeodiversity and of rock volume show long-term covariation through geological time. Rock volume per million years increases towards the present day (Raup 1976; Smith 2001, 2007*a*), as does geological complete-ness of palaeontological sampling (Peters & Heim 2010). Further, there are substantial rises and falls in rock volume per million years within any selected time span, resulting from sea-level change (Smith 2001, 2007*a*; Peters 2005). Generally, the apparent diversity of marine animals follows both of these signals (Smith 2001), rising from the Cambrian onwards and rising and falling in line with sea-level rises and falls, except in the past 100 Ma, when sea-levels have generally fallen while marine diversity has risen. Terrestrial diversity likewise correlates with rock volume (Kalmar & Currie 2010), although causes of variations in rock volume for continental habitats are complex.

The close covariation of rock volume and palaeodiversity (Smith 2001, 2007*a*; Smith & McGowan 2007, 2008), and counts of geological formations (Peters & Foote 2001, 2002; Fröbisch 2008; Barrett *et al.* 2009; Butler *et al.* 2009; Wall *et al.* 2009; Benson *et al.* 2010) or sequence strati-graphic rock packets (Peters 2005, 2008; Peters & Heim 2010) can be interpreted in various ways, namely (1) rock bias; (2) common cause; and (3) redundancy.

The *rock bias hypothesis* is that the fossil record is wholly or partly a result of sampling bias, primar-ily determined by rock volume and accessibility (e.g. Raup 1972; Peters & Foote 2001, 2002; Smith 2001, 2007*a*; Smith & McGowan 2007; Fröbisch 2008; Barrett *et al.* 2009). In simple terms, the more rock, the more fossils (combining aspects of all three biasing factors – rock volume, accessibility, and human effort). Low volumes of rock (low outcrop area or low numbers of for-mations) means fossils cannot be found and so low fossil diversity is interpreted as sampling failure. The implication is that more intensive search of these formations, or ideally the discovery of new formations or areas of a poorly sampled time inter-val then ought to yield fossils at a faster rate than those from a well-sampled time bin.

The *common cause hypothesis* (the 'biologically driven' model of Smith 2001; Peters 2005, 2008) is that measures of rock volume and fossil diversity or abundance are correlated because they are driven by a third, common cause, such as sea-level change. It could be that biodiversity is driven by rock volume, so that marine life is much more diverse at times of rapid sedimentation (wide continental shelves; high habitat heterogeneity; high productivity, with much organic matter swept in from the land) than at times of low sedimentation (narrow continental shelves; low habitat heterogeneity; low productivity). The expanding and contracting marine shelf then drives expansions and contractions in marine bio-diversity, which is dominated by the portion that occupies continental shelves, a kind of species-area effect (Smith 2001). Sea-level has also been posited as a driver of terrestrial diversity (e.g. Smith 2001; Mannion & Upchurch 2010), with high sea-levels corresponding either to low terrestrial diversity because of smaller land areas and fewer opportu-nities for preservation of habitats (Markwick 1998; Smith 2001) or to high terrestrial diversity because of more islands, leading to more endemism, as well as more chances for skeletons to fall into aquatic habitats and for coastal habitats to be swamped by the sea (Haubold 1990; Mannion *et al.* 2011). However, these suggestions are not sup-ported by evidence from the Cretaceous terrestrial record (Fara 2002), and, although terrestrial diver-sity correlates with rock volume (Kalmar & Currie 2010), it is not clear how the volume of terrestrial clastic sediment could relate to sea-level (Butler *et al.* 2011).

The *redundancy hypothesis*, proposed here, is that in many cases the supposed sampling proxy and the fossil record are to a greater or lesser extent redundant with each other. This is especially true when the fossil record is patchy, perhaps for birds or pterosaurs – each fossil find adds a new fos-siliferous formation to the roster. In such cases (e.g. Butler *et al.* 2009), the correlation between fossil count and formation count will be nearly perfect. The simplest 'correction', where number of fossil taxa is divided by number of formations will yield a flat line, as found by Lloyd *et al.* (2008) when they subsampled numbers of dinosaur species by dinosaur localities through 12 Mesozoic time bins. But such a confrontation of two redundant signals says nothing about sampling or the true unbiased pattern.

We now present re-analyses of two recently published studies, one on anomodonts and one on pterosaurs, to explore situations where bias (or 'megabias') was claimed, but where the supposed sampling proxy is really redundant with the palae-ontological diversity signal.

Anomodonts: redundancy of data and sampling proxy

As noted earlier, it is important to consider the kind of formation count employed, whether the strict, wider, or comprehensive fossiliferous formation count (FFC). We explore the effects of using these three variants in a recent study of anomodont diver-sity through the Permian and Triassic (Fröbisch 2008). Fröbisch (2008, 2009) divided the fossil record of anomodonts into a variety of time bins: land vertebrate faunachrons (LVFs), standard marine stages, and million-year divisions. The LVFs were his first determination of ages, and the others were derived from them, so the test here focuses on the LVFs. All 128 anomodont species were assigned to the 13 LVFs according to data from Fröbisch's (2008) paper, and the numbers of anomodont-bearing formations for each time bin were listed (Table 1). In addition, the numbers of named stratigraphic units with *any* tetrapod fossils were also listed, based on unpublished data from Benton *et al.* (2004), Sahney & Benton (2008), Sahney *et al.* (2010), and other sources. As Fröbisch (2008) found, the time series of anomodont species diversity correlates significantly with the time series of anomodont-bearing formations, but not with the time series of all tetrapod-bearing formations, with which anomodont diversity shows a negative, but not significant, correlation (Fig. 3; Table 2).

In order to test further whether the strict FFC could ever be an independent sampling proxy for 'diversity of X', a series of randomized trials was carried out to see whether it would be possible to break the strong correlation of time series of anomodont species diversity with number of anomodont-bearing formations. The numbers of recorded formations per time bin (Table 1) were used as a starting point. For each of the 84 anomodont-bearing formations reported in Fröbisch (2008), a total species diversity from 0–5 was generated using random numbers – the minimum of 0 allows for no finds in a formation, and 5 is the maximum reported, for the Gamkan 1 time interval, in Fröbisch (2008). These were summed for each time bin according to the reported numbers of formations, and forty simulations were performed, sufficient to assess the statistical significance of the

results at the 95% level for a two-tailed test (95% probability = 1 in 20; × 2 for a two-tailed test). Remarkably, the simulated anomodont totals and the actual numbers of anomodont-bearing formations were significantly correlated in all but three cases (Table 2). In fact these randomized trials gave rank correlation values in a tight range around a mean and mode of 0.60, identical to the actual value obtained by Fröbisch (2008), and with significance values equal to or slightly better than in the real example (Table 2). This suggests not only that random data *can* produce results similar to those found with real data, but that this will occur most of the time. Hence, it is probably impossible ever to find a non-significant relationship between time series of sparsely occurring fossils and their strict FFC, and so the discovery of such a correlation in such cases says nothing about sampling, quality of the fossil record, or megabiases.

In order to avoid any spurious correlations arising from autocorrelation within the time series, the data were detrended by taking first differences of anomodont diversity and formation numbers, and these confirmed the significant correlation of changes in anomodont diversity and changes in strict FFC, but an absence of correlation of changes in anomodont diversity and changes in the comprehensive FFC (Table 2). The same is true also for generalized differencing (GD), a more comprehensive technique that incorporates detrending and differencing but modulates the differences by the strength of the correlation between successive data points (McKinney 1990). The GD results show a highly significant correlation between numbers of

Table 1. *Comparison of diversity of anomodont fossils through 13 time bins, spanning from the Middle Permian (bottom) to Late Permian*

Stage	Lucas LVF	Karoo LVF	Myr	Anomodont species	Anomodont-bearing fms	All fmns
Nor(l)	Revueltian		2	2	2	16
Crn(u)	Adamanian		3.5	2	2	17
Crn(m)	Otischalkian		2.5	2	3	17
Lad(u)-Crn(l)	Berdvankian		5	3	6	31
Ans(u)-Lad(l)	Perovkan	*Cynognathus C*	4	23	8	40
Ole(u)-Ans(l)	Nonesian	*Cynognathus A, B*	6.5	13	13	59
Ind-Ole(l)	Lootsbergian	*Lystrosaurus*	2.5	9	8	49
Chx(m-u)	Platbergian	*Dicynodon*	3	34	12	8
Wuc(u)-Chx(l)	Steilkransian	*Cistecephalus*	1.5	27	6	16
Wuc(m)	Hoedemakeran	*Tropidostoma*	2.5	16	8	15
Cap(u)-Wuc(l)	Gamkan II	*Pristerognathus*	2	7	4	11
Cap(l-m)	Gamkan I	*Tapinocephalus*	4.5	13	3	11
Wor	Kapteinskraalian	*Eodicynodon*	2	5	2	15

Standard stratigraphic stages, and two systems of Land Vertebrate Faunachrons (LVF) are given, as well as durations of these intervals, in Myr (millions of years). The Dicynodont-bearing formations (fmns) are those from Fröbisch (2008) that yielded anomodont fossils, whereas 'All formations' are all named stratigraphic units that have yielded any kind of fossil tetrapod remains. Stage-name abbreviations: Ans, Anisian; Cap, Capitanian; Chx, Changhsingian; Crn, Carnian; Ind, Induan; Lad, Ladinian; Nor, Norian; Ole, Olenekian; Wor, Wordian; Wuc, Wuchiapingian.

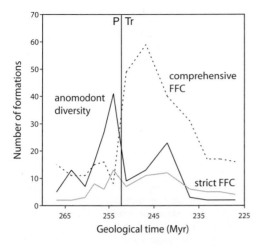

Fig. 3. Diversity of anomodont species through the Permo-Triassic, showing the close matching of the time series for anomodont diversity (black line) and anomodont-bearing formations (grey line; strict FFC). The time series for *all* tetrapod-bearing formations (dashed line; comprehensive FFC) follows a rather different course. The Permo-Triassic (P-Tr) boundary is indicated by a vertical line. Based on data in Fröbisch (2008, 2009), and unpublished.

anomodont species and the strict FFC, but a highly insignificant negative correlation with the comprehensive FFC.

As a non-independent metric, these results suggest that anomodont FFC cannot be used as a sampling proxy or correcting factor for the time series of anomodont genera because the two signals are essentially identical. Much more fruitful would be to compare the diversities of anomodonts with total numbers of tetrapod-bearing formations (Table 2), and to seek to understand why, for example, anomodonts were apparently highly diverse during *Tapinocephalus*, *Cistecephalus*,

Dicynodon, and *Cynognathus* C Zone times when the comprehensive FFC was relatively low (Fig. 3).

Pterosaurs: redundancy and the Lagerstätten effect

The fossil record of pterosaurs is an extreme example of episodicity (Barrett *et al.* 2008; Butler *et al.* 2009), where a dozen or so Lagerstätten, such as the Late Triassic Zorzino Limestone of North Italy, the Late Jurassic Solnhofen Limestone of Germany, the Early Cretaceous Santana Formation of Brazil and the Jehol Group of China, and the Late Cretaceous Niobrara Chalk Formation of North America account for more than half the species and genera of pterosaurs ever recorded. This gives the time series of pterosaur species occurrences a spiky appearance (Fig. 4a), each peak representing one or more Lagerstätten.

When the data are detrended and normalized, the strong correlation between pterosaur-bearing formations and pterosaur diversity remains with first differences (FD), but becomes a highly non-significant negative correlation with generalized differencing (GD; Table 3). This mixed finding suggests that much of the correlation between ptero-saur palaeodiversity and pterosaur-bearing formations may relate to the overall trend (GD), but the surviving correlation with FD may be a trivial result because the count of pterosaur-bearing formations (PBF) is not a comprehensive FFC (it includes formations from which named pterosaurs were found as well as others that yielded pterosaur fragments), and so both curves are intimately linked, each documenting the same episodic preservation of fossils, and it cannot be said that one metric explains the other. The key question is whether a patchy fossil record such as this is simply tied to the Lagerstätten, or might also reflect some wider dependence on rock volume or accessibility.

Table 2. *Correlations of anomodont species diversity through time with different proxies for formation numbers, showing rank-order correlations for the raw data and for first differences (FD) and generalized differences (GD)*

	Spearman's ρ	P
Anomodont-bearing formations	0.60	0.034*
Randomized species numbers ($n = 40$)	0.53–0.66	0.015* − 0.062
Mode	0.60	0.029*
Mean	0.60	0.033*
FD anomodont-bearing formations	0.58	0.041*
GD anomodont-bearing formations	0.76	0.006**
All tetrapod-bearing formations	−0.22	0.472
FD all tetrapod-bearing formations	−0.13	0.646
GD all tetrapod-bearing formations	−0.12	0.716

Probabilities (P) for each correlation measure are given, and these are marked as significant ($P < 0.5*$) and highly significant ($P < 0.005**$).

Table 3. *Correlation of phylogenetically corrected diversity estimate (PDE) for species of pterosaurs (actual records plus ghost ranges), from Butler et al. (2009) with number of pterosaur-bearing formations (PBFs) from Butler et al. (2009), counted as raw data, $\log_{10}$ of raw values, to create a normal distribution of the data, and first differences (FD) to detrend the data, and divided by substage duration (∂t) to standardize for time. Comparisons are also made of raw pterosaur counts (TDE) and PDE with dinosaur-bearing collections (DBC) and dinosaur-bearing formations (DBF), as raw data and $\log_{10}$-tranformed data, both from Butler et al. (2011), to approximate a comprehensive FFC*

	Pearson's r	P	Spearman's ρ	P	Kendall's τ	P
PDE v. PBF	0.61	0**	0.56	0**	0.46	0**
$\log_{10}$ (PDE v. PBF)	0.49	0**	0.56	0**	0.46	0**
FD (PDE v. PBF)	0.40	0**	0.34	0**	0.30	0.002**
GD (PDE v. PBF)	−0.05	0.645	−0.09	0.429	−0.07	0.406
FD/∂t (PDE v. PBF)	0.97	0**	0.35	0**	0.29	0.002**
TDE v. DBC	0.32	0.113	0.64	0**	0.51	0**
TDE v. DBF	0.49	0.012*	0.54	0.004**	0.41	0.005*
$\log_{10}$ (TDE v. DBC)	0.60	0.001**	0.65	0**	0.48	0.001**
$\log_{10}$ (TDE v. DBF)	0.59	0.002**	0.55	0.004**	0.42	0.004**
PDE v. DBC	0.08	0.699	0.37	0.061	0.31	0.032*
PDE v. DBF	0.20	0.319	0.21	0.314	0.17	0.238
$\log_{10}$ (PDE v. DBC)	0.21	0.313	0.36	0.068	0.31	0.034*
$\log_{10}$ (PDE v. DBF)	0.14	0.499	0.20	0.333	0.17	0.246

Probabilities (*P*) for each correlation measure are given, and these are marked as significant (*P* < 0.5*) and highly significant (*P* < 0.005**).

If the measure of rock volume had been a comprehensive one, such as 'all fossiliferous Mesozoic rock units' or 'all Mesozoic rock units with vertebrate fossils', then the sampling measure would be less evidently redundant with the pterosaur fossil record. In this study, we chose counts of all dinosaur-bearing formations and all dinosaur collections from the PaleoDB, as given by Butler *et al.* (2011), as proxies for comprehensive FFCs (Fig. 5). Correlations are strong with the raw diversity measure (TDE), but limited with the phylogenetically corrected measure (PDE) (Table 3; Fig. 5). For the raw data, all three correlation measures yielded largely significant results, although the strict correlation (Pearson) provides generally less significant results than the rank-based measures (Spearman, Kendall). This suggests, perhaps surprisingly, that there is a sampling signal linked to the wider availability of suitable rocks through the Mesozoic, lying behind the dominant sampling signal from the small number of crucial Lagerstätten. Interestingly, linear correlations were very poor (Pearson's *r*), rank-order correlations were poor (Spearman's *ρ*), and yet the phylogenetically corrected pterosaur numbers and dinosaur-bearing collections showed some evidence that rises and falls were in phase with each other (Kendall's *τ*). This study confirms that the pterosaur fossil record is dominated by ten or twelve Lagerstätten, as already shown, but that the Mesozoic record of fossiliferous units (whether DBF or DBC) apparently follows the pattern of occurrence of those

Lagerstätten, and so covaries with the pterosaur palaeodiversity curve to some extent as well (Fig. 5); in other words, and unexpectedly, the concentration of Lagerstätten in the Late Triassic, late Jurassic, and mid Cretaceous matches times of high numbers of dinosaur collections in the PaleoDB.

To return to the Butler *et al.* (2009) paper, it is not, however, clear what the modelled pterosaur diversity curve, with wider FFC removed, actually documents (Fig. 4a): it is hardly a 'true' or corrected global signal of pterosaurian palaeodiversity because Lagerstätten and rock volume have nothing to do with each other. In other words, if a particular fossil record is dominated by Lagerstätten, rock volume and diversity need not correlate. If there had been a clear correlation between number of formations and palaeodiversity, so that each spike in diversity really averaged out across several formations or localities, then this would not be a record dominated by Lagerstätten. Both could occur at the same time, with rock volume and palaeodiversity rising and falling together and, on top of that, times of particularly high diversity might be made even more pronounced by the presence of one or more Lagerstätten. However, these are two separate things – Lagerstätten are rich in fossils, not rich in rock.

If this is the case, then a fossil record dominated by Lagerstätten, such as that of pterosaurs, is largely determined by intimate details of how each extraordinarily rich deposit is exploited – the 'Jehol peak' (Aptian–Albian), for example, was zero a

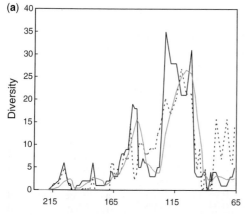

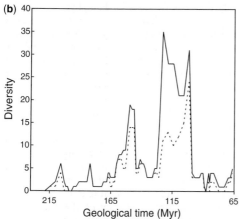

Fig. 4. Diversity of pterosaur species through the Mesozoic. (**a**) Raw species count (solid black line), modelled species diversity, with the effect of number of formations removed (dashed line), and five-point moving average (grey line). (**b**) Raw species diversity according to a 2008 data base (solid black line), and based on the number of pterosaurian taxa named before 1990 (dashed line). Data from Barrett *et al.* (2008) and Butler *et al.* (2009).

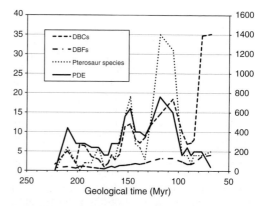

Fig. 5. Covariance of pterosaur diversity and wider sampling measures, the number of dinosaur collections in the PaleoDB (DBC) and the number of dinosaur-bearing formations (DBF). The DBC measure in particular shows a spiky pattern, with highs in the Late Jurassic (Kimmeridgian, Tithonian), mid Cretaceous (Aptian, Albian), and latest Cretaceous (Campanian, Maastrichtian). The first two of these peaks correspond to times of pteroaur-bearing Lagerstätten, and this is highlighted especially in the raw pterosaur palaeodiversity measure (TDE), but much less so in the phylogenetically corrected measure, including ghost ranges (PDE). Correlation measures are shown in Table 3.

few years ago (Fig. 4b), and now consists of 20 pterosaur species, and soon might reach 25 or 30, or it might fall if [sp.] taxonomic revisions reveal some synonymy of species names. The Jehol peak is founded on the fossil contents of two formations, the Yixian and Jiufotang, in NE China. The raw data, and the modelled sampling time series, are equally dependent on the current state of research, and so neither can 'correct' the other. This can be illustrated by stripping the raw data back to the position at the end of 1989 – all post-1990 finds are excluded from the Barrett *et al.* (2008) data base, and the high peaks in particular are substantially reduced (Fig. 4b). This removal of the past 20 years of research effort effectively halves the total

number of pterosaur species known (from 118 today to 66 at the end of 1989), but 20 of the 52 species removed come from the Yixian and Jiufotang formations of China, eight from the Crato and Santana formations of Brazil, and the remainder scattered throughout other less productive localities. The great majority of new finds reported since 1990 are from previously known formations, and most are from a small number of Lagerstätten. The shape of the diversity time series, whether based on raw or modelled data, in a Lagerstätten-driven signal such as the pterosaur fossil record, is dependent more on intensity of collecting in known Lagerstätten rather than the number and distribution of those fossil-bearing formations. Further, at any time a new Lagerstätte may be found or exploited, as the Jehol Group formations were in the 1990s, and the addition of numerous fossil taxa corresponds to only a comparatively trivial addition to the FFC, the basis of the modelling approach employed by Butler *et al.* (2009).

Is there any meaningful way to turn such a Lagerstätten-driven fossil record into a sampling-free distribution? One might apply various techniques to reduce the spikiness of the plot, such as reading only the residuals after formation numbers have been considered, or rarefying, or shareholder quorum sampling (Alroy 2010), although the latter two are likely to return a flat line. The modelling approach (Smith & McGowan 2007) is to rank the species and number of formations time series, calculate their straight-line relationship ($y = 1.1076x - 4.1948$, in this case; Butler *et al.* 2009), then apply a correction by subtracting this modelled diversity estimate (MDE) from the observed diversity estimate (TDE) for the time bin, that is assuming that true diversity is constant and observed diversity is driven entirely by sampling. The modelled diversity estimates track the raw data closely for much of the signal (Fig. 4a), but lie below or above in places, so suggesting the influence of factors other than sampling on the signal (Butler *et al.* 2009). However, as noted, this method has not accounted for sampling, and it has probably removed much real diversity signal.

A second approach might be to seek to smooth the spikiness of the species diversity time series. This has already been done to some extent in the lumping approach taken to the stratigraphic substages, similar to the method employed in Barrett *et al.*'s (2009) dinosaur study. For example, the Yixian Formation is dated by Barrett *et al.* (2008) as 'late Barremian to Aptian', the Jiufotang Formation as 'Aptian', the Crato Formation as 'late Aptian to early Albian', and the Santana Formation as 'Aptian–Albian'. These age designations correctly reflect current uncertainties, and they vary from 2 to 6 substages in duration. Each valid pterosaurian species from those formations was then scored from 2 to 6 times, depending on the age uncertainty, but this bears no relation to the actual age, which might eventually turn out to fall entirely within a single substage for each of the formations. A similar 'smearing effect' could be achieved by adopting a 5-point moving average, for example (Fig. 4a), but there is no justification for either approach. Alternatively, a 'tightrope' could be drawn, linking the high peaks of pterosaurian diversity, based on the assumption that the Lagerstätten reveal something about the true diversity of pterosaurs at any time. This approach at least avoids the problem of all sampling standardization techniques in that they penalize the best fossil records in favour of the poorest fossil records in a time series. However, any such corrections are transient, dependent on minute details of the study of a small number of geological formations, and impossible to interpret, representing as they do sporadic and geographically restricted samples. Probably both approaches are best avoided with such sporadic fossil records as that of pterosaurs in that any 'corrections' add levels of uncertainty and hypothesis to an already uncertain and patchy fossil record. In conclusion, the pterosaur record is patchy – we know that – and for phylogenetic interpretation we can identify weaker and stronger episodes, but statistical manipulations probably add little information.

Collector curves and age v. clade metrics

In both studies, the authors (Fröbisch 2008; Butler *et al.* 2009) argued that they had demonstrated that their fossil records were biased. In doing so, they rejected the common cause hypothesis, and did not consider the redundancy hypothesis advanced here.

Certainly, we would argue that Fröbisch (2008) did not demonstrate 'an obvious rock record bias affecting the diversity curve of anomodonts during at least parts of the Permian and Triassic', nor that Butler *et al.* (2009) showed that the pterosaur fossil record is 'controlled by geological and taphonomic megabiases rather than macroevolutionary processes'. Although they are almost certainly right, their method did not demonstrate what was claimed.

The key point of the redundancy hypothesis is that it rejects the possibility of using the rock volume measure as a sampling proxy. It does not address whether the record is biased or not. Our point is that the studies by Fröbisch (2008) and Butler *et al.* (2009) tell us nothing about whether the fossil records of anomodonts or pterosaurs are good or bad – other investigations are needed to assess that. Nor are we arguing that these two papers are uniquely uninformative – such assumptions have been made in many other papers, all of which require careful reconsideration along the lines we suggest.

Our key concern is that, in cases such as these, the authors show a 'corrected' diversity curve, as if the error has been removed (e.g. Fröbisch 2008; Barrett *et al.* 2009; Butler *et al.* 2009). And yet, having failed to distinguish the empirical fossil record signal from the sampling signal (strict FFC data for anomodonts; wider FFC data for pterosaurs), the 'corrected' curves might represent something closer to the true diversity pattern than the uncorrected curves, but equally they might not. In other words, modifying the raw data with information from any measure of FFC, gives a different pattern of diversity through time, but it is unclear what is represented.

Something more can be said about the quality of the fossil record of anomodonts and pterosaurs. Even though the formations proxy approach has said nothing about the quality of these respective fossil records, there are at least two established methods that provide some insights, namely collector curves and age v. clade metrics. Collector curves (= species discovery curves) for anomodonts and pterosaurs show rather different trajectories for each group (Fig. 6). The anomodonts show a steady accretion of new species from 1850 to the present day, close to the pattern of species discovery detected for North American fossil mammals and for trilobites. The pterosaurs, on the other hand, show a rapid accumulation of valid species in the nineteenth century, relatively faster than for any of

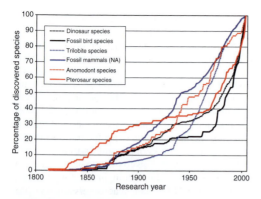

Fig. 6. Collector curves, or species discovery curves, for several fossil groups, including anomodonts and pterosaurs. All discovery curves are shown as percentages, even though final totals, in 2003, are very different: trilobites ($n = 4126$), early tetrapods ($n = 515$), dinosaurs ($n = 694$), fossil birds ($n = 221$), fossil mammals of North America ($n = 3340$), anomodonts ($n = 124$), and pterosaurs ($n = 130$). The 50% line marks the 'half life' of the discovery curve, the date by which half the currently valid taxa had accumulated. Data from these sources: trilobites (Tarver *et al.* 2007), dinosaurs (Benton 2008*a*), fossil birds (Fountaine *et al.* 2005), fossil mammals (Alroy 2002), anomodonts (Fröbisch 2008, 2009), and pterosaurs (Butler *et al.* 2009; Dyke *et al.* 2009).

the other test groups (although note the relatively small sample size for pterosaurs), and somewhat akin to the species discovery record of fossil birds, although beginning much earlier. Fossil birds, dinosaurs, and pterosaurs share a pattern of steeply rising rates of species discovery since 1970, a pattern not seen in the other groups. None of the species discovery curves (Fig. 6) shows a convincing asymptote, although North American fossil mammals come closest, followed by anomodonts. These observations suggest that the potential anomodont fossil record is probably better explored than the potential pterosaur fossil record, confirming the expectation of most palaeontologists. The collector curve approach cannot shed any light on how well the fossil records of these taxa reflect reality (Fig. 1).

An alternative method of exploring quality and sampling, and one that does have the potential to compare the fossil record with reality, is to confront age (stratigraphic) and clade (phylogenetic) data (Norell & Novacek 1992; Benton & Storrs 1994; Benton *et al.* 2000): good congruence between the two indicates that the phylogeny is reasonably accurate and that the fossil record is good enough to document fossils in the right order, whereas low congruence could mean that either the phylogeny or the fossil record, or both, are at fault. Age v. clade congruence metrics for anomodonts and

pterosaurs are good, but not exceptional. For example, 13 cladograms of synapsids and therapsids, the larger clades including anomodonts, have Stratigraphic Consistency Indices (SCI; Huelsenbeck 1994) of 0.60–0.86 (mean, 0.74), Relative Completeness Indices (RCI; Benton & Storrs 1994) of 66.7–97.9 (mean, 80.4), and Gap Excess Ratios (GER; Wills 1999) of 0.50–0.96 (mean, 0.82), all well above the mean values for a sample of 1000 cladograms of all taxa (Benton *et al.* 2000), including plants, invertebrates, and vertebrates (SCI, 0.55; RCI, 31.13; GER, 0.56). In all cases, values of 1.00 (SCI, GER) or 100 (RCI) indicate perfect congruence. In his comparison of anomodont cladograms, Angielczyk (2002) found a range of RCI values from 28.2–63.2 (mean, 41.5), and GER values from 0.68–0.86 (mean, 076), again well above the global means, and so suggesting that, at the scale of genera and stratigraphic stages or substages, the anomodont fossil record shows better-than-average congruence with phylogeny. In the case of pterosaurs, Dyke *et al.* (2009) found RCI of 39.4 and −102.1, SCI of 0.58 and 0.62, and GER of 0.85 and 0.82 for two cladograms of pterosaurs; apart from the devastatingly low RCI value for the second cladogram, which reflects an enormous amount of ghost range, the values are well above average, suggesting generally excellent congruence between the fossil record of pterosaurian genera and phylogeny.

Exploring fossil record incompleteness

The position reached so far is not that the fossil record is good or bad, but rather that many of the global-scale methods used recently to explore the bias and incompleteness of the fossil record fail in their core aim. The key question in the minds of palaeontologists is whether the fossil record is adequate to make a particular macroevolutionary or palaeobiological study, or not. Two major subsidiary issues are (1) testing the validity of the bias, common cause, and redundancy hypotheses, and (2) seeking to correct the empirical fossil record time series to generate a truer signal.

We consider these two issues first, and then outline four approaches for exploring error and bias in the fossil record, (1) regional exploration of geological completeness; (2) regional and local exploration of sampling completeness; (3) phylogenetic and gap-counting methods; and (4) model-based comparison of sampling bias and other explanatory variables.

Common cause or bias?

Peters (2005, 2008) advanced the common cause hypothesis as the best explanation for the pervasive correlation of fossil and rock volume signals in the marine realm, and Smith (2007*a*) argued against it and for the bias hypothesis. So far, a decisive test has not been attempted on the global-scale data.

In a more focused study, Butler *et al.* (2011) attempted to test between the bias and common cause hypotheses in the evolution of dinosaurs; they showed how the various signals are mixed and indeed how difficult it is to devise a conclusive test. They found strong correlation among all metrics, namely between the fossil record signal of dinosaurian species diversity through time with measures of sampling (dinosaur-bearing collections from the PaleoDB, and dinosaur localities), and with measures of sea-level (two sea-level curves, estimated non-marine surface areas). The results became clearer after detrending, when the strong linkage between diversity and the sampling measures was confirmed, but not with the sea-level measures. They rejected the 'terrestrial common cause hypothesis' and they considered 'variation in sampling to be the preferred null hypothesis for short-term diversity variation in the Mesozoic terrestrial realm.' This is still suggestive, however, as Butler *et al.* (2011) acknowledge, when they add that 'The long-term trend towards increased sampling and dinosaur taxic diversity through the Mesozoic may result from a genuine increase in dinosaur diversity through this time period, increased opportunities to sample dinosaurs in younger rocks, or a combination of these two factors.' When numerous metrics covary in different ways and to varying degrees, it is hard to reject the influence of one or other factor.

A further issue with this study is that one 'terrestrial common cause' model has been rejected, but others almost certainly exist: is the terrestrial rock record and biodiversity driven by sea-level, continental area, mountain building, rifting, or climate change? All could affect rock volume and biodiversity. Most likely, habitable areas on land are determined by a combination of these factors, and terrestrial biodiversity may be influenced by a combination of such physical environmental drivers as these, as well as opportunism and the evolution of novel adaptations (e.g. the evolution of the ability to fly in pterosaurs and birds), and so it would be hard to capture such complexity in a comparison of diversity and physical signals.

Correcting the fossil record for sampling

Attempts have been made to correct the raw fossil record data with evidence from sampling proxies, namely outcrop areas (e.g. Smith 2001; Smith & McGowan 2007; Wall *et al.* 2009) or formation counts (e.g. Barrett *et al.* 2009; Benson *et al.* 2010; Butler *et al.* 2010). In these cases, the method calculates a modelled diversity estimate that represents the diversity expected if observed diversity variations result solely from the correcting factor (outcrop area or formation count). Diversity residuals (i.e. the differences between modelled diversity values and actual diversity values) following correction for sampling then provide the supposedly sampling-free signal that may represent genuine biological signal, or may be explained in other ways. Note, however, that the 'sampling-free' residuals left after sampling standardization by outcrop areas differ in the studies by Smith & McGowan (2007) and Wall *et al.* (2009), as a result of the different palaeodiversity estimates and the different metrics of outcrop area used by both teams. It is not clear then which of these two 'corrected' curves is likely to be more informative about the true palaeodiversity signal, or whether in fact either of them is closer to the truth than the empirical data.

We have presented evidence above that number of formations is a poor global sampling proxy because of huge variations in the scale and definition of formations, and because formations document sediment and fossil heterogeneity and so are not independent of the signal they seek to correct. Further, map (= outcrop) area may also be suspect as a global sampling proxy because it does not always correspond to exposure area, perhaps a closer measure of rock availability (Dunhill 2011, in press). Therefore, these proxies on their own may be inadequate as simple correction metrics, and yet they might be elaborated to assess rock volume, accessibility, and human effort by the use of alternative metrics such as counts of collections or localities, considerations of fossil quality, and Lagerstätten. In all these cases, however, the risk of circularity (two-way causation), in other words the partial to complete redundancy of empirical and sampling signals, as noted above for formation counts and palaeontological interest units (Raup 1977; Smith 2007*a*), must be considered. It is not wise to term any of these sampling proxies simply 'sampling', as if the complex interdependence of data and sampling signal does not exist.

As argued below, the use of more subtle sampling measures, such as the number of localities or fossil collections, used as a basis for sampling standardization by several authors (e.g. Alroy *et al.* 2001, 2008; Crampton *et al.* 2003; Benton *et al.* 2004; Alroy 2010; Butler *et al.* 2011), may offer a better approach. Other sampling measures might include number of specimens (whether raw numbers or relative numbers), dispersion of sampling sites (Barnosky *et al.* 2005), and quality of specimens (completeness of preservation), but all of these are unlikely to be practical for global-scale studies incorporating diverse taxa. They are perhaps best employed at regional or clade scale, where rarefaction and other sample standardization methods may be used (e.g. Raup 1975; Benton *et al.* 2004; Barnosky *et al.* 2005).

It should be noted that rarefaction, although commonly used to standardize sample sizes, makes the assumption that 'even sampling is fair sampling', which requires that organismal abundance is constant through time, which is unlikely. In addition, there may also be a problem with community composition – if a diverse community is dominated by one or more particularly abundant taxa, it will be undersampled in terms of diversity. Overall, subsampling global data by rarefaction risks seriously 'dampening' the results, reducing peaks in palaeodiversity to a flat line (Bush *et al.* 2004; Marshall 2010). Alroy's (2010) shareholder quorum sampling method seeks to reduce this problem, ensuring that uncommon taxa are more fairly represented, but the method can only retrieve 'most of the common taxa and a stochastic assortment of the rare ones', and so some of the global signal-damping effects of rarefaction are retained. Finally, it may be that rarefaction, and equivalent techniques, really lead to a 'relative' diversity estimate, rather than an absolute one, being equivalent to, for example, 'diversity per X samples'. It is not clear how this relates to actual palaeodiversity.

This then casts doubt on the usefulness of global palaeodiversity curves 'corrected' by the use of sampling standardization (e.g. Alroy *et al.* 2001, 2008; Alroy 2010). The correction techniques themselves have been questioned (Bush *et al.* 2004), and these approaches correct only the collections included in the study, and do not consider missing collections (Smith 2007*a*). Further, the empirical curves, and the 'corrected' versions have evolved substantially over the ten-year span of these studies, as more data have been added to the PaleoDB. The 'corrected' curves differ from the empirical curve (Sepkoski 1993; Benton 1995) not only in suggesting that diversity in the sea reached modern levels in the Palaeozoic, but also in highlighting elevated diversity spikes in the Devonian, Permian, Late Cretaceous, and Palaeogene. These could be novel discoveries that require explanation, or they could reflect uneven data entry into the PaleoDB, or they may have been generated in part from the data manipulations.

Testing and correcting for bias

Regional exploration of geological completeness. Peters and colleagues (Peters 2005, 2008; Peters & Heim 2010) have pioneered a new approach to investigating the completeness of the rock record by focusing on gap-bound sediment packages.

Some 19 000 such units spanning the Phanerozoic, and encompassing all recorded surface and subsurface rock sections from the United States and Canada, are compiled in their macrostratigraphic database 'Macrostrat' (http://macrostrat.geology. wisc.edu). These sediment packages are not subject to human whim or dependent on habitat or fossil heterogeneity, as are geological formations. In their analysis of these data, Peters & Heim (2010) identify a long-term increase in rock record completeness through the Phanerozoic, with many rises and falls, an especially high peak in the latest Cretaceous, and a dip to early Mesozoic completeness levels in the Neogene. The Cretaceous peak and Neogene dip correspond to a similar phenomenon reported by Wills (2007) in assessing congruence indices through geological time, perhaps indicating a real pattern of rock record completeness.

These stratigraphic data allow detailed estimates of rock volume through time, as well as estimates of completeness of representation of fossiliferous units taken from the PaleoDB. The Mesozoic and Cenozoic are better sampled than the Palaeozoic; on average, Cenozoic time intervals have a geological completeness that is approximately 40% greater than mean Palaeozoic completeness (Peters & Heim 2010).

This approach to assessing geological completeness, limited to North America at present, has the benefit of representing sedimentary rock volume in a more comprehensive and accurate manner than counts of geological formations or map areas. The implications for assessing fossil record bias through time may also be important. For example, sampling standardization of PaleoDB data using rarefaction and equivalent techniques, omission of taxa with extant members, and other data processing approaches (Alroy *et al.* 2001, 2008; Alroy 2010) produce corrected curves for marine diversification through time that confirm Raup's (1972) bias simulation model, namely that most of the apparently low diversity levels in the Palaeozoic and Mesozoic represent sampling failure, and that the rise in diversity over the past 100 Ma is not real. The Macrostrat data, on the other hand, seem to imply that sampling cannot be solely or even largely responsible for the observed increase in marine generic diversity in the past 100 Ma (Sepkoski 1997), especially in view of the steep dip in sampling proportions in the Neogene according to the three criteria assessed by Peters & Heim (2010).

Regional and local exploration of sampling completeness. Areas of outcropping sedimentary rock may not yield any fossils at all, and large expanses of homogeneous outcrop might very well yield the same fossil assemblage throughout, or might show increasing diversity with area. It may, therefore, be preferable to use a measure of sampling directly related to the diversity data under scrutiny. Direct measures of sampling might be the volume of material collected (e.g. number of specimens), the number of individual find spots within a location, or an assessment of the effort involved. Although relatively straightforward to apply to small-scale field studies, it is difficult to apply such direct measures to large-scale palaeodiversity studies, and at best it may be possible to compare global or continental datasets to the number (Fara 2002; Lloyd et al. 2008) or area (Barnosky et al. 2005) of recorded fossil-bearing localities.

Direct sampling measures are most applicable to small-scale studies, where palaeontologists might sample a fixed volume (Mander et al. 2008) or area (Barras & Twitchett 2007) of sediment at defined sampling horizons (Little 1996), or spend a fixed amount of time sampling at each locality to ensure parity between samples. Mander et al. (2008) provide an example of controlling for sampling bias in their palaeoecological study of the Late Triassic mass extinction event in the SW UK, where fixed samples of 1.6 kg of sediment were collected for diversity and abundance analysis at intervals of 1 m. However, it has been noted that bulk-sampling methods are not always effective at recording rare species, and it may be necessary to integrate field samples with data from museum collections and the literature to gain a more reliable picture of palaeodiversity (Harnik 2009).

It could be contended that poorly sampled time intervals might sometimes correspond to times when fossils are of poor quality. Especially among complex organisms that are rarely preserved, such as vertebrates, it could be worth assessing whether some time bins have yielded more complete skeletons than others, and whether mean specimen completeness correlates with apparent diversity. If this were the case, then specimen quality might provide a guide to sampling.

Fossil quality has been considered in previous studies of dinosaurs (e.g. Benton 2008a, b; Mannion & Upchurch 2010) where the quality of type specimens was found to have improved through research time. In their detailed study of sauropodomorph fossils, Mannion & Upchurch (2010) found that mean skeletal completeness and mean character completeness varied between time bins, but roughly halved from the Triassic and Early Jurassic to the Late Cretaceous. Mannion & Upchurch (2010, p. 291) note that 'The ... results suggest that sea-level has, in some fashion, controlled the quality of the sauropodomorph fossil record, but only through part of the group's evolutionary history, with high sea-level correlated with low average completeness scores, and low sea-level with high completeness scores in the Jurassic – Early Cretaceous.' It is equally likely that the apparent sporadic covariation of sauropodomorph specimen quality with sea-level does not indicate a causal link at all: note that many of the completeness measures are based on relatively small sample sizes ($n = 4-24$ taxa), so a single locality can dominate the findings within a time bin. Perhaps such studies of fossil specimen quality based on modest numbers of specimens and localities cannot address sampling at the global scale.

In a basinwide study of the Late Permian and Early to Middle Triassic red beds of the South Urals basins (Benton et al. 2004), some 289 localities, assigned to 13 stratigraphic divisions in succession, have yielded 675 identified tetrapod fossils. These were assigned to four 'quality' categories, namely (1) single isolated fragments, (2) several individual elements of a taxon, (3) one or more skulls, and (4) one or more skeletons. All the noted materials, even the fragments, could be identified at least to family level, and so bone scrap is excluded. Across the whole study, the numbers of specimens in each category were 313, 288, 63, and 11 respectively. As reported before (Benton et al. 2004), the sampling measures of number of localities and number of specimens per time bin covary (Fig. 7a), but these do not covary with either diversity of genera or families. The 'quality' measure (number of good specimens/total number of specimens), where 'good specimens' are the complete skulls or skeletons, shows a different pattern (Fig. 7b) from either locality or specimen numbers. Ignoring the first value, which is based on a very low sample size, fossil quality in the Permian is out of synchrony with generic and familial diversity (Fig. 7b), but seems to be in line with generic diversity in particular in the Triassic. However, the measure of fossil specimen quality does not appear to covary with number of localities or specimens (Fig. 7a, b). The change in behaviour of the specimen quality measure across the Permo-Triassic boundary is probably not a sample-size artefact: Permian samples range from 11–63 specimens per time bin, excluding the first time bin (mean, 38.6) and Triassic sample sizes range from 17–147 (mean, 68). Certainly, in the late Early and Middle Triassic (time divisions 11–13), numbers of localities and specimens appear to peak in time bin 11 with number of families, but not number of genera (Fig. 7a). Further, the specimen quality index peaks in time bin 12 with number of genera (Fig. 7b). None of the three sampling measures, including specimen quality, could be said, however, to show a convincing covariation with apparent diversity, so suggesting that much of the palaeodiversity signal is probably real.

In a similar study of echinoids, Smith (2007b) showed that the Triassic fossil record was much poorer than that of the Jurassic. He noted that 27%

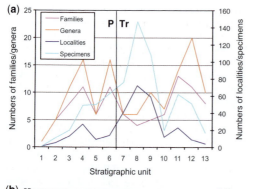

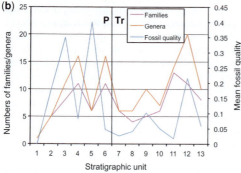

Fig. 7. Diversity and sampling through the Permo-Triassic continental redbeds of the South Urals basin, Russia. (**a**) Diversity of families and genera covary, as do the sampling measures of numbers of localities and specimens, but the diversity measures do not covary with the sampling measures. (**b**) Mean fossil quality (number of 'good' specimens: all specimens) does not covary with diversity of genera or families in the Permian, but appears to do so in the Triassic. Data from Benton *et al.* (2004).

Triassic species were based on relatively complete material (whole tests or whole tests plus spines), compared to 69% from the Lower Jurassic. Further, among Triassic species, 60% had been established on isolated spines or dissociated inter-ambulacral plates, whereas only 21% of Jurassic species are named on such incomplete material. The relatively poorer Triassic record was confirmed also by slower accumulation of Triassic taxa in a comparison of collector curves, longer implied gaps from a study of molecular trees, and more ghost lineages. Smith (2007*b*) explained these differences by a combination of more limited marine rocks in the Triassic when compared to the Jurassic, as well as to evolutionary changes among the echinoids, which acquired more robust tests in the Jurassic. Such a clear demonstration of relative differences in sampling quality of members of a single clade between two time units then points to the possibility of exact numerical correction when

comparing the palaeodiversity signals of Triassic and Jurassic echinoids.

Phylogenetic and gap-counting methods. The debate about whether number of formations is a covariate or a determinant of palaeodiversity could continue on its circular way unless additional data can be introduced. A possible source of such information might be ghost ranges and Lazarus taxa. A ghost range, or ghost lineage, is the minimum gap implied by a cladogram where the oldest fossils of two sister lineages differ in age (Norell 1992), and a Lazarus taxon (Flessa & Jablonski 1983) is a break in the record of a lineage that exists below and above a particular sampling horizon. In both cases, providing the cladogram is correct in the first case, and providing the taxa below and above the gap are the same in the second case, these can both provide independent evidence for a failure in sampling.

This has been noted before (Paul 1998). For example, Smith (2001, p. 364) pointed out, 'The only realistic way to distinguish between sampling and biologically driven patterns is to gather phylogenetic information. The key here is the recognition of ghost lineages and the stratigraphical distribution of pseudoextinctions.' Pseudoextinctions are false extinctions marking artificial truncations of lineages, which, when corrected indicate Lazarus gaps. The correct identification of such gaps can change perceptions of evolutionary pattern: for example, when Modesto *et al.* (2001) revised the cladogram of procolophonids, a clade of small reptiles from the Permian and Triassic, they found that several ghost lineages spanned the Permo-Triassic boundary, so showing that the group was not so severely affected by the Permo-Triassic mass extinction as had been thought previously.

Here, we concentrate on ghost ranges. A number of gap analyses have been carried out (e.g. Paul 1982, 1998; Flessa & Jablonski 1983; Benton 1987; Fara & Benton 2000; Smith 2001, 2007*a*, *b*; Fara 2002), and we cannot add to what was said in those papers, other than to urge caution. It has so far been generally assumed that gaps in ranges occur when fossils are not found as a result of missing rocks and missing sampling. This is doubtless commonly the case, but there is a risk that Lazarus gap analysis might still involve a measure of circularity in that the method cannot distinguish poor sampling from low abundance and diversity. A lineage that showed genuine rises and falls in abundance or breadth of geographical distribution might show Lazarus gaps even if sampling is constant throughout, but this would be an evolutionary, not a sampling, signal. This was the contention by Wignall & Benton (1999) for times of low diversity and high gap counts following mass extinctions.

A further weakness of gap analysis is that the detection of Lazarus taxa becomes harder as the gap duration increases (Benton 1987), and of course the method cannot detect gaps before and after the currently known stratigraphic range. These criticisms are true also of ghost ranges, and one might very well expect that many ghost ranges do indeed arise from low diversity and abundance of lineages and clades soon after they became established and before they had diversified fully.

Nevertheless, it is worth perhaps exploring ghost range distribution in time as an independent guide to sampling (Paul 1998). The assumptions would be (1) that ghost ranges might be distributed in time in negative proportion to putative sampling proxies such as comprehensive counts of formations or localities, and (2) that raw counts of taxa should correlate better with those putative sampling proxies than phylogenetially corrected counts of taxa. The rationale of this last suggestion emerges from a comparison of three counts of palaeodiversity, the taxon diversity estimate (TDE), a raw count of numbers of taxa reported per time bin, the ghost range diversity estimate (GDE), based on a cladogram plotted against time, and the phylogenetic diversity estimate (PDE), the sum of TDE and GDE (Barrett *et al.* 2009; Mannion & Upchurch 2010; Mannion *et al.* 2011).

We present four examples, the first two based on relatively small sample sizes, the second two on large examples, and these show broadly that phylogenetic gaps can indicate sampling failure.

Triassic archosaurian diversity. A recent comprehensive cladistic analysis of the relationships of Triassic archosaurs (Brusatte *et al.* 2010, fig. 8) offers a summary phylogeny plotted against time, and highlighting ghost ranges. Lazarus taxa are not shown. From this (Fig. 8; Table 4), ghost ranges were summed for substages, and compared with numbers of archosaur-bearing and tetrapod-bearing formations for those same substages (data from Benton *et al.* 2004; Sahney & Benton 2008; and unpublished). Note that formations were assigned to substages based on independent stratigraphic evidence in each case, and with no interpolation. Two counts of archosaur-bearing formations were considered, first the 'strict' count, taken only from the taxa included in the Brusatte *et al.* (2010) cladogram, and then the 'all archosaur' count, based on all archosaurs known from the Triassic. The sums of these three counts vary substantially: 37 actual archosaur (strict) FFC, 94 all-archosaur (wider) FFC, and 292 all-tetrapod (comprehensive) FFC.

To compare gaps in the stratigraphic record with ghost ranges, a measure of the 'absence of formations' is required. As a visual approximation (Fig. 8), the inverse of the number of formations

per time bin was taken, by subtracting the actual number from the maximum possible number of formations (maximum number of formations per time bin were: actual archosaur-bearing formations, 7 in the upper Carnian and in the lower Norian; all archosaur-bearing formations, 14 in the middle Norian; all tetrapod-bearing formations, 26 in the upper Olenekian). There is no obvious visual matching (Fig. 8) of times of significant ghost range, such as the Anisian to Carnian interval, with times of lower sampling (the Ladinian and lower Carnian coincide, but the later Triassic epochs do not).

In comparisons of GDE (Table 5), the strict and comprehensive FFC gave non-significant negative associations between ghost ranges and formation counts, whereas the wider FFC correlated negatively highly significantly with ghost ranges. However, this strong correlation disappears when first differences are considered (Table 5), so the strong correlation might be an artefact of parallel trends of increasing numbers through the Triassic combined with small data sets. Comparison of TDE and PDE (Table 5) shows that TDE correlates with the strict FFC, but only at $P < 0.1$ with the wider FFC, and not at all with the comprehensive FFC. These relationships disappear with the PDE, which shows both negative and positive non-significant correlations with the sampling counts.

These results are equivocal, confirming the proposition that PDE correlates much less well with formation counts than TDE, but highlighting the odd result that ghost range counts (GDE) also correlate strongly with the wider FFC. The Triassic archosaur fossil record is not simply dependent on rock volume (no correlation with the comprehensive FFC), and it is unresolved how well ghost ranges predict sampling.

Mesozoic bird diversity. When the simple cladogram of Mesozoic birds from Chiappe & Witmer (2002) is compared with minimum and maximum estimates of bird-bearing formations, there is no correlation with ghost ranges, whether using the raw data or detrended data (Table 6). The minimum estimate of locality numbers (= strict FFC) consists of just the localities that yielded the bird taxa included in Chiappe & Witmer's (2002) cladogram, whereas the comprehensive FFC comes from all records of Mesozoic birds, as documented by Fountaine *et al.* (2005). The fossil record of Mesozoic birds certainly includes very many ghost ranges (55 stage-level ghost ranges and only 29 stage-level records), and formation numbers, whether as a strict or wider FFC, might be thought to have been a suitable predictor of ghost ranges, but this is not the case whether for total or detrended data (Table 6).

When the raw palaeodiversity (TDE) is compared, however, it shows a remarkably strong

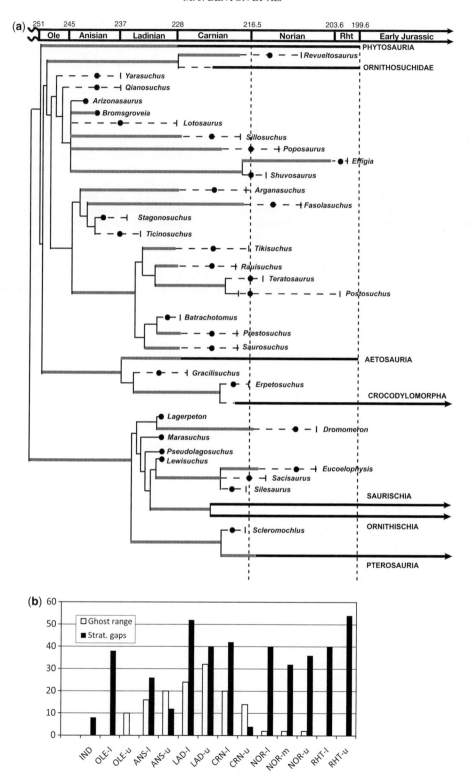

Fig. 8.

Table 4. *Comparison of phylogenetically implied gaps and formation numbers for Triassic archosaurs*

Substage	Duration (My)	Ghost range	Archosaur formations	Stratigraphic gaps	All-tetrapod formations	Stratigraphic gaps
Ind	1.5	0	5	9	32	4
Ole(l)	1.5	0	3	11	17	19
Ole(u)	1.5	5	7	7	36	0
Ans(l)	3.5	8	6	8	23	13
Ans(u)	3.5	10	6	8	30	6
Lad(l)	1.5	12	3	11	10	26
Lad(u)	1.5	16	2	12	16	20
Crn(l)	3.35	10	1	13	15	21
Crn(u)	3.5	7	13	1	34	2
Nor(l)	6.5	1	10	4	16	20
Nor(m)	6.5	1	14	0	20	16
Nor(u)	6.5	1	13	1	18	18
Rht(l)	3	0	8	6	16	20
Rht(u)	3	0	3	11	9	27
Totals	46.85	71	94	102	292	212

Data are tabulated from a recent cladistic analysis in Brusatte *et al.* (2010, fig. 8), from which phylogenetically implied gaps ('Ghost range') are drawn, and then compared with the inverse of the number of formations, as a measure of absence of information ('Stratigraphic gaps' = maximum number of formations in a time bin [14] minus actual number). Comparisons are made with the strict FFC ('Archosaur formations') and the comprehensive FFC ('All-tetrapod formations').
Abbreviations of stratigraphic stage names as in Table 1, plus Rht, Rhaetian.

Table 5. *Correlations of archosaurian ghost ranges (GDE, Ghost range diversity estimate) with counts of restricted and all archosaur-bearing and all tetrapod-bearing formations (strict, wider, and comprehensive FFCs) for the Brusatte* et al. *(2010) study of Triassic archosaur phylogeny, showing rank-order correlations for the raw data and for first differences (FD). Taxon diversity estimates (TDE) and phylogenetic diversity estimates (PDE = TDE + GDE) are also compared with the three formation counts*

	Spearman's ρ	P
GDE v. Strict FFC	−0.41	0.237
GDE v. Wider FFC	−0.86	0.001**
GDE v. Comprehensive FFC	−0.34	0.337
GDE v. FD strict FFC	0.20	0.577
GDE v. FD wider FFC	0.15	0.681
GDE v. FD comprehensive FFC	0.27	0.451
TDE v. Strict FFC	0.79	0.006
TDE v. Wider FFC	0.57	0.088
TDE v. Comprehensive FFC	−0.02	0.947
PDE v. Strict FFC	0.40	0.249
PDE v. Wider FFC	−0.27	0.443
PDE v. Comprehensive FFC	−0.24	0.498

Fig. 8. Phylogeny of basal archosaurs (**a**), showing dates of the major lineages and ghost ranges. (**b**) Histograms across the bottom show number of ghost ranges and a measure of the 'absence of formations' (= maximum number minus actual number; Strat. gaps, stratigraphic gaps) for each time bin. Abbreviations: ANS, Anisian; CRN, Carnian; IND, Induan; l, lower; LAD, Ladinian; NOR, Norian; OLE, Olenekian; RHT, Rhaetian; u, upper. A, based on data in Brusatte *et al.* (2011).

Table 6. *Correlations of Cretaceous bird ghost ranges (GDE) with bird-bearing FFCs, read as a strict FFC, representing only those formations with the named bird taxa (from Chiappe & Witmer 2002) and wider FFC figures (from Fountaine et al. 2005), showing rank-order correlations for the raw data and for first differences (FD). Correlations between these measures and TDE and PDE are also given. Data are calculated from Berriasian to Campanian only, to avoid the edge effects of wide variation in number of ghost ranges in the first time bin (Tithonian), and necessary absence of ghost ranges in the last (Maastrichtian)*

	Spearman's ρ	P
GDE v. Strict FFC	−0.44	0.180
GDE v. Wider FFC	−0.22	0.518
GDE v. FD strict FFC	−0.43	0.191
GDE v. FD wider FFC	−0.17	0.619
TDE v. Strict FFC	0.90	0.000**
TDE v. Comprehensive FFC	0.65	0.032*
PDE v. Strict FFC	0.38	0.248
PDE v. Comprehensive FFC	0.47	0.147

correlation with a strict count of formations with fossil birds and a weaker correlation with the wider formation count (Table 6), but these two signals are doubtless essentially redundant with each other, as in the pterosaur case above. These correlations disappear for the phylogenetically corrected diversity estimate (PDE; Table 6).

In this rather extreme case, with high proportions of ghost ranges (relative completeness index = −0.527), these minimum estimates of phylogenetically determined gap may provide a guide to sampling that is not achievable through the various strict and wider FFCs.

Dinosaurs. In an attempt to go beyond such small-scale studies, an analysis of the dinosaurian fossil record was conducted. This consists of the 420 species included in the formal dinosaur super-tree of Lloyd et al. (2008), plotted against time, using stratigraphic data to establish stage-level divisions of the Mesozoic. Dinosaurian distribution data comes from the Paleobiology database (http://paleodb.org/; download of all non-avian body fossil data on 29th June, 2010). We compared the GDE:PDE ratio, diversity (GDE) and phylogenetically corrected diversity (PDE = TDE + GDE) to three sampling proxies: (1) number of dinosaur-bearing formations (DBFs), (2) number of dinosaur-bearing localities (DBLs), and (3) the palaeoarea of a spherical polygon described by drawing a convex hull around the DBLs (Fig. 9).

As with previous dinosaur studies (e.g. Lloyd et al. 2008; Barrett et al. 2009; Butler et al. 2011)

we find strong correlation between all of our sampling proxies and species diversity (Fig. 9). However, we note a consistent weakening of this relationship when PDE is used instead of TDE (Table 7), despite a strong correlation between GDE and sampling. The sampling proxies, dinosaur-bearing formations and dinosaur-bearing localities, doubtless mix some redundancy (many formations/localities yield a single species) with genuine sampling signal, as discussed above, and so the strong correlation between sampling proxy and palaeodiversity could reflect a mix of true sampling signal and redundancy. A better sampling proxy in these cases might be the total number of formations that have yielded any kind of vertebrate fossil, or that are of the correct facies to do so: this would allow inclusion of localities and formations that have been searched, but failed to produce dinosaur specimens.

In seeking to understand whether the relative proportion of ghost ranges might provide a more reliable guide to sampling than the traditional geological measures, the weak negative relationship between the GDE:PDE ratio and the formation/locality counts (only barely significant at $P < 0.05$ in the DBF case) is suggestive and indeed is strengthened when generalized differencing is used (McKinney 1990), where all three proxies show a strong negative correlation (Table 7). Consequently, despite being only a minimal correction it does appear that for dinosaurs at least the proportion of phylogenetically-inferred to sampled lineages is a good predictor of sampling.

Data and R code for all analyses are available from GTL.

Baleen whales. In a further large-scale study, taxonomic and phylogenetic diversity estimates of mysticete whales were considered. The phylogeny is based on Marx (2010, fig. 2), with *Cetotherium megalophysum* excluded owing to a lack of precise stratigraphic information. The formations that produced the taxa included in the tree of Marx (2010) (strict FFC), as well as the total number of cetacean-bearing formations (wider FFC), and all marine fossiliferous formations (comprehensive FFC) assignable to stage level, which were downloaded from the PaleoDB on 1st June, 2010, were all compared to the two diversity estimates using Spearman rank correlation (Table 7). While both the strict and wider raw FFCs showed a significant positive correlation with the raw taxonomic diversity estimate, this correlation disappeared when phylogenetic diversity was considered instead. Furthermore, the comprehensive FFC did not correlate with either taxonomic or phylogenetically adjusted diversity. When generalized differences were used instead of the raw data, the correlation of the taxonomic diversity estimate with all three formation counts was

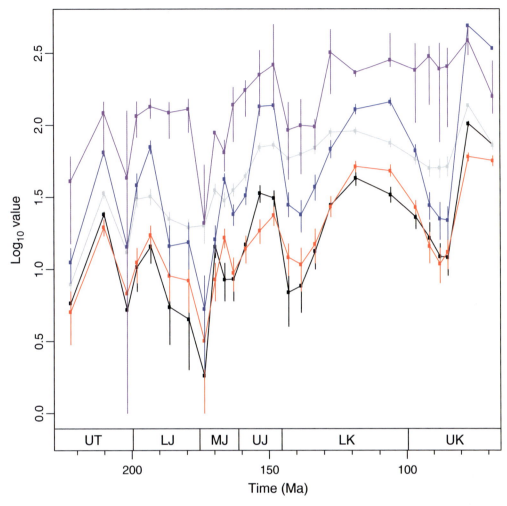

Fig. 9. Dinosaur diversity and sampling. Taxonomic Diversity Estimate (TDE; black), Phylogenetic Diversity Estimate (PDE; grey, based on Lloyd *et al.* 2008), Dinosaur-bearing Formations (DBFs; red), Dinosaur-bearing Localities (DBLs; blue) and palaeoarea of a spherical polygon encompassing the DBLs (purple). Values are logged to allow plotting on same scale. NB: Palaeoarea measure is further modified to allow plotting on same scale as values are orders of magnitude larger than for other variables. Vertical lines indicate 95% confidence interval reflecting 1000 randomizations of dating uncertainty. Stratigraphic divisions: UT, Upper Triassic; LJ, Lower Jurassic; MJ, Middle Jurassic; UJ, Upper Jurassic; LK, Lower Cretaceous; UK, Upper Cretaceous.

weakened, and indeed obliterated in the case of the strict FFC. By contrast, the correlation of the phylogenetic diversity estimate with the rock record was strengthened, but also turned negative in all cases. However, none of the correlations were statistically significant following differencing.

Perspective. In light of the need to distinguish between the bias and common cause models (e.g. Peters 2005, 2008; Smith & McGowan 2007), the observation that phylogenetic diversity estimates seem to decrease or remove existing correlations

between taxonomic estimates and a range of different measures of sampling gives rise to two possible interpretations. First, if it were assumed that the phylogenetic trees used are a reasonable representation of biological reality, the weakening of the diversity-sampling correlation might indicate that observed taxonomic diversity is largely driven by bias, with the phylogenetically adjusted estimate offering an improved and, presumably, truer picture of biological reality. However, if the cladograms we used to perform this correction were in some way flawed, 'correcting' diversity based on

Table 7. *Correlations of dinosaur and mysticete taxonomic and phylogenetic diversities with different measures of sampling*

Taxon	Correlation	Spearman's ρ	P
Dinosauria	GDE:PDE v. DBF	-0.39	0.047*
	GDE:PDE v. DBL	-0.37	0.060
	GDE:PDE v. Palaeoarea	-0.30	0.138
	TDE v. DBF	0.91	0.000**
	TDE v. DBL	0.88	0.000**
	TDE v. Palaeoarea	0.73	0.000**
	PDE v. DBF	0.80	0.000**
	PDE v. DBL	0.76	0.000**
	PDE v. Palaeoarea	0.70	0.000**
	GDE v. DBF	0.71	0.000**
	GDE v. DBL	0.66	0.000**
	GDE v. Palaeoarea	0.66	0.000**
	GD GDE:PDE v. DBF	-0.64	0.000**
	GD GDE:PDE v. DBL	-0.65	0.000**
	GD GDE:PDE v. Palaeoarea	-0.51	0.010*
	GD TDE v. DBF	0.80	0.000**
	GD TDE v. DBL	0.82	0.000**
	GD TDE v. Palaeoarea	0.73	0.000**
	GD PDE v. DBF	0.67	0.000**
	GD PDE v. DBL	0.78	0.000**
	GD PDE v. Palaeoarea	0.63	0.001**
	GD GDE v. DBF	0.55	0.006*
	GD GDE v. DBL	0.64	0.000**
	GD GDE v. Palaeoarea	0.48	0.016*
Mysticeti	TDE v. strict FFC	0.64	0.035*
	PDE v. strict FFC	0.30	0.366
	TDE v. wider FFC	0.71	0.015*
	PDE v. wider FFC	0.17	0.610
	TDE v. comprehensive FFC	0.40	0.227
	PDE v. comprehensive FFC	-0.23	0.503
	GD TDE v. strict FFC	-0.01	0.973
	GD PDE v. strict FFC	-0.60	0.067
	GD TDE v. wider FFC	0.50	0.144
	GD PDE v. wider FFC	-0.53	0.105
	GD TDE v. comprehensive FFC	0.39	0.248
	GD PDE v. comprehensive FFC	-0.33	0.330

Abbreviations: DBF, dinosaur-bearing formations; DBL, dinosaur-bearing localities; FFC, fossiliferous formation count; GD, generalized differences (McKinney 1990); PDE, phylogenetic diversity estimate; TDE, taxonomic diversity estimate; strict FFC, number of formations from which the taxa in the tree were recovered; wider FFC, total number of cetacean-bearing formations, as downloaded from the PaleoDB; comprehensive FFC, total number of marine fossiliferous formations as downloaded from the PaleoDB.
Probabilities (P) for each correlation measure are given, and these are marked as significant ($P < 0.05$*) and highly significant ($P < 0.005$**).

their topology might result in the addition of more noise than signal. In this case, the observed correlation between the sampling proxies and diversity might either be the result of an actual bias, or of a common cause – in either case, the addition of a large number of spurious ghost ranges could obliterate any statistically significant relationship. In addition, it is also worth noting that any phylogenetic correction fundamentally relies on the assumption that cladograms ignoring the potential presence of ancestral taxa in the fossil record are an adequate

model of evolution. However, treating genuine ancestor-descendant pairs as sister taxa may lead to the inference of ghost lineages where none exists, and hence the over-inflation of taxon estimates per time bin, which could bias phylogenetic diversity corrections even if the topology of the cladogram itself were accurate.

Finally, cladograms may also suffer from other problems, including the one-sidedness of the correction they provide (for obvious reasons, no ranges leading upwards in time can be inferred from them), and non-random taxon sampling, particularly when the cladogram was constructed to analyse the relationships of a particular subgroup of the taxon in question (Lane *et al.* 2005). These factors certainly have the potential to bias diversity estimates, and, if the taxa included in the tree present a non-random or very small sample of the taxon of interest, could even lead to completely spurious diversity patterns.

Model-based comparison of bias and other explanatory variables. While a large number of studies have investigated the impact of geological or human bias on measures of palaeodiversity, relatively few have tried to contrast the latter with the explanatory power of potentially biologically relevant variables that might account for some, or most of the observed diversity signal (e.g. Smith & McGowan 2007; Barrett *et al.* 2009; Benson *et al.* 2010). While it may well be possible that genuine biological signals in the fossil record are overwhelmed by geological biases, this assumption needs to be tested explicitly. Correcting palaeodiversity data using some form of sampling proxy (e.g. Smith & McGowan 2007; Barrett *et al.* 2009) and then attempting to interpret the residuals in a biologically meaningful way may be counterproductive in this sense, as it runs the risk of throwing the baby out with the bathwater: if the presumed sampling proxy is either redundant with diversity or the result of a common cause, removing it from the data may obliterate the actual (biological) signal

of interest, leaving little more than a flat line or random noise to be interpreted by the researcher.

One way to address this issue may be to consider both potential sampling proxies and evolutionarily meaningful variables at the same time, giving them an equal chance to explain the data of interest (e.g. Mayhew *et al.* 2007; Marx & Uhen 2010). If in such an analysis the explanatory power of sampling proxies outperforms that of the proposed biological model, the case for a large-scale bias in the data is corroborated. If, on the other hand, the biological model outperforms the bias hypothesis, a common cause or sampling proxy/diversity data redundancy explanation may be considered. Finally, the best model might also include aspects of both sampling bias and a biological signal. One example of this approach was recently implemented by Mayhew *et al.* (2007), who tested for, and found, a significant association of Phanerozoic diversity with temperature, while simultaneously assessing the effect of sampling probability on their results. Similarly, Marx & Uhen (2010) applied a series of models including food abundance, climate change, and a sampling proxy (number of fossiliferous marine formations) to late Oligocene to Pleistocene neocete whale diversity (Fig. 10; Table 8), and assessed their respective goodness-of-fit using the second-order Akaike's Information Criterion (AICc) and Akaike weights (Sugiura 1978; Burnham & Anderson 2002).

It is clear that in both cases the models chosen were far from exhaustive in their exploration of

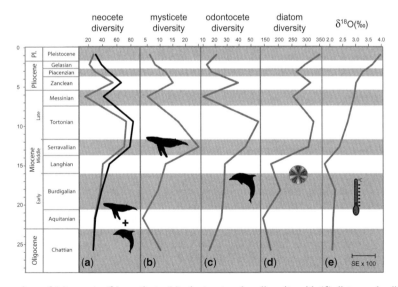

Fig. 10. Comparison of (**a**) neocete; (**b**) mysticete; (**c**) odontocete palaeodiversity with (**d**) diatom paleodiversity and (**e**) global $\delta^{18}O$ values (E), from Marx & Uhen (2010). Cetacean diversity is shown as sampled in bin data as downloaded from the Paleobiology Database (grey) and as a ranged through estimate (black). Based on data in Marx & Uhen (2010).

Table 8. *Estimated best-fit model parameters for the neocete, mysticete and odontocete datasets, as reported by Marx & Uhen (2010)*

	Neoceti sampled in bin		Neoceti ranged through		Mysticeti sampled in bin		Odontoceti sampled in bin	
	Estimate	S.E.	Estimate	S.E.	Estimate	S.E.	Estimate	S.E.
Intercept	6.694	1.465	6.649	1.111	2.214	0.179	5.566	1.179
st. dur.	−0.047	0.189	−0.068	0.167	0.096	0.023	0.049	0.166
Diatom	0.029	0.005	0.028	0.003	0.015	<0.001	0.020	0.003
$\delta^{18}O$	−2.881	0.577	−2.253	0.422	−1.077	0.081	−2.147	0.351
Rock	−	−	−0.013	0.006	−	−	−	−

Explanation of terms: $\delta^{18}O$, oxygen isotope records used as proxy for climate change; diatom, diatom species diversity (Neptune database); rock, total number of fossiliferous marine formations as downloaded from the Paleobiology Database; st. dur., geological stage duration; the latter was included as a non-optional predictor in all models on order to account for the potentially biasing effects of unequal Caenozoic stage durations. Based on data from Marx & Uhen (2010).

potential predictors, and other methods of simultaneously assessing the relative impact of bias and biology may be envisaged. Nevertheless, they make the point that combining sampling proxies and potential evolutionary drivers represents a more inclusive and, most likely, fairer way of assessing palaeodiversity than analysing either of these variables in isolation.

Conclusions

While it is evident that the fossil record is incomplete, some recent approaches to identifying bias, or 'megabias', have been flawed. The sampling proxies, such as number of formations containing particular fossils, or map areas from particular parts of the world, may not suffice as independent evidence for sampling failure. The two signals, rock volume and palaeodiversity, often covary closely, but this need not indicate that the former drives the latter. In fact, as already suggested (Peters 2005), both may be driven by an external 'common cause' such as sea-level change or, in the case of terrestrial organisms, by rates of uplift and by the weather, and consequent variation in volumes of sedimentary rock accumulation.

Further, as argued here, much of the covariance of rock volume metrics and palaeodiversity is likely a result of redundancy of the signals – the number of formations containing dinosaurs is tightly linked to the number of dinosaur species because finds are sporadic and interdependent (Benton 2008*a*). Removing the formation count from the species count produces a flat line because all signals, both geological and biological, have been removed. This observation of redundancy is a criticism of the assumption that measures of rock volume are independent proxies for sampling, and it says nothing about the quality of the fossil

record of dinosaurs (and other similar terrestrial taxa), which is undoubtedly patchy and incomplete.

We suggest four reasonable approaches to exploring sampling of the fossil record that avoid the problems of recent global-scale numerical explorations of covariance. First, regional-scale explorations of sampling may work because sampling metrics can be more detailed and can explore aspects of both rock volume and human effort. Further, explorations of rock volume that avoid the confusions of imprecise measurements of map areas that may not relate to rock availability (Dunhill 2011, in press) and the human quirks of geological formations (that scale over at least eight orders of magnitude), may provide independent estimators of sampling potential. A third approach may be to explore gaps (Lazarus gaps) and ghost ranges, which are both independently determined measures of known fossil absences. Our initial studies here are only moderately promising, however. A fourth approach, and perhaps the best of all because it does not assume primacy of either the fossil record or the sampling metrics, is to compare multiple models with a palaeodiversity curve, some models reflecting changes in the environment and others reflecting sampling, and yet others combining environmental change and sampling. The benefit of this approach is that there are no prior assumptions, and it assesses a variety of models for goodness of fit; the weakness is that the real drivers of palaeodiversity in any particular case may elude measurement and so may be missed.

Our proposal is that palaeontologists should be less obsessed about the poor quality of the fossil record, and that global-scale, single-hit analyses may never address the issue of whether the fossil record is good or bad, whether it is driven primarily by macroevolution or megabiases. Each time bin, each geographical region, and each clade is sampled differently, and so a global answer can probably never be found. Paul (1998) noted how palaeontologists seem to over-react and make a special issue out of fossil record incompleteness when compared to other biologists and earth scientists, who are comfortable that their data are not perfect, and who use standard methods to explore quality and confidence issues appropriate to each study.

We thank A. Smith and A. J. McGowan for organizing the meeting, and for their invitation to participate. This work was funded in part by NERC grant NE/C518973/1 to MJB, NERC doctoral training grant NE/H525111/1 to AMD, GTL was supported by NERC grant NE/F016905/1 to A. Smith, J. Young and P. Pearson, and FGM by a University of Otago Doctoral Scholarship. We thank A. McGowan, A. Smith, M. Wills, and an anonymous reviewer for reading the manuscript at various stages, and for helpful comments.

References

ADRAIN, J. M. & WESTROP, S. R. 2000. An empirical assessment of taxic paleobiology. *Science*, **289**, 110–112.

ALROY, J. 2002. How many named species are valid? *Proceedings of the National Academy of Sciences, USA*, **99**, 3706–3711.

ALROY, J. 2010. The shifting balance of diversity among major marine animal groups. *Science*, **329**, 1191–1194.

ALROY, J., MARSHALL, C. R. ET AL. 2001. Effects of sampling standardization on estimates of Phanerozoic marine diversification. *Proceedings of the National Academy of Sciences, USA*, **98**, 6261–6266.

ALROY, J., ABERHAN, M. ET AL. 2008. Phanerozoic trends in the global diversity of marine invertebrates. *Science*, **321**, 97–100.

ANGIELCZYK, K. D. 2002. A character-based method for measuring the fit of a cladogram to the fossil record. *Systematic Biology*, **51**, 176–191.

BARNOSKY, A. D., CARRASCO, M. A. & DAVIS, E. B. 2005. The impact of the species-area relationship on estimates of paleodiversity. *PLoS Biology*, **3**, 1356–1361.

BARRAS, C. G. & TWITCHETT, R. J. 2007. Response of the marine infauna to Triassic–Jurassic environmental change: ichnological data from Southern England. *Palaeogeography, Palaeoecology, Palaeoclimatology*, **244**, 223–241.

BARRETT, P. M., BUTLER, R. J., EDWARDS, N. P. & MILNER, A. R. 2008. Pterosaur distribution in time and space: an atlas. *Zitteliana B*, **26**, 61–107.

BARRETT, P. M., MCGOWAN, A. J. & PAGE, V. 2009. Dinosaur diversity and the rock record. *Proceedings of the Royal Society, B*, **276**, 2667–2674.

BENSON, R. J., BUTLER, R. J., LINDGREN, J. & SMITH, A. S. 2010. Palaeodiversity of Mesozoic marine reptiles: mass extinction and temporal heterogeneity in geologic megabiases affecting vertebrates. *Proceedings of the Royal Society, B*, **277**, 829–834.

BENTON, M. J. 1987. Mass extinctions among families of non-marine tetrapods: the data. *Mémoires de la Société Géologique, France*, **150**, 21–32.

BENTON, M. J. 1995. Diversification and extinction in the history of life. *Science*, **268**, 52–58.

BENTON, M. J. 1998. The quality of the fossil record of the vertebrates. *In*: DONOVAN, S. K. & PAUL, C. R. C. (eds) *The Adequacy of the Fossil Record*. Wiley, New York, 269–303.

BENTON, M. J. 2001. Finding the tree of life: matching phylogenetic trees to the fossil record through the 20th century. *Proceedings of the Royal Society, B*, **268**, 2123–2130.

BENTON, M. J. 2008a. Fossil quality and naming dinosaurs. *Biology Letters*, **4**, 729–732.

BENTON, M. J. 2008b. How to find a dinosaur, and the role of synonymy in biodiversity studies. *Paleobiology*, **34**, 516–533.

BENTON, M. J. 2009. The Red Queen and the Court Jester: species diversity and the role of biotic and abiotic factors through time. *Science*, **323**, 728–732.

BENTON, M. J. & HITCHIN, R. 1997. Congruence between phylogenetic and stratigraphic data on the history of life. *Proceedings of the Royal Society, B*, **264**, 885–890.

BENTON, M. J. & STORRS, G. W. 1994. Testing the quality of the fossil record: paleontological knowledge is improving. *Geology*, **22**, 111–114.

BENTON, M. J., WILLS, M. & HITCHIN, R. 2000. Quality of the fossil record through time. *Nature*, **403**, 534–537.

BENTON, M. J., TVERDOKHLEBOV, V. P. & SURKOV, M. V. 2004. Ecosystem remodelling among vertebrates at the Permian–Triassic boundary in Russia. *Nature*, **432**, 97–100.

BENTON, M. J., ZHOU, Z., ORR, P., ZHANG, F. & KEARNS, S. 2008. The remarkable fossils from the Early Cretaceous Jehol Biota of China and how they have changed our knowledge of Mesozoic life. *Proceedings of the Geologists' Association*, **119**, 209–228.

BERNARD, E. L., RUTA, M., TARVER, J. E. & BENTON, M. J. 2010. The fossil record of early tetrapods: worker effort and the end-Permian mass extinction. *Acta Palaeontologica Polonica*, **55**, 213–228.

BRUSATTE, S. L., BENTON, M. J., DESOJO, J. B. & LANGER, M. C. 2010. The higher-level phylogeny of Archosauria (Tetrapoda: Diapsida). *Journal of Systematic Palaeontology*, **8**, 3–47.

BRUSATTE, S. L., BENTON, M. J., LLOYD, G. T., RUTA, M. & WANG, S. C. 2011. Macroevolutionary patterns in the evolutionary radiation of archosaurs (Tetrapoda: Diapsida). *Earth and Environmental Science Transactions of the Royal Society of Edinburgh*, **101**, 285–299.

BURNHAM, K. P. & ANDERSON, D. R. 2002. *Model Selection and Multimodel Inference: A Practical Information-theoretic Approach*. Springer-Verlag, New York.

BUSH, A. M., MARKEY, M. J. & MARSHALL, C. R. 2004. Removing bias from diversity curves: the effects of spatially organized biodiversity on sampling-standardization. *Paleobiology*, **30**, 666–686.

BUTLER, R. J., BARRETT, P. M., NOWBATH, S. & UPCHURCH, P. 2009. Estimating the effects of the rock record on pterosaur diversity patterns: implications for hypotheses of bird/pterosaur competitive replacement. *Paleobiology*, **35**, 432–446.

BUTLER, R. J., BENSON, R. J., CARRANO, W. T., MANNION, P. D. & UPCHURCH, P. 2011. Sea level, dinosaur diversity and sampling biases: investigating the 'common cause' hypothesis in the terrestrial realm. *Proceedings of the Royal Society, B*, **278**, 1165–1170.

CARPENTER, K. 1997. Morrison formation. *In*: CURRIE, P. J. & PADIAN, K. (eds) *Encyclopedia of Dinosaurs.* University of California Press, Berkeley, California, 5.

CHIAPPE, L. M. & WITMER, L. M. (eds) 2002. *Mesozoic Birds: Above the Heads of Dinosaurs.* University of California Press, Berkeley, California.

COX, B. M., SUMBLER, M. G. & IVIMEY-COOK, H. C. 1999. *A formational framework for the Lower Jurassic of England and Wales (onshore area).* British Geological Survey Research Report **RR/99/01**, 1–28.

CRAMPTON, J. S., BEU, A. G., COOPER, R. A., JONES, C. M., MARSHALL, B. & MAXWELL, P. A. 2003. Estimating the rock volume bias in palaeodiversity studies. *Science*, **301**, 358–360.

DARWIN, C. 1859. *On the Origin of Species by Means of Natural Selection.* John Murray, London.

DODSON, P. 1990. Counting dinosaurs, how many kinds were there? *Proceedings of the National Academy of Sciences, USA*, **87**, 7608–7612.

DONOVAN, S. & PAUL, C. R. C. (eds) 1998. *The Adequacy of the Fossil Record.* Wiley, New York.

DUNHILL, A. M. 2011. Using remote sensing and a GIS to quantify rock exposure in England and Wales: implications for paleodiversity studies. *Geology*, **39**, 111–114.

DUNHILL, A. M. in press. Problems with using rock outcrop area as a paleontological sampling proxy: comparing rock outcrop and exposure area in California, New York State, Australia, and England and Wales. *Paleobiology*, **38**, in press.

DYKE, G. J., MCGOWAN, A. J., NUDDS, R. L. & SMITH, D. 2009. The shape of pterosaur evolution: evidence from the fossil record. *Journal of Evolutionary Biology*, **22**, 890–898.

ERWIN, D. H. 2009. Climate as a driver of evolutionary change. *Current Biology*, **19**, R575–R583.

FARA, E. 2002. Sea-level variations and the quality of the continental fossil record. *Journal of the Geological Society, London*, **159**, 489–491.

FARA, E. & BENTON, M. J. 2000. The fossil record of Cretaceous tetrapods. *Palaios*, **15**, 161–165.

FLESSA, K. W. & JABLONSKI, D. 1983. Extinction is here to stay. *Paleobiology*, **9**, 315–321.

FOOTE, M. & RAUP, D. M. 1996. Fossil preservation and the stratigraphic ranges of taxa. *Paleobiology*, **22**, 121–140.

FOOTE, M. & SEPKOSKI, J. J. 1999. Absolute measures of the completeness of the fossil record. *Nature*, **398**, 415–417.

FOREY, P., FORTEY, R. A., KENRICK, P. & SMITH, A. B. 2004. Taxonomy and fossils: a critical appraisal. *Philosophical Transactions of the Royal Society, B*, **359**, 639–653.

FOUNTAINE, T., BENTON, M. J., DYKE, G. J. & NUDDS, R. L. 2005. The quality of the fossil record of Mesozoic birds. *Proceedings of the Royal Society, B*, **272**, 289–294.

FRÖBISCH, J. 2008. Global taxonomic diversity of anomodonts (Tetrapoda, Therapsida) and the terrestrial rock record across the Permo-Triassic boundary. *PLoS One*, **3**, e3733.

FRÖBISCH, J. 2009. Composition and similarity of global anomodont-bearing faunas. *Earth-Science Reviews*, **95**, 119–157.

GUISAN, A. & THUILLER, W. 2005. Predicting species distribution: offering more than simple habitat models. *Ecology Letters*, **8**, 993–1009.

GUISAN, A. & ZIMMERMANN, N. E. 2000. Predictive habitat distribution models in ecology. *Ecological Modelling*, **135**, 147–186.

HARNIK, P. G. 2009. Unveiling rare diversity by integrating museum, literature, and field data. *Paleobiology*, **35**, 190–208.

HAUBOLD, H. 1990. Dinosaurs and fluctuating sea levels during the Mesozoic. *Historical Biology*, **4**, 75–106.

HENDY, A. J. W. 2009. The influence of lithification on Cenozoic marine biodiversity trends. *Paleobiology*, **35**, 51–62.

HUELSENBECK, J. P. 1994. Comparing the stratigraphic record to estimates of phylogeny. *Paleobiology*, **20**, 470–483.

KALMAR, A. & CURRIE, D. J. 2010. The completeness of the continental fossil record and its impact on patterns of diversification. *Paleobiology*, **36**, 51–60.

KIESSLING, W. 2005. Habitat effects and sampling bias on Phanerozoic reef distribution. *Facies*, **51**, 24–32.

LANE, A., JANIS, C. M. & SEPKOSKI, J. J. JR. 2005. Estimating paleodiversities: a test of the taxic and phylogenetic methods. *Paleobiology*, **31**, 21–34.

LITTLE, C. T. S. 1996. The Pliensbachian–Toarcian (Lower Jurassic) extinction event. *Geological Society of America, Special Paper*, **307**, 505–512.

LLOYD, G. T., DAVIS, K. E. *ET AL.* 2008. Dinosaurs and the Cretaceous Terrestrial Revolution. *Proceedings of the Royal Society, B*, **275**, 2483–2490.

LLOYD, G. T., SMITH, A. B. & YOUNG, J. R. 2011. Quantifying the deep-sea rock and fossil record bias using coccolithophores. *In*: MCGOWAN, A. J. & SMITH, A. B. (eds) *Comparing the Geological and Fossil Records: Implications for Biodiversity Studies.* Geological Society, London, Special Publications, **358**, 167–177.

MANDER, L., TWITCHETT, R. J. & BENTON, M. J. 2008. Palaeoecology of the Late Triassic extinction event in the SW UK. *Journal of the Geological Society*, **165**, 319–332.

MANNION, P. D. & UPCHURCH, P. 2010. Completeness metrics and the quality of the sauropodomorph fossil record through geological and historical time. *Paleobiology*, **36**, 283–302.

MANNION, P. D., UPCHURCH, P., CARRANO, W. T. & BARRETT, P. M. 2011. Testing the effect of the rock record on diversity: a multidisciplinary approach to elucidating the generic richness of sauropodomorph dinosaurs through time. *Biological Reviews*, **86**, 157–181.

MARKWICK, P. J. 1998. Fossil crocodilians as indicators of Late Cretaceous and Cenozoic climates: implications for using palaeontological data in reconstructing palaeoclimate. *Palaeogeography, Palaeoecology, Palaeoclimatology*, **137**, 205–271.

MARSHALL, C. R. 1990. Confidence intervals on stratigraphic ranges. *Paleobiology*, **16**, 1–10.

MARSHALL, C. R. 1997. Confidence intervals on stratigraphic ranges with non-random distributions of fossil horizons. *Paleobiology*, **23**, 165–173.

MARSHALL, C. R. 2010. Marine biodiversity dynamics over deep time. *Science*, **329**, 1156–1157.

MARX, F. G. 2009. Marine mammals through time: when less is more in studying palaeodiversity. *Proceedings of the Royal Society, B*, **276**, 887–892.

MARX, F. G. 2010. The more the merrier? A large cladistic analysis of mysticetes, and comments on the transition from teeth to baleen. *Journal of Mammalian Evolution*, **28**, 77–200.

MARX, F. G. & UHEN, M. D. 2010. Climate, critters, and cetaceans: Cenozoic drivers of the evolution of modern whales. *Science*, **327**, 993–996.

MAXWELL, W. D. & BENTON, M. J. 1990. Historical tests of the absolute completeness of the fossil record of tetrapods. *Paleobiology*, **16**, 322–335.

MAY, R. M. 1990. How many species? *Philosophical Transactions of the Royal Society of London, Series B*, **330**, 293–304.

MAYHEW, P. J., JENKINS, G. B. & BENTON, T. G. 2007. A long-term association between global temperature and biodiversity, origination and extinction in the fossil record. *Proceedings of the Royal Society B*, **275**, 47–53.

MCGOWAN, A. & SMITH, A. 2008. Are global Phanerozoic diversity curves truly global? A study of the relationship between regional rock records and global Phanerozoic marine diversity. *Paleobiology*, **34**, 80–103.

MCKINNEY, M. L. 1990. Classifying and analysing evolutionary trends. *In*: MCNAMARA, K. J. (ed.) *Evolutionary Trends*. Belhaven Press, London, 28–58.

MODESTO, S., SUES, H.-D. & DAMIANI, R. 2001. A new Triassic procolophonoid reptile and its implications for procolophonoid survivorship during the Permo-Triassic extinction event. *Proceedings of the Royal Society, B*, **268**, 2047–2052.

NICHOL, D. 1977. The number of living animals likely to be fossilized. *Florida Scientist*, **40**, 135–139.

NORELL, M. 1992. Taxic origin and temporal diversity: the effect of phylogeny. *In*: NOVACEK, M. & WHEELER, Q. (eds) *Extinction and Phylogeny*. Columbia University Press, New York, 89–118.

NORELL, M. A. & NOVACEK, M. J. 1992. The fossil record and evolution: comparing cladistic and paleontologic evidence for vertebrate history. *Science*, **255**, 1690–1693.

PAUL, C. R. C. 1982. The adequacy of the fossil record. *In*: JOYSEY, K. A. & FRIDAY, A. E. (eds) *Problems of Phylogenetic Reconstruction*. Academic Press, London, 75–117.

PAUL, C. R. C. 1998. Adequacy, completeness and the fossil record. *In*: DONOVAN, S. K. & PAUL, C. R. C. (eds) *The Adequacy of the Fossil Record*. Wiley, New York, 1–22.

PETERS, S. E. 2005. Geologic constraints on the macroevolutionary history of marine animals. *Proceedings of the National Academy of Sciences, USA*, **102**, 12 326–12 331.

PETERS, S. E. 2006. Macrostratigraphy of North America. *Journal of Geology*, **114**, 391–412.

PETERS, S. E. 2008. Environmental determinants of extinction selectivity. *Nature*, **454**, 626–629.

PETERS, S. E. & FOOTE, M. 2001. Biodiversity in the Phanerozoic: a reinterpretation. *Paleobiology*, **27**, 583–601.

PETERS, S. E. & FOOTE, M. 2002. Determinants of extinction in the fossil record. *Nature*, **416**, 420–424.

PETERS, S. E. & HEIM, N. A. 2010. The geological completeness of paleontological sampling in North America. *Paleobiology*, **36**, 61–79.

PRESTON, F. W. 1948. The commonness, and rarity, of species. *Ecology*, **29**, 254–283.

PURNELL, M. A. & DONOGHUE, P. C. J. 2005. Between death and data: biases in interpretation of the fossil record of conodonts. *Special Papers in Palaeontology*, **73**, 7–25.

RAUP, D. M. 1972. Taxonomic diversity during the Phanerozoic. *Science*, **177**, 1065–1071.

RAUP, D. M. 1975. Taxonomic diversity estimates under rarefaction. *Paleobiology*, **1**, 333–342.

RAUP, D. M. 1976. Species diversity in the Phanerozoic: a tabulation. *Paleobiology*, **2**, 279–288.

RAUP, D. M. 1977. Systematists follow the fossils. *Paleobiology*, **3**, 328–329.

RONOV, A. B. 1978. The earth's sedimentary shell. *International Geology Review*, **24**, 1313–1363.

RONOV, A. B. 1994. Phanerozoic transgressions and regressions on the continents: a quantitative approach based on areas flooded by the sea and areas of marine and continental deposition. *American Journal of Science*, **294**, 777–801.

RUSSELL, D. A. 1995. China and the lost worlds of the dinosaurian era. *Historical Biology*, **10**, 3–12.

SAHNEY, S. & BENTON, M. J. 2008. Recovery from the most profound mass extinction of all time. *Proceedings of the Royal Society, B*, **275**, 759–765.

SAHNEY, S., BENTON, M. J. & FERRY, P. A. 2010. Links between global taxonomic diversity, ecological diversity, and the expansion of vertebrates on land. *Biology Letters*, **6**, 544–547.

SEPKOSKI, J. J. JR. 1993. Ten years in the library: how changes in taxonomic data bases affect perception of macroevolutionary pattern. *Paleobiology*, **19**, 43–51.

SEPKOSKI, J. J. JR. 1997. Biodiversity: past, present, and future. *Journal of Paleontology*, **71**, 533–539.

SEPKOSKI, J. J. JR, BAMBACH, R. K., RAUP, D. M. & VALENTINE, J. W. 1981. Phanerozoic marine diversity and the fossil record. *Nature*, **293**, 435–437.

SHEEHAN, P. M. 1977. Species diversity in the Phanerozoic: a reflection of labor by systematists? *Paleobiology*, **3**, 325–328.

SIMMS, M. J., CHIDLAW, N., MORTON, N. & PAGE, K. N. 2004. *British Lower Jurassic Stratigraphy*. Geological Conservation Review Series, **30**. JNCC, London.

SMITH, A. B. 2001. Large-scale heterogeneity of the fossil record: implications for Phanerozoic diversity studies. *Philosophical Transactions of the Royal Society, Series B*, **356**, 351–367.

SMITH, A. B. 2007a. Marine diversity through the Phanerozoic: problems and prospects. *Journal of the Geological Society, London*, **164**, 731–745.

SMITH, A. B. 2007b. Intrinsic versus extrinsic biases in the fossil record: contrasting the fossil record of echinoids in the Triassic and early Jurassic using sampling data, phylogenetic analysis, and molecular clocks. *Paleobiology*, **33**, 310–323.

SMITH, A. B. & MCGOWAN, A. J. 2007. The shape of the Phanerozoic marine palaeodiversity curve: how much can be predicted from the sedimentary rock record of Western Europe? *Palaeontology*, **50**, 1–10.

SMITH, A. B. & MCGOWAN, A. J. 2008. Temporal patterns of barren intervals in the Phanerozoic. *Paleobiology*, **34**, 155–161.

STANLEY, S. M. 2007. An analysis of the history of marine animal diversity. *Paleobiology*, **33** (Suppl. 4), 1–55.

SUGIURA, N. 1978. Further analysis of the data by Akaike's information criterion and the finite corrections. *Communications in Statistics, Series A*, **7**, 13–26.

TARVER, J. E., BRADDY, S. J. & BENTON, M. J. 2007. The effects of sampling bias on Palaeozoic faunas and implications for macroevolutionary studies. *Palaeontology*, **50**, 177–184.

TARVER, J. E., DONOGHUE, P. C. J. & BENTON, M. J. 2011. Is evolutionary history repeatedly rewritten in light of new fossil discoveries? *Proceedings of the Royal Society, B*, **278**, 599–604.

UPCHURCH, P. & BARRETT, P. M. 2005. Phylogenetic and taxic perspectives on sauropod diversity. *In*: CURRY-ROGERS, K. A. & WILSON, J. A. (eds) *The Sauropods: Evolution and Paleobiology*. University of California Press, Berkeley, California, 104–124.

UPCHURCH, P., MANNION, P. D., BENSON, R. B. J., BUTLER, R. J. & CARRANO, M. T. 2011. Geological and anthropogenic controls on the sampling of the terrestrial fossil record: a case study from the Dinosauria. *In*: SMITH, A. B. & MCGOWAN, A. J. (eds) *Comparing the Geological and Fossil Records: Implications for Biodiversity Studies*. Geological Society, London, Special Publications, **358**, 209–240.

VALENTINE, J. W. 1969. Patterns of taxonomic and ecological structure of the shelf benthos during Phanerozoic time. *Palaeontology*, **12**, 684–709.

VERMEIJ, G. J. & GROSBERG, R. K. 2010. The great divergence: when did diversity on land exceed that in the sea? *Integrative and Comparative Biology*, **50**, 675–682.

WALL, P., IVANY, L. & WILKINSON, B. 2009. Revisiting Raup: exploring the influence of outcrop area on diversity in light of modern sample-standardization techniques. *Paleobiology*, **35**, 146–167.

WANG, S. C. & DODSON, P. 2006. Estimating the diversity of dinosaurs. *Proceedings of the National Academy of Sciences, USA*, **103**, 13 601–13 605.

WEISHAMPEL, D. B. 1996. Fossils, phylogeny, and discovery: a cladistic study of the history of tree topologies and ghost lineage durations. *Journal of Vertebrate Paleontology*, **16**, 191–197.

WICKSTRÖM, L. M. & DONOGHUE, P. C. J. 2005. Cladograms, phylogenies and the veracity of the conodont fossil record. *Special Papers in Palaeontology*, **73**, 185–218.

WIGNALL, P. B. & BENTON, M. J. 1999. Lazarus taxa and fossil abundance at times of biotic crisis. *Journal of the Geological Society, London*, **156**, 453–456.

WILLIAMS, H. S. 1901. The discrimination of time-values in geology. *Journal of Geology*, **9**, 570–585.

WILLS, M. A. 1999. Congruence between phylogeny and stratigraphy: randomization tests. *Systematic Biology*, **48**, 559–580.

WILLS, M. A. 2007. Fossil ghost ranges are most common in some of the oldest and some of the youngest strata. *Proceedings of the Royal Society, B*, **274**, 2421–2427.

Macrostratigraphy and macroevolution in marine environments: testing the common-cause hypothesis

SHANAN E. PETERS* & NOEL A. HEIM

Department of Geoscience, University of Wisconsin-Madison, 1215 W. Dayton St., Madison, WI 53706, USA

Corresponding author (e-mail: peters@geology.wisc.edu)

Abstract: Quantitative patterns in the sedimentary rock record predict many different macroevolutionary patterns in the fossil record, but the reasons for this predictability remain uncertain. There are two competing, but non-mutually exclusive, hypotheses: (1) similarities reflect a sampling bias imposed by variable and incomplete sampling of fossils, and (2) similarities reflect environmental perturbations that influence both the patterns of sedimentation and macroevolution (i.e., common-cause). Macrostratigraphy, which is based on the quantitative analysis of hiatus-bound rock packages, permits variation in the rock record to be expressed in terms of rock quantity and, more importantly, spatiotemporal continuity. In combination with spatially-explicit fossil occurrence data in the Paleobiology Database, it is now possible to more rigorously test alternative hypotheses for similarities in the rock and fossil records and to start distinguishing between geologically-controlled sampling bias and the common-cause hypothesis. Here we summarize results from measuring the intersection of Macrostrat and the Paleobiology Database. Our results suggest that patterns in the fossil record are not dominated by large-scale stratigraphic biases. Instead, they suggest that linkages between multiple Earth systems are driving both spatiotemporal patterns of sedimentation and macroevolution.

Darwin (1872, p. 289) famously described the stratigraphic record as 'a history of the world imperfectly kept.' This pessimism towards the stratigraphic record of evolution was articulated for the specific purpose of explaining why palaeontologists had failed to sample the innumerable intermediate forms that Darwin expected in his particular formulation of the theory of evolution by natural selection. Although this view of life is demonstrably misguided and macroevolutionary theory, such as punctuated equilibrium (Eldredge & Gould 1972), now accounts for the abundant non-continuous patterns of evolution that are observed in the fossil record, many palaeobiologists remain skeptical about the fidelity of the fossil record. In particular, there is concern that observed patterns of diversity, origination, and extinction in the fossil record are artifacts of the variable quantity and quality of the Phanerozoic sedimentary rock record. This bias-oriented perspective is rooted in the pervasive and strong positive correlations that have been documented between many different tabulations of rock quantity and diversity (Gregory 1955; Raup 1976; Smith 2001; Smith *et al.* 2001; Peters & Foote 2002; Crampton *et al.* 2003; McGowan & Smith 2008; Wall *et al.* 2009). Although past diversity can only be sampled from the incomplete rock and fossil records, it may also be the case that variability in the rock record both reflects and controls fundamental changes in Earth systems that in turn directly and/or indirectly influence the macroevolutionary history of life. This alternative interpretation of the observed co-variation between the fossil and rock records is known as the common-cause hypothesis; that is, there is a common set of processes that have similarly shaped both stratigraphic patterns and biological evolution.

The common-cause hypothesis has been discussed, if not formalized, by geologists for many decades. Sloss (1963) recognized that the large-scale variability and nature of the preserved stratigraphic record on the craton primarily reflects tectonically-driven cycles in continental flooding, and not the destruction of sedimentary rocks by erosion and metamorphism (see also Melott & Bambach 2011; Meyers & Peters 2011). Similarly, Newell (1952) proposed that mass extinctions, the large die-offs that he first described as such (Bambach 2006), were driven by sea-level fluctuations that resulted in the geographical spread of unconformities (e.g. Simberloff 1974). The common-cause hypothesis has not, however, been tested with quantitative data until very recently.

Initial tests of the common-cause hypothesis against the null hypothesis of stratigraphic and preservation-induced sampling bias have been made using an integration of Macrostrat (http://macrostrat.org) and the Paleobiology Database

From: McGowan, A. J. & Smith, A. B. (eds) *Comparing the Geological and Fossil Records: Implications for Biodiversity Studies*. Geological Society, London, Special Publications, **358**, 95–104.
DOI: 10.1144/SP358.7   0305-8719/11/$15.00 © The Geological Society of London 2011.

(PaleoDB; http://paleodb.org). Macrostrat contains a summarization of the rock record for several different geographical regions, including (as of February 2011) North America, the circum-Caribbean, New Zealand, and the deep-sea. Macrostrat is continually growing and improving, but here we use data consisting of 15 773 formation-level lithostratigraphic units from a spatial array of 831 stratigraphic columns in the USA, Canada, Caribbean, eastern Mexico and northern South America (Fig. 1).

The PaleoDB is global in scope and contains fossil occurrences that are reported in the primary literature and by individual field investigators. Because both the PaleoDB and Macrostrat contain information on geographical location and basic lithostratigraphic data (Fig. 1a, b), PaleoDB fossil collections located in the study area are matched to specific units in Macrostrat on the basis of time and lithostratigraphy (see Peters & Heim 2010 for full descriptions of the matching procedures). In essence, the fossil record, at least as represented in the PaleoDB, has been returned to its original stratigraphic context, at least as captured by Macrostrat. With this unique integration of large-scale stratigraphic and palaeobiologic data, the common-cause hypothesis can be more rigorously tested against the null hypothesis of stratigraphic bias. Here we explore a few of these tests using the analytical framework of macrostratigraphy.

Macrostratigraphy

Macrostratigraphy (Peters 2006a; Hannisdal & Peters 2010) is a recent development in quantitative stratigraphy that takes advantage of spatially explicit stratigraphic data, which can be summarized quantitatively using a variety of metrics, most of which relate to the spatiotemporal extent and continuity of the sedimentary rock record. Fundamental to the approach is that any geographical location, the rock record can be divided into a succession of sediment packages that are bound by temporal gaps. These gap-bound rock packages can be defined on any criteria, including the hiatuses that occur at unconformities or lithological contacts, which may or may not correspond to temporal hiatuses. For example, if the dynamics of carbonate environments need to be quantified separately from those of siliciclastic environments, a sandstone unit separating two limestone units at a location would constitute a gap, even if sedimentation were in fact temporally continuous through the whole succession and there were no hiatus (Peters 2008). In the analyses presented here, marine sedimentary packages are based on hiatuses in all types of marine sedimentation that are resolvable at a temporal resolution of c. 1–3 Ma.

The advantage of partitioning the stratigraphic record into gap-bound packages is that time series for the number of sedimentary packages, rates of package initiation and rates of package truncation can be derived, yielding macrostratigraphic quantities that are quantitatively identical to, and also analogous to, macroevolutionary quantities, such as taxonomic richness and rates of origination and extinction (Peters 2006a, b; Hannisdal & Peters 2010). Rock package initiation and truncation rates are quantities that reflect the spatial rates of translation, as well as expansion and contraction,

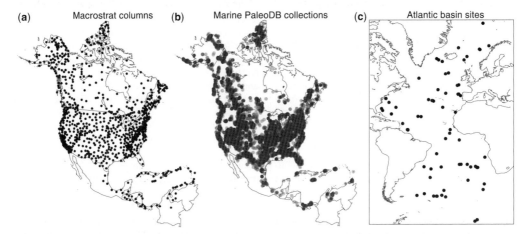

Fig. 1. Maps of continental coverage area and the Atlantic Ocean basin. (**a**) Map showing the 831 locations of stratigraphic columns used here. (**b**) Map of the marine PaleoDB collections. PaleoDB points are transparent so that dark areas indicate many stacked collections. (**c**) Map showing the locations of the ODP/DSDP/IODP core locations used for the deep-sea. All cores were drilled from the sediment surface down to or very near basaltic seafloor.

respectively, of sedimentary environments through time. With paired measures of the geological and fossil records, quantitative tools can be used to interrogate how these systems respond to each other and to covarying environmental changes.

For all of the analyses presented here, only marine packages and marine fossil genera are used. Future work will address the non-marine rock and fossil records (Rook et al. 2010). The temporal gaps that define marine packages are hiatuses at unconformities and gaps in marine sedimentation that occur at environmental transitions between marine and non-marine environments. The average duration of the marine packages used here is 26 Ma, which corresponds to c. 2nd order stratigraphic sequences (Vail et al. 1977) and, remarkably, to the average duration of marine invertebrate genera in the PaleoDB (e.g. 29 Ma in Anthozoa, 24 Ma in Rhychonellata).

Common Cause v. sampling bias

A frequent criticism of the fidelity of the fossil record is that it is biased by temporal heterogeneity in the quantity and quality of preserved sedimentary rock. A simple characterization of the bias hypothesis predicts that observed diversity should be high for intervals of time with a large quantity of preserved rock (e.g. rock volume, outcrop area) and low for time intervals with less preserved rock because rock quantity determines the number of fossils recovered and, therefore, diversity estimates (Raup 1979; Smith 2001; Smith et al. 2001; Peters & Foote 2001, 2002; Crampton et al. 2003). Another version of the bias hypothesis was formalized in a sequence stratigraphic framework by Holland (1995), who demonstrated that taxon first appearance datums (FADs) will be artificially concentrated at sequence bases and maximum flooding surface whereas last appearance datums (LADs) will cluster at sequence tops and maximum flooding surfaces. In both cases, it is the presence of a temporal gap in the sampling of taxon ranges that promotes the artificial clustering of FADs and LADs at discrete and predictable stratigraphic levels.

Holland's (1995) unconformity bias hypothesis is explicitly testable using macrostratigraphic data because the recognition of packages is based on hiatuses in sedimentation and, therefore, potential gaps in the sampling of taxon ranges. Thus, both the 'gappiness' and the 'rockiness' of the sedimentary rock record can be simultaneously quantified using the principles of macrostratigraphy. Contrary to the predictions of the sampling bias hypothesis, changes in the mean durations of the hiatuses that bound marine sedimentary rock packages do not predict changes in the magnitude of origination or extinction in the fossil record (Fig. 2). If sampling bias were the dominant driver of patterns of extinction and origination in the fossil record, then, all else being equal, long gaps in sampling should be associated with bigger artifactual rate pulses on either side of that gap. This is, however, not what we observe. Instead, it is the spatial extent of the unconformities that define package boundaries that predicts a significant amount of the variability in rates of genus extinction and origination (Fig. 3). The spatial extent of an unconformity can be related to the magnitude of the causal environmental change (e.g. retraction of a large epicontinental sea).

The common-cause hypothesis is an alternative to the bias hypothesis that invokes either direct or indirect links between the physical environmental changes that are reflected in the spatial extent of

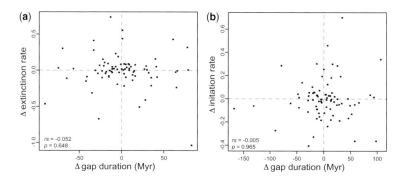

Fig. 2. Macroevolutionary rates for marine organisms v. hiatus durations in marine sedimentation. (**a**) First differences in change in genus origination rate v. first differences in mean duration (millions of years, Ma) of hiatuses that terminate in the previous time interval. (**b**) First differences in change in extinction rate v. mean duration of hiatuses initiating in the following interval. Rate excursions are not correlated with the durations of the hiatuses that define package boundaries but they are correlated with the spatial extent of those hiatuses (Fig. 3). This result is not consistent with an unconformity-induced sampling bias.

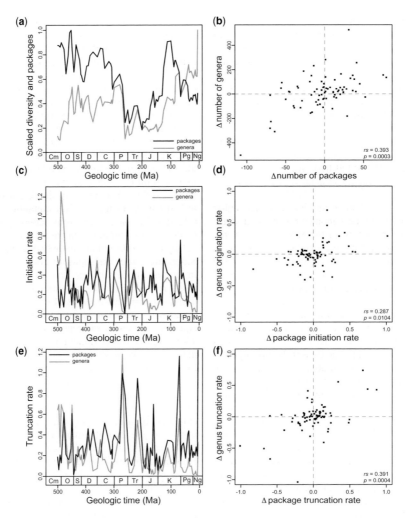

Fig. 3. Macrostratigraphy v. macroevolution. (**a**) Time series of the number of marine genera and packages. Points are plotted at interval bases. (**b**) First differences of the data presented in (a) with the Spearman rank-order correlation coefficient (ρ) and p-value in the bottom left. (**c**) Per capita, per interval regional origination rates for marine genera and per capita, per interval rates of marine package initiation. Plotting conventions same as for (a). (**d**) First differences of the data presented in (c). Plotting conventions same as for (b). (**e**) Per capita, per interval regional extinction rates for marine genera and per capita, per interval rates of marine package truncation. Plotting conventions same as for (a). (**f**) First differences of the data presented in (e). Plotting conventions same as for (b). Time-scale abbreviations are as follows: Cm, Cambrian; O, Ordovician; S, Silurian; D, Devonian; C, Carboniferous; P, Permian; Tr, Triassic; J, Jurassic; K, Cretaceous; Pg, Palaeogene; Ng, Neogene.

unconformities, variability in rock quantity and biological evolution. Of course, it is likely that both common-cause mechanisms and stratigraphically-controlled sampling biases have influenced the fossil record of marine animals. Isolating and quantifying the relative magnitude of sampling and common-cause are non-trivial. Crampton *et al.* (2003), found that the aerial extent of outcrop in a region is a good predictor of sampling effort and taxonomic richness in the fossil record. However, they could not attribute the area-diversity relationship

strictly to bias. In modelling expected extinction rates based on observed outcrop area, Smith & McGowan (2007) found that many small fluctuations in the number of marine genera could be explained by outcrop area, but that mass extinctions could not.

Previous work (Peters 2005, 2006*b*, 2008; also Heim & Peters 2011) has focused on macroevolutionary rates because they are less sensitive to variations in sample size than tallies of taxa and because changes in taxonomic richness reflect shifts in the balance between rates of origination and extinction

(Foote 2000); a given change in diversity may be driven by a change in origination, extinction, or some combination of the two. One of the most compelling results that supports the common-cause hypothesis, and that is not consistent with a simple sampling bias, is the asymmetry in the ability of macrostratigraphic quantities to predict genus origination v. genus extinction, a result to which we return here.

Asymmetry in the relationship between macroevolution and macrostratigraphy

If the fossil record of extinction and origination were dominated by an unconformity-related stratigraphic bias, then both origination and extinction should be similarly positively correlated with variability in the rock record (Holland 1995) and the magnitude and timing of intervening hiatuses should predict apparent turnover pulses. This is, however, not observed when comparing macroevolutionary and macrostratigraphic rates (Figs 2 & 3; Table 1). Extinction rates are significantly positively correlated with sediment truncation rates (Fig. 3e, f), but origination rates do not strongly covary with sediment initiation rates (Fig. 3c, d). The discrepancy between origination and extinction suggests that macroevolutionary rates are not determined by a straight-forward sampling bias that is due to the clustering of taxon FADs and LADs at unconformities (Holland 1995; Fig. 2).

When the timing of the FADs and LADs for genera occurring in the study area (Fig. 1) are based on their global fossil occurrences in the PaleoDB, rather than only those occurrences within the target regions (Fig. 3), the asymmetry between origination and extinction is strengthened (Table 1). Rates of origination based on the globally-determined genus FADs are not correlated with macrostratigraphic rates of initiation, whereas the correlation between

extinction and truncation for the global fossil data remains virtually unchanged (Table 1).

Because the PaleoDB is a continuously growing database of published fossil occurrences, it is also possible that incomplete sampling has influenced the observed asymmetry in the predictability of rates of origination and extinction. As published collection records are entered into the PaleoDB, new taxa are added and the stratigraphic ranges of previously entered taxa may be extended. Using the creation date field for collections in the PaleoDB, the correlations between macroevolutionary patterns and macrostratigraphic quantities were calculated for each year of the PaleoDB between 1998 to February 2011 (Fig. 4). The strengths of the correlations between taxonomic turnover rates and macrostratigraphic rates increased sharply between 1998 and 2000, as the PaleoDB rapidly matured. Since then, the correlations appear to be approaching an asymptote and are not changing rapidly or strongly systematically. This result suggests that our knowledge of the relationship between macroevolution and macrostratigraphy is relatively mature and that the relative strengths of the correlations are not likely to change markedly by the addition of new fossil data in the next several years. Ongoing improvements and additions to Macrostrat may, however, have a large quantitative effect, but that effect is anticipated to strengthen, not weaken, any statistical similarities, and to do so in a way that similarly influences both extinction and origination.

Distribution of fossils within hiatus-bound sediment packages

The stronger correlation between sediment truncation and generic extinction suggests that there are systematic differences in the distribution of fossils, FADs and LADs within sedimentary rock packages. To test this hypothesis, the relative locations of

Table 1. *Macrostrat–macroevolution results for three different data sets. Spearman rank-order correlation coefficients for first differences on time series of genus diversity, origination and extinction compared to the analogous macrostratigraphic quantities. The Sepkoski (2002) global genus compendium is compared to the macrostratigraphy of North America (Peters 2005). The Paleobiology Database data are in comparison to the macrostratigraphic data for the same region (Figs 1 & 3) and for the same set of genera with global, rather than regional, FADs and LADs. A global compendium of Planktonic foraminifera genera are also compared to the macrostratigraphy of the Atlantic Ocean basin (Peters et al. 2010)*

	USA, Sepkoski's global genus compendium	Paleobiology Database (regional)	Paleobiology Database (global)	Atlantic Basin, Planktonic Foraminifera
Diversity	0.51	0.39	0.24**	0.46
Origination	0.36	0.29	−0.01*	0.27
Extinction	0.50	0.39	0.42	0.55

All p-values <0.01, except *p = 0.93, **p = 0.03.

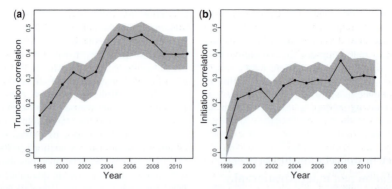

Fig. 4. Evolution of the macroevolution–macrostratigraphy correlations with increased sampling in the PaleoDB. New palaeontological data has been continuously entered into the PaleoDB since 1998, thus adding new taxa and extending the ranges of existing genera. (**a**) Time series of the Spearman rank-order correlation coefficient based on first differences between marine genus origination rate and marine package initiation rates. The gray field is a 95% bootstrap confidence interval and the slope of the linear regression through the data is given in the upper left. (**b**) Correlations between genus extinction and package truncation as the sampling in the PaleoDB increased. Plotting conventions are the same as in (a).

PaleoDB fossil collections within each of the marine packages in Macrostrat was determined (Peters & Heim 2011). All packages that span more than one geological stage and that contain more than one lithostratigraphic unit were scaled to unit duration and then divided into 100 equal-duration increments. That is, time was uniformly distributed within the constituent units of the package. Finally, the number of PaleoDB collections in each of the time increments was tabulated and then scaled to the maximum value in that package. This process was performed on all fossiliferous packages and an average occupancy curve for fossil collections was calculated for various time intervals (Fig. 5a).

To generate an expected window of collection occupancy (i.e., a null distribution), fossil collections were randomized within packages multiple times. The randomization envelope is not flat, indicating that there is a non-uniform distribution of lithostratigraphic units within marine packages, possibly due to changing environmental heterogeneity during the evolution of *c.* 2nd-order sedimentary sequences, a hypothesis which we hope to soon test with field data. It is also, however, possible that this modal null distribution reflects the inherent binning scheme used here. Regardless of the reason for the modal null, this provides the expected pattern under a model of randomly distributed fossil collections.

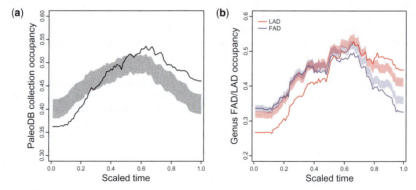

Fig. 5. Stratigraphic occupancy of marine fossils. (**a**) Mean Phanerozoic distribution of fossil collections within hiatus-bound sedimentary packages. Shaded regions are 95% confidence limits around the expected mean occupancy based on 1000 bootstrap randomizations of fossil collections within hiatus-bound sedimentary packages. (**b**) Mean stratigraphic distribution of generic first and last appearance datums in marine sediment packages. Shaded regions encompass all outcomes for 1000 randomizations of FADs and LADs among collections within packages. Null distributions differ because packages that contain FAD- and LAD-defining collections are a partially overlapping subset of all fossil-bearing packages.

These results indicate that PaleoDB fossil collections occur more frequently than expected in the top 40% of package durations and less frequently than expected in the bottom 25% of package durations. This result is difficult to explain without more refined environmental and lithological data, but one hypothesis is that fossil distribution is related to changing environmental characteristics of marine shelves during the evolution of large-scale, c. 2nd-order transgressive–regressive cycles. Alternatively, the pattern could be driven by changes in mean sediment accumulation rate, such that rates are initially high during the accommodation formation phase of 2nd-order successions, and then decrease as the basin fills, allowing more frequent beds that contain rich concentrations of skeletal remains to be formed at the top of packages (Kidwell 1989, 1991, 1993; Banerjee & Kidwell 1991; Abbott 1997; Abbott & Carter 1997).

The stratigraphic distribution of generic FADs and LADs was calculated using the same procedure described above for PaleoDB fossil collections (Fig. 5b). LAD occupancy is not markedly surprising in light of the asymmetrical relationship between origination and extinction and macrostratigraphy; LADs are in fact preferentially located in the top of packages, and they are found at package bases less frequently than expected due to chance alone. The FAD results are in some ways more interesting. FADs are more uniformly distributed within packages, but are much less common near package tops than expected under a model of random distribution.

The fact that FADs occur only rarely at package tops is curious given that the majority of fossil collections are found in the top 40% of packages. If a simple sampling bias were a dominant component of the reason for the rock-fossil similarities documented here, then recovering more fossils should result in more FADs and more LADs. The discrepancy between FAD and LAD distribution within packages indicates that there is a tendency for FADs and LADs to reflect some other attribute of packages than simply the preservation of fossil occurrences that might identify range end-points. Instead, it is likely that differences in the distribution of lineage origination and extinction within package durations reflect a macroevolutionary response to the palaeoenvironmental changes that occur during the formation and maturation of 2nd-order sedimentary sequences. This is, we believe, *prima facie* evidence for a complex common-cause relationship between sedimentation and macroevolution that involves process-response lags and biological responses to the evolution of shelf environments during large-scale draining and flooding of epicontinental seas.

The alternative hypothesis for the asymmetry in the FAD–LAD distribution with gap-bound packages is that the biological meaning of genus first and last appearances are disjunct. If, for example, taxonomists tend to artificially truncate continuous lineages at unconformities and then artificially lump lineages on the other side of unconformities, then it is conceivable that the result shown in Figure 5 could be an artifact of taxonomy. Although we strongly suspect that this is not the case, a more rigorously phylogenetic approach to the identification of lineage range end points would be highly advantageous to testing the common-cause hypothesis.

Deep-sea macrostratigraphy

Marine sedimentation on the continents is controlled by the extent to which the continents are flooded by shallow seas. Thus, eustatic sea-level and continental freeboard are the dominant mechanisms involved. Sedimentation in the deep-sea is, however governed by a completely different set of rules. Here we compare deep-sea sedimentation patterns to macroevolutionary patterns in surface-dwelling plankton.

Data on the lithology, thickness, and temporal distribution of deep-sea sediments were recovered from published Ocean Drilling Program, Deep Sea Drilling Program and Integrated Ocean Drilling Program scientific reports. Hiatuses in sedimentation at each site were identified wherever one or more nannofossil zones were missing entirely or not represented by a measurable thickness of sediment. A total of 249 hiatus-bound sediment packages composed of 3692 lithological units were recognized from 73 suitably drilled Atlantic sites (Fig. 1c). (Suitable sites have continuous core recovered from the sediment–water interface down to or very near basaltic basement rock.)

Planktonic foraminifera are abundant in the global ocean and play an important role in understanding the history of the oceans, but relatively little is know about the factors that have governed their macroevolutionary history. First differences in planktic foraminiferal diversity are significantly positively correlated with changes in the number of deep-sea sediment packages (Table 1; Peters *et al.* 2010). The correlations remain when the correlation imposed by the variable durations of calcareous nannofossil zones is removed by partial correlation. The only systematic departure from the statistical relationship between foraminifera diversity and the number of packages occurs during the late Eocene through the Oligocene, a time which also represents the most dramatic climate transition in the past 200 Ma (Zachos *et al.* 2008).

First differences in planktic foraminifera per-capita species/genus extinction rates are also significantly positively correlated with first differences in rates of deep-sea sediment package truncation. Rates of planktic foraminifera species/genus origination, by contrast, are less strongly positively correlated with rates of package initiation, though the correlation is significant. Thus, similarities between changes in planktic foraminiferal diversity and changes in the number of sediment packages reflect more the congruence in rates of sediment truncation and lineage extinction than they do rates of origination and package initiation.

These new results for the deep ocean are remarkably similar to those for the continental shelves. Like the shelf results, these new results for the deep-sea indicate that the macroevolutionary history of planktic foraminifera is linked causally to the oceanographic factors that govern spatiotemporal patterns of sedimentation. In the deep-sea, such changes are related to oceanographic factors, including surface water nutrient loads, the strength and pathways of ocean circulation and bottom currents (Keller & Barron 1983; Lyell 2003), and the age and source of deep water and the position of the calcium carbonate compensation depth (Van Andel 1975). The mechanisms that govern

sedimentation in the deep-sea are radically different from those that govern sedimentation on the shelves, and yet the quantitative linkages between macroevolution and sedimentation are similar. This too, is compelling evidence for a causal, common-cause link between patterns of sedimentation and macroevolution, though many outstanding questions remain about the relative roles of sampling, taxonomic practice, and common biological response to environmental perturbations that manifest in the rock record.

Discussion

The weight of evidence currently suggests that many patterns in the macroevolutionary history of the marine fossil record are not best explained by sampling artifacts in fossil preservation alone. Sampling bias at unconformities predicts that taxonomic origination rates and extinction rates should be covary in similar ways with the sampling bias induced by sampling failures at unconformities. Remarkably, the correlations between origination and initiation are much weaker than extinction and

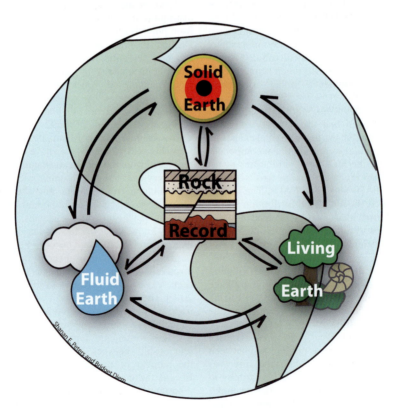

Fig. 6. Our view of the Earth systems process connections that are operating in our new formulation of the 'common-cause' hypothesis.

truncation across three different fossil data compilations (Table 1): Sepkoski's global genus compendium and the macrostratigraphy of North America (Peters 2005), the PaleoDB and macrostratigraphy of North America (Heim & Peters 2011), and global planktonic foraminifera and the macrostratigraphy of the Atlantic Ocean basin (Peters *et al.* 2010). Furthermore, the stratigraphic distribution of marine fossil collections within hiatus-bound packages, which is a direct measure of sampling effort, is not consistent with the hypothesis that sampling bias dominates macroevolutionary patterns. Fossil collections occur predominantly in the top 40% of packages, but genus FADs occur less frequently than expected in the top of packages and LADs even more frequent in the top of packages than expected.

These results suggest that the extinction of marine organisms is likely to be causally linked to many of the environmental changes that occur during contractions in the area of marine sedimentation, whereas origination is not linked in a similar way to the expansions in the area of sedimentation. Perhaps the origination of new lineages exhibits a lagged response to the environmental changes that occur when packages initiate, thereby dampening any process-response linkages in interval-to-interval changes in patterns of sedimentation and origination. Alternatively, originations may simply be more biologically deterministic and extinctions more environmentally deterministic. Further exploration of the specific drivers of taxonomic origination, extinction and marine sedimentation are needed. It is also absolutely imperative that differences in sampling intensity be explicitly taken into consideration and accounted for quantitatively. Our analyses accept the fossil record at face-value, and although there is some evidence to suggest that the fossil record of extinction has relatively high fidelity (Foote 2003), incomplete sampling distorts macroevolutionary patterns and, therefore, is expected to obscure, rather than to enhance, correlation with such macroevolutionary quantities as rates of truncation and initiation.

One of the goals going forward in stratigraphic palaeobiology should be to interrogate the exact mechanisms that are involved in driving similar sedimentary and biological responses. Simple changes in habitable area, the most direct common cause mechanism for marine shelf organisms, are unlikely to exert a strong control on the macroevolutionary history of marine life. Instead, common cause mechanisms are likely to involve numerous direct and indirect linkages and feedbacks between climate, tectonics, sedimentation, and biological evolution (Fig. 6). In the deep-sea, these linkages include seawater chemistry, ocean circulation, and surface water productivity, all of which respond to, and interact with, climate and tectonics. On the continents these linkages include the draining and flooding of continents due to changes in climate and plate tectonics, which itself can perturb the climate cycle and alter global climate. Under this view of the common cause hypothesis, the sedimentary record is more akin to an integrative record of the direct and indirect linkages between Earth systems than it is a direct driver of macroevolutionary change. Testing this next-generation formulation of the common cause hypothesis will be exciting because it demands a much higher level of quantitative integration of biological, geochemical, and sedimentary data on large spatial and temporal scales than has ever before been accomplished. Assembling the requisite data, rigorously controlling for variability in sampling in ways that do not discriminate against a process-connection to variable quantities of fossils, and then integrating these data quantitatively will be rewarding.

We thank A. Smith and A. J. McGowan for the invitation to participate in the 2010 Lyell Symposium and to contribute to this volume. M. Aberhan and A. Smith provided very helpful reviews. Acknowledgement is made to the Donors of the American Chemical Society Petroleum Research Fund, the United States Geological Survey, and the National Science Foundation EAR 0819931 for financial support. Both authors contributed equally to this work. This is official Paleobiology Database publication 131.

References

ABBOTT, S. T. 1997. Mid-cycle condensed shellbeds from mid-Pleistocene cyclothems, New Zealand: implications for sequence architecture. *Sedimentology*, **44**, 805–824.

ABBOTT, S. T. & CARTER, R. M. 1997. Macrofossil associations from mid-Pleistocene cyclothems, Castlecliff section, New Zealand: implications for sequence stratigraphy. *Palaios*, **12**, 188–210.

BAMBACH, R. K. 2006. Phanerozoic biodiversity mass extinctions. *Annual Review of Earth and Planetary Science*, **34**, 127–155.

BANERJEE, I. & KIDWELL, S. M. 1991. Significance of molluscan shell beds in sequence stratigraphy: an example from the Lower Cretaceous Mannville Group of Canada. *Sedimentology*, **38**, 913–934.

CRAMPTON, J. S., BEU, A. G., COOPER, R. A. & JONES, C. M. 2003. Estimating the rock volume bias in paleobiodiversity studies. *Science*, **301**, 358–360.

DARWIN, C. 1872. *The Origin of Species by Means of Natural Selection, or the Preservation of Favoured Races in the Struggle for Life*. 6th ed. John Murray, London.

ELDREDGE, N. & GOULD, S. J. 1972. Punctuated equilibria: an alternative to phyletic gradualism. *In*: SCHOPF, T. J. M. (ed.) *Models in Paleobiology*. Freeman, Cooper & Co., San Francisco, 82–115.

FOOTE, M. 2000. Origination and extinction components of taxonomic diversity: general problems. *In*: ERWIN, D. H. & WING, S. L. (eds) *Deep Time: Paleobiology's*

Perspective. Paleobiology, **26**, (Suppl. to no. 4). Lawrence, Kansas, 74–102.

FOOTE, M. 2003. Origination and extinction through the Phanerozoic. *Journal of Geology*, **111**, 125–148.

GREGORY, J. 1955. Vertebrates in the geologic time scale. *In*: POLDERVAART, A. (ed.) *Crust of the Earth*. Geological Society of America Special Paper no. 62. Geological Society of America, Boulder, CO, 593–608.

HANNISDAL, B. & PETERS, S. E. 2010. On the relationship between macrostratigraphy and geological processes: quantitative information capture and sampling robustness. *Journal of Geology*, **118**, 111–130.

HEIM, N. A. & PETERS, S. E. 2011. Covariation in macrostratigraphic and macroevolutionary patterns in the marine record of North America. *Geological Society of America Bulletin*, **123**, 620–630.

HOLLAND, S. M. 1995. The stratigraphic distribution of fossils. *Paleobiology*, **21**, 92–109.

KELLER, G. & BARRON, J. A. 1983, Paleoceanographic implications of Miocene deep-sea hiatuses. *Geological Society of America Bulletin*, **94**, 590–613.

KIDWELL, S. M. 1989. Stratigraphic condensation of marine transgressive records: origin of major shell deposits in the Miocene of Maryland. *Journal of Geology*, **97**, 1–24.

KIDWELL, S. M. 1991. The stratigraphy of shell concentrations. *In*: STEHLI, F. G., JONES, D. S., ALLISON, P. A. & BRIGGS, D. E. G. (eds) *Taphonomy: Releasing the Data Locked in the Fossil Record*. Plenum Press, New York, 211–290.

KIDWELL, S. M. 1993. Influence of subsidence on the anatomy of marine siliciclastic sequences and on the distribution of shell and bone beds. *Journal of the Geological Society, London*, **150**, 165–167.

LYELL, M. 2003. Neogene carbonate burial in the Pacific Ocean. *Paleoceanography*, **18**, 1059, doi: 10.1029/2002PA000777.

MCGOWAN, A. J. & SMITH, A. B. 2008. Are global Phanerozoic marine diversity curves truly global? A study of the relationship between regional rock records and global Phanerozoic marine diversity. *Paleobiology*, **34**, 80–103.

MELOTT, A. L. & BAMBACH, R. K. 2011. A ubiquitous ~62-Myr periodic fluctuation superimposed on general trends in fossil biodiversity. II. Evolutionary dynamics associated with periodic fluctuation in marine diversity. *Paleobiology*, **37**(3), 383–408.

MEYERS, S. R. & PETERS, S. E. 2011. A 56 million year rhythm in North American sedimentation during the Phanerozoic. *Earth and Planetary Science Letters*, **303**, 174–180.

NEWELL, N. D. 1952. Periodicity of invertebrate evolution. *Journal of Paleontology*, **26**, 371–385.

PETERS, S. E. 2005. Geologic constraints on the macroevolutionary history of marine animals. *Proceedings of the National Academy of Sciences*, **102**, 12 326–12 331.

PETERS, S. E. 2006a. Macrostratigraphy of North America. *The Journal of Geology*, **114**, 391–412.

PETERS, S. E. 2006b. Genus extinction, origination, and the durations of sedimentary hiatuses. *Paleobiology*, **32**, 387–407.

PETERS, S. E. 2008. Environmental determinants of extinction selectivity in the fossil record. *Nature*, **454**, 626–629.

PETERS, S. E. & FOOTE, M. 2001. Biodiversity in the Phanerozoic: a reinterpretation. *Paleobiology*, **27**, 583–601.

PETERS, S. E. & FOOTE, M. 2002. Determinants of extinction in the fossil record. *Nature*, **416**, 420–424.

PETERS, S. E. & HEIM, N. A. 2010. The geological completeness of paleontological sampling in North America. *Paleobiology*, **36**, 61–79.

PETERS, S. E. & HEIM, N. A. 2011. The stratigraphic distribution of marine fossils in North America. *Geology*, **39**, 259–262, doi: 10.1130/G31442.1.

PETERS, S. E., KELLY, D. C. & FRAASS, A. 2010. *Deep-sea macrostratigraphy and the macroevolution of planktic foraminifera*. GSA Abstracts with Programs, **42**, 138.

RAUP, D. M. 1976. Species diversity in the Phanerozoic: an interpretation. *Paleobiology*, **2**, 289–297.

RAUP, D. M. 1979. Biases in the fossil record of species and genera. *Bulletin of the Carnegie Museum of Natural History*, **13**, 85–91.

ROOK, D. L., HEIM, N. A. & PETERS, S. E. 2010. *What are we missing? Geological completeness of paleontological sampling in the terrestrial Cenozoic of North America*. Society of Vertebrate Paleontology Annual Meeting Abstracts with Program, 185A.

SEPKOSKI, J. J., JR. 2002. A compendium of fossil marine animal genera. *Bulletins of American Paleontology*, **363**, 1–560.

SIMBERLOFF, D. S. 1974. Permo-Triassic extinctions: effects of area on biotic equilibrium. *Journal of Geology*, **82**, 267–274.

SLOSS, L. L. 1963. Sequences in the cratonic interior of North America. *Geological Society of America Bulletin*, **74**, 93–113.

SMITH, A. B. 2001. Large-scale heterogeneity of the fossil record: implications for Phanerozoic biodiversity studies. *Philosophical Transactions of the Royal Society, London*, **B 356**, 351–367.

SMITH, A. B. & MCGOWAN, A. J. 2007. The shape of the Phanerozoic marine palaeodiversity curve: how much can be predicted from the sedimentary rock record of Western Europe? *Palaeontology*, **50**, 765–774.

SMITH, A. B., GALE, A. S. & MONKS, N. E. A. 2001. Sea-level change and rock-record bias in the Cretaceous: a problem for extinction and biodiversity studies. *Paleobiology*, **27**, 241–253.

VAIL, P. R., MITCHUM, R. & THOMPSON, S. 1977. Seismic stratigraphy and global changes of sea level, Part 4: Global cycles of relative changes of sea level. *In*: PAYTON, C. E. (ed.) *Seismic Stratigraphy: Applications to Hydrocarbon Exploration*. AAPG Memoir no. 26. American Association of Petroleum Geologists, Tulsa, 83–97.

VAN ANDEL, T. H. 1975. Mesozoic/Cenozoic calcite compensation depth and the global distribution of calcareous sediments. *Earth and Planetary Science Letters*, **26**, 187–194.

WALL, P. D., IVANY, L. C. & WILKINSON, B. H. 2009. Revisiting Raup: exploring the influence of outcrop area on diversity in light of modern sample-standardization techniques. *Paleobiology*, **35**, 146–167.

ZACHOS, J. C., DICKENS, G. R. & ZEEBE, R. E. 2008. An early Cenozoic perspective on greenhouse gas warming and carbon-cycle dynamics. *Nature*, **451**, 279–283.

The fossil record and spatial structuring of environments and biodiversity in the Cenozoic of New Zealand

JAMES S. CRAMPTON[1]*, MICHAEL FOOTE[2], ROGER A. COOPER[1], ALAN G. BEU[1] & SHANAN E. PETERS[3]

[1]GNS Science, PO Box 30368, Lower Hutt, New Zealand

[2]Department of the Geophysical Sciences, The University of Chicago, 5734 South Ellis Avenue, Chicago, Illinois 60637, USA

[3]Department of Geology & Geophysics, University of Wisconsin-Madison, 1215 W. Dayton St., Madison, Wisconsin 53706, USA

*Corresponding author (e-mail: j.crampton@gns.cri.nz)

Abstract: There is increasing evidence to suggest that drivers of bias in the fossil record have also affected actual biodiversity history, so that controls of artefact and true pattern are confounded. Here we examine the role of spatial structuring of the environment as one component of this common cause hypothesis. Our results are based on sampling standardized analyses of the post-Middle Eocene record of shelf molluscs from New Zealand. We find that spatial structuring of the environment directly influenced the quality of the fossil record. Contrary to our expectations, however, we find no evidence to suggest that spatial structuring of the environment was a strong or direct driver of taxic rates, net diversity, or spatial structuring in mollusc faunas at the scale of analysis. Stage-to-stage variation in sampling standardized diversity over the past 40 Ma exhibits two superficially independent dynamics: (a) changes in net diversity were associated primarily with changes in origination rate; and (b) an unknown common cause related extinction rate to the quality of the fossil record and, indirectly, to spatial structuring of the environment. We suggest that tectonic drivers, manifest as second-order sequence stratigraphic cycles, are likely to have been a key element of this common cause.

Our perceptions of biodiversity history are influenced strongly by secular biases in the quality of the fossil record (e.g. Raup 1976; Peters & Foote 2001). Many authors have argued against a literal reading of the record, and various techniques to minimize the impacts of sampling biases have been developed (e.g. Connolly & Miller 2001; Foote 2003; Alroy 2010). In addition, however, a number of studies have proposed that the relationship between sampling biases and diversity dynamics may be complex. These studies have suggested that there are latent common-cause factors, such as plate tectonics and sea-level change, which have affected simultaneously the quality of the fossil record and *true* macroevolutionary history (e.g. Newell 1952; Sepkoski 1976; Smith 2001; Peters & Foote 2002; Foote 2003; Peters 2005, 2006a; Wall *et al.* 2009). Demonstrating potential common causes, separating their effects from simple preservation and sampling biases, and interpreting them mechanistically, are challenging problems that must be solved if palaeobiologists are to resolve true patterns and processes of macroevolutionary change through geological time.

The aim of the present paper is to examine the role of spatial structuring – of the environment and of biodiversity – in driving both macroevolutionary dynamics and biases in the fossil record (Fig. 1). Specifically, we are interested in the extent to which partitioning of marine environments may have influenced biodiversity and the preservation of that diversity. In part, the study is founded on the expectation that the geographical range and evolution–extinction dynamics of organisms occupying a particular habitat will be related to the area and fragmentation of that habitat (e.g. Rosenzweig 1995; Maurer & Nott 1998). At the same time, we anticipate that the quality of the marine fossil record will be influenced by, for example, the areal extent and distribution of depositional v. non-depositional or erosional environments, factors that are also likely to affect habitat fragmentation and diversity. To test the role of spatial structuring in the common cause hypothesis, we use the exemplary, well-studied and relatively complete fossil record of post-Middle Eocene, level-bottom, shelf molluscs from New Zealand (e.g. Crampton *et al.* 2003, 2006a). This record is approximately

From: McGOWAN, A. J. & SMITH, A. B. (eds) *Comparing the Geological and Fossil Records: Implications for Biodiversity Studies*. Geological Society, London, Special Publications, **358**, 105–122. DOI: 10.1144/SP358.8 0305-8719/11/$15.00 © The Geological Society of London 2011.

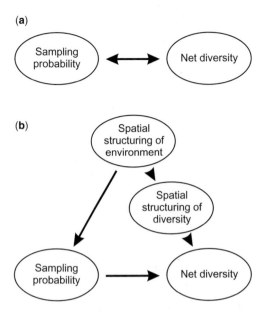

Fig. 1. Schematic showing the common-cause hypothesis and the relationships being tested here. (**a**) The apparent relationship between quality of the fossil record – here sampling probability – and diversity. A positive association is observed when diversity data are not corrected for uneven sampling through time. The common-cause hypothesis posits that this relationship results from both sampling related bias and also from latent factors that simultaneously influenced true diversity and the quality of the fossil record. (**b**) The potential common-causal agent that is tested here: spatial structuring of the environment. Arrows indicate direction of causality in the observed positive association.

40% complete at the species level for the Neogene, as estimated using two very different approaches (Cooper *et al.* 2006; Crampton *et al.* 2006*a*).

In this paper, therefore, we derive an index for spatial heterogeneity of the environment and adapt established indices of spatial differentiation in the New Zealand Cenozoic mollusc fauna. We then compare time series of these measures to time series of sampling probability (a proxy for the quality of the fossil record) and sampling standardized estimates of overall species richness, origination rate and extinction rate. All analyses and computations reported here were undertaken using R (R Development Core Team 2010).

Materials and methods

Material

This study is based on the rich fossil record of Cenozoic molluscs from New Zealand (including the Chatham Islands). Our analyses are restricted to gastropods, bivalves and scaphopods that are inferred to have inhabited level bottom environments at shelf depths (0–200 m). We have excluded pelagic, littoral, estuarine, bathyal and abyssal taxa because they are relatively poorly represented in the fossil record. The study was undertaken at the species level. Results are based primarily on two datasets that have been described elsewhere. The first, the synoptic dataset, comprises stratigraphic ranges and palaeoecological data for 5241 species, 1949 of which are undescribed. This dataset is taxonomically highly vetted and consistent, and was compiled by just two palaeontologists (A. G. Beu, P. A. Maxwell) over a period of many years (e.g. see Crampton *et al.* 2006*a*).

The second dataset is derived from the Fossil Record File Electronic Database (FRED), comprises individual occurrence records from over 8000 collection lists of Cenozoic molluscs, and contains data of varying vintage and quality. The FRED data have been subjected to several iterations of cleaning and vetting. In particular:

(1) All records from identifiers of unknown or doubtful expertise were eliminated.
(2) All uncertain identifications were eliminated.
(3) Collections that could not be identified to a single time bin (see below) were eliminated.
(4) All taxonomic names were checked against a list of synonyms and, where necessary, updated.
(5) Geographically and stratigraphically restricted, and biogeographically distinctive, records from northernmost New Zealand were excluded in order to avoid possible biasing effects, although inclusion of these data appears to make no significant difference to interpretations (results not presented here).

Following these adjustments, and including only level bottom shelf taxa, the dataset contains 30 114 occurrences of 1954 species from 6106 collections.

In order to use both these data compilations, and to derive maximum benefit from the taxonomically less reliable but otherwise valuable occurrence data in FRED, all our analyses are based on 1557 shelf taxa that are shared between the two datasets. These shared taxa are represented by 19 946 occurrences in 3482 collections in the FRED dataset. Because pre-Middle Eocene data are sparse, interpretations are restricted to post-40 Ma (late Middle Eocene) faunas (Fig. 2 and see below).

Measuring relationships between time series

Throughout this study, comparisons between different time series employ first differences between successive time bins, denoted with the Δ symbol, and

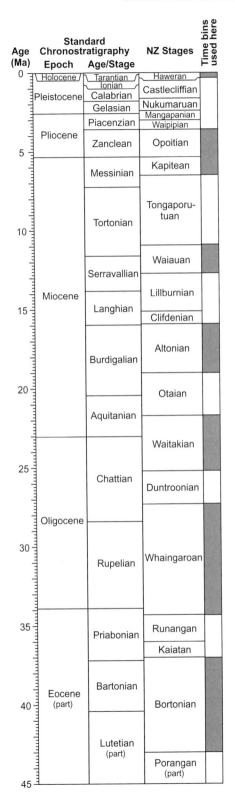

Age (Ma)	Standard Chronostratigraphy		NZ Stages	Time bins used here
	Epoch	Age/Stage		
0	Holocene	Tarantian	Haweran	
		Ionian	Castlecliffian	
	Pleistocene	Calabrian	Nukumaruan	
		Gelasian	Mangapanian	
		Piacenzian	Waipipian	
	Pliocene	Zanclean	Opoitian	
5				
		Messinian	Kapitean	
		Tortonian	Tongaporu-tuan	
10				
			Waiauan	
		Serravallian		
	Miocene		Lillburnian	
15		Langhian	Clifdenian	
			Altonian	
		Burdigalian		
20			Otaian	
		Aquitanian		
			Waitakian	
25		Chattian	Duntroonian	
	Oligocene			
30		Rupelian	Whaingaroan	
35		Priabonian	Runangan	
			Kaiatan	
	Eocene (part)	Bartonian	Bortonian	
40				
		Lutetian (part)	Porangan (part)	
45				

Spearman's rank-order correlation coefficient (r_s). First-differencing is a simple treatment for minimizing autocorrelation in the data (e.g. McKinney 1990) and it means that comparisons between time series focus on short-term, time bin-to-time bin patterns of variation. The first difference is calculated simply as the value for one time bin minus the value for the preceding bin. Use of the non-parametric Spearman's correlation coefficient is a statistically conservative approach that makes few assumptions regarding underlying distributions of data. In cases where we consider pair-wise comparisons of several time series, we have corrected for inflation of type I error rates by adjusting individual significance levels using the false discovery rate procedure (Curran-Everett 2000).

Where relevant, we have explored relationships between different time series further using bivariate and multiple linear regression (again, based on first differences). In particular, we investigated models containing up to two predictor variables that were fitted individually and as additive or multiplicative (interactive) terms. We assessed the goodness-of-fit of simple and more complex regression models using F-tests and analyses of variance based on the residual sums of squares of competing nested models. In cases where assignment of predictor and response variables is not clear, we have tested all possible options for consistency of interpretation.

Estimation of diversity

Raw and sampling standardized estimates of mollusc diversity were generated using species occurrence records from the FRED dataset and the sampled within-bin counting protocol (e.g. Alroy *et al.* 2001, reasons for preferring this protocol are outlined in Crampton *et al.* 2006*b*; Alroy *et al.* 2008). The need for sampling standardization, and methods for achieving this, have been discussed at length elsewhere (e.g. Raup 1975; Alroy 1996, 2000, 2010; Miller & Foote 1996; Alroy *et al.* 2001, 2008). Here we have employed three different approaches to sampling standardization that make rather different assumptions about underlying diversity structure (e.g. Bush *et al.* 2004): by-lists unweighted standardization (LUW, Smith *et al.* 1985), by-lists occurrences weighted standardization (OW, Alroy 1996), and shareholder quorum standardization (SQ, Alroy 2010) (see Appendix A for details). Temporal patterns of diversity change are very similar across the family of analyses and

Fig. 2. New Zealand Eocene to Recent time-scale (after Cooper 2004; Hollis *et al.* 2010), showing the time bin groupings of stages used here.

correlations are high (Appendix A), although the relative magnitudes of diversity peaks, the strength of a long-term trend, and error bars do vary. Given overall consistency between methods, for simplicity we report results based on the well-tested OW method, but note that none of our key interpretations change if we use results from either the LUW or SQ methods. Throughout the text, all comparisons between time series involve sampling standardized diversity unless specified with the prefix 'raw'.

Estimation of taxic rates

We estimated origination and extinction rates from biostratigraphic ranges as recorded in the synoptic dataset. To circumvent problems caused by incomplete sampling, we used the inverse survivorship modelling approach of Foote (2003, 2005). (The synoptic data, which we think provide the best representation of species' ranges, consist only of first and last appearances. Sampling standardization is therefore not an option, cf. Alroy 2008.) For living species, in order to avoid pull-of-the-Recent problems, we used the youngest occurrences as recorded in the FRED dataset (see Appendix A for further details). Rates given here are per-time bin, per-capita rates and assume pulsed turnover. Again, throughout the text, all comparisons between time series involve sampling standardized rates unless specified with the prefix 'raw'.

Estimation of spatial structuring of biodiversity

In order to quantify New Zealand-wide biogeographical structuring of molluscan diversity, we need a measure that is analogous to beta diversity (β) at the ecological scale. Beta diversity is the 'extent of differentiation of communities along habitat gradients' (Whittaker 1972, p. 214). In effect, we seek a measure of overall beta diversity averaged across all shelf habitat gradients; henceforth we refer to this as 'spatial turnover' of faunas. Unlike gamma (regional, γ) and alpha (community, α) diversities that are measured quantities, beta diversity is a derived metric. A plethora of concepts have been proposed for beta and its calculation has been the subject of extensive and vigorous debate (e.g. Ellison 2010 and associated forum papers). In particular, debate has revolved around the use of multiplicative v. additive partitions of diversity (e.g. Layou 2007; Holland 2010a, for discussion in a palaeontological context), and the conversion of some common beta indices to 'numbers equivalents'. In the present study, we lack species abundance proportions and are limited to measures that are appropriate for presence–absence data (Koleff *et al.* 2003 list 24 such measures). Our aim is not

to contribute to debate about the merits of different approaches, but merely to derive some robust estimate of spatial turnover.

To this end, we have experimented with two measures of spatial turnover based on rather different, multiplicative formulations of beta diversity, namely Whittaker's (1960) original index, β_w, and Simpson's (1943) β_{sim} index (see Appendix A for details). Both are expressed as measures of dissimilarity, so that values of 0 correspond to perfect similarity between faunas and high positive values indicate maximum dissimilarity. Although the two indices yield time series that are significantly correlated, β_{sim} has the distinct advantage of being insensitive to sample size bias and variations in richness; henceforth we base all our interpretations on results from this method. We also calculated map sheet occupancy, an estimate of the proportion of potentially available geographical range actually inhabited by a species (Foote *et al.* 2007), as an independent proxy for spatial structuring of the biota; as expected, we find this to be inversely correlated with spatial turnover even though the two metrics target somewhat different attributes of faunal distribution (see Appendix A). We do not consider map sheet occupancy further here.

Estimation of sampling probability

Per-time bin sampling probability (R) is the probability that any species ranging through an entire bin has been sampled and recorded at least once from that bin. We use this as a metric of the quality of the fossil record. We estimated sampling probability using the standard gap statistic of Paul (1982), as modified by Foote & Raup (1996), calculated using the FRED occurrence data (see also Crampton *et al.* 2006a). Uncertainties were calculated as binomial standard errors.

Estimation of spatial structuring of the environment

In order to examine the relationship between spatial structuring of biodiversity and spatial structuring of the environment, we need to quantify temporal changes in the degree of gross fragmentation or partitioning of the environment. (Facies variation *per se*, although of interest, is likely to be expressed at much finer spatial and temporal scales than the partitioning of interest here; see Appendix A.) To measure environmental partitioning, we used data from the lithostratigraphic sections presented in King *et al.* (1999), as compiled into a digital database by Peters (see http://macrostrat.geology.wisc.edu and Peters 2006b). In this database, each of the 403 stratigraphic sections, arranged within 39 transects, has been digitized and coded according to lithology, age and lithostratigraphic unit. The depositional environment of each unit was inferred separately using the compilation of Crampton *et al.* (2003), corrected for omissions and updated with new interpretations. For consistency with the molluscan diversity data, we considered only units that were deposited entirely or partly in shelfal marine environments.

To quantify environmental partitioning, we computed an index of spatial dispersion for deposition of shelfal facies (d_s), deposition of non-shelfal facies (dominantly deep marine, not considered here), and non-deposition or erosion (unconformities, d_u) (see Appendix A for details). For d_s, high values of this index indicate that shelfal units were deposited over much of the available and preserved area, and low values indicate patchy deposition relative to available and preserved area – that is, high environmental partitioning (Fig. 3). For d_u, high values indicate that non-deposition or erosion in shelfal environments was widespread, and low values correspond to geographical localization of non-deposition/erosion. We stress that, to the extent possible, d_u was computed in such a way as to encode changes that occurred in the shelfal environment and/or to reflect only short-term excursions out of this setting. Thus, for example, relatively brief subaerial exposure and minor erosion of shelfal habitats, lasting less than one time bin, have been captured. In contrast, periods of erosion/non-deposition that lasted through several time bins, or removed strata deposited during several preceding time bins, or were restricted apparently to deep marine environments, have been ignored. We also stress that our measures of spatial dispersion were calculated by comparing the observed spatial extent of facies to a random expectation derived from the actual spatial distribution of preserved sections in each time bin (see Appendix A); these measures are therefore insensitive to uneven representation of the rock record through time. Given the comparatively small number of sections available to us, this last point is an important reason why we could not simply use established coefficients of spatial structuring that have been developed in population ecology (e.g. Green 1966), which assume more comprehensive landscape coverage than available with our stratigraphic data.

In addition to our measures of (non-)depositional dispersion, following Peters (2005, 2006a) we also calculated per-time bin rates of initiation and truncation of shelfal sedimentary packages (macrostratigraphic 'origination' and 'extinction' rates, p_{strat} and q_{strat}, respectively). For these calculations, boundaries of shelfal sedimentary packages were defined by the limits of unconformities or intervals of non-shelf units. In other words, moving up a section, any transition from non-deposition, erosion, or non-shelf facies into shelf facies was coded as an initiation, and any transition out of shelf facies was counted as a truncation. Conformable transitions between different lithologies within the shelf, for example limestone to mudstone, were not counted as initiations or truncations.

Time-scale

Geological ages are given in terms of the New Zealand Cenozoic stages (Cooper 2004; Hollis *et al.* 2010). In order to minimize potential biases caused by unequal stage durations, short stages were combined to yield 13 Late Eocene to Pleistocene time bins (Fig. 2). The youngest of these, the Haweran Stage, is short and was retained simply as an edge-bounding bin. Ignoring the Haweran, the time bins have a mean duration of 3.5 Ma and a standard deviation of 1.4 Ma.

All the quantities discussed above are expressed per-time bin. Whereas some of these measures might be expected to correlate with time bin duration (e.g. sampling probability, see Crampton *et al.* 2006a), others should be independent of duration (e.g. origination and extinction rates under the assumption of pulsed turnover and our measures of environmental heterogeneity). In order to test for possible confounding effects of unequal time bin duration, we calculated correlation coefficients between all our measures and bin duration (excluding values for the short Haweran Stage). In addition, for all relationships of interest, we used multiple regression to test for sensitivity to the inclusion of time bin duration as a second predictor variable. In all but two cases, correlations between time bin duration and our key measures are low ($r_s < 0.17$) and non-significant, and inclusion of bin duration does not significantly improve regression fits between predictor and response variables. The two exceptions are β_{sim} and macrostratigraphic initiation rate (correlations against time bin duration, $r_s = -0.570$, $P = 0.067$ and $r_s = 0.557$, $P = 0.034$, respectively). Implications of these time bin duration effects are discussed in the Results.

We note that Crampton *et al.* (2006a) found a significant, positive correlation between sampling probability and time bin duration ($r_s = 0.442$, $P < 0.05$) for the New Zealand mollusc data, whereas here we do not ($r_s = 0.068$, $P = 0.842$). This difference results from the fact that in the present study we have grouped the seven shortest stages into time bins of relatively uniform duration, whereas Crampton *et al.* (2006a) did not.

Results

The key time series used in this study are shown in Figure 4. Bivariate scatter plots for important pairs

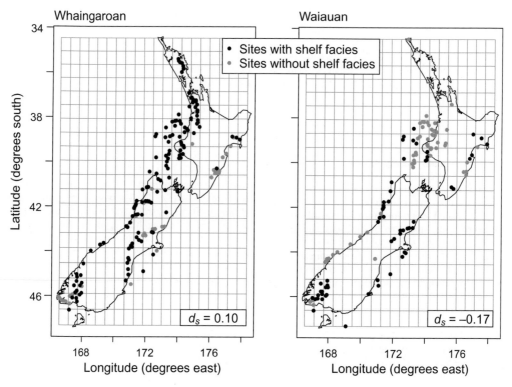

Fig. 3. Maps illustrating our measures of spatial structuring of the environment, in this case dispersion of deposition of shelfal facies (d_s) for two time bins with contrasting values. For clarity in these plots, sites of deposition of non-shelf facies and sites of non-deposition or erosion (unconformities) are combined; in our numerical analyses we separated these two categories. For the Whaingaroan, shelf facies are more dispersed relative to random expectation for the given distribution of measured sections (i.e. d_s is positive), whereas for the Waiauan, shelf facies are more clumped (i.e. d_s is negative). The 20 × 20 grid cells are equal-area but appear unequal in the latitude–longitude projection shown here. Note that some non-shelf facies or unconformities are present also at sites with shelf facies, but are not indicated here because such occurrences do not impact on the calculation or perception of d_s (see Appendix A for details and for a description of how we defined unconformities).

of metrics are shown in Figure 5. In the following, we consider three families of comparisons: (a) relationships between sampling probability and raw and sampling standardized diversity parameters; (b) relationships between taxic rates and other diversity parameters; and (c) relationships between spatial structuring of the environment and all other parameters. In passing, we also comment briefly on relationships between taxic rates and rates of macrostratigraphic initiation and truncation. Partly because of sparse data and partly because of poor constraints that result from being at the beginning of the time series (Foote 2001), the oldest interval considered here, the Bortonian stage, has a highly uncertain origination rate, with a standard error of 7.7 compared with a per-capita rate of 15.3. Therefore, this point is omitted from comparisons involving origination rate.

In general, palaeontological occurrence data are known to be affected by a lithification bias related to the combined effects of enhanced recovery of taxa from unlithified rocks and uneven temporal distribution of such rocks (e.g. Hendy 2009; Sessa *et al.* 2009). Ignoring the Haweran Stage (see above), 50% of the FRED collection lists used here have lithification recorded, 1% of these are unlithified, and most of the unlithified collections (35) are from the second youngest time bin, for which they represent 4% of collections with known lithification state. Whereas this situation might be expected to impart some bias, we find that removing these collections from analyses based on the FRED data makes no measurable difference (results not shown). We infer, therefore, that our key conclusions are unlikely to be influenced by a lithification bias.

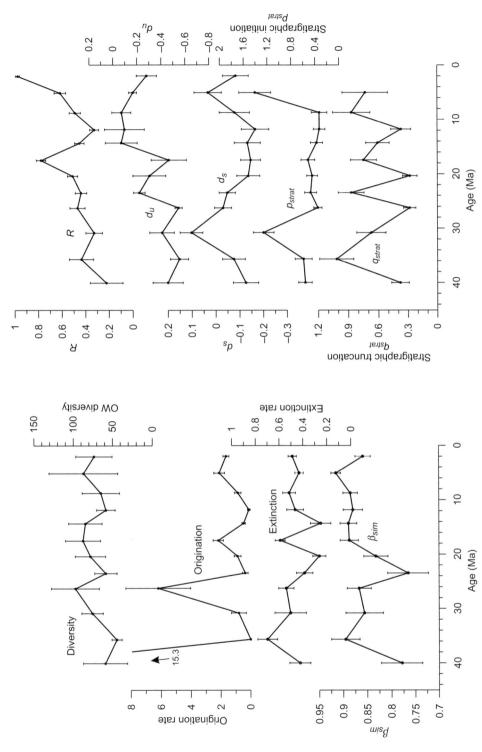

Fig. 4. Time series of key diversity and environmental metrics used in this study. All error bars are ±1 standard error. For explanations of the various measures, see Methods and Appendix A.

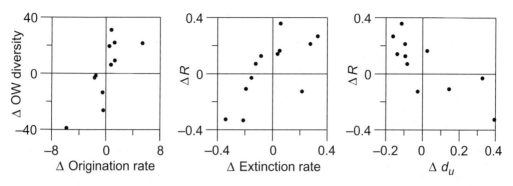

Fig. 5. Bivariate scatter plots for key pairs of metrics discussed in the text; the plots are based on first differences.

Relationships between sampling probability and raw sampling standardized diversity parameters

There are strong and significant positive correlations between raw diversity and sampling probability, and between raw extinction rate and sampling probability (Table 1). Sampling standardization removes the correlation between diversity and sampling probability, but the extinction rate-sampling probability correlation remains. Multiple regression suggests that the association between extinction and sampling probability is not affected by additive or interactive effects with origination or spatial turnover of faunas (results not shown).

Relationships between sampling standardized diversity, β_{sim}, and taxic rates

Temporal variations in diversity are strongly positively correlated with variations in origination rate

(Table 2). The time series of extinction rate, by contrast, is uncorrelated with either origination rate or diversity. These results are robust irrespective of the measure of diversity used (results not shown). Furthermore, correlations between lagged time series (either first differences or raw data) are all apparently low and non-significant, although we acknowledge that our time series may be too short to test this rigorously. For example, origination rate is not correlated positively with extinction rate in the previous time bin and extinction rate is not correlated with diversity in the previous bin (results not shown), as might be expected if simple diversity-dependent dynamics were operating at the temporal scale of the time bins used here (cf. Alroy 2008).

There is a marginally significant and positive association between β_{sim} and origination rate, suggesting that origination rate may be higher at times when there is more spatial differentiation among faunas. This correlation is not significant once the effects of multiple comparisons are taken into account (Table 2). In addition, β_{sim} is possibly

Table 1. *Correlations between diversity parameters and sampling probability*

	r_s	P
Δ Raw diversity v. Δ R	0.839*	0.001
Δ OW diversity v. Δ R	0.056	0.869
Δ Raw origination v. Δ R	0.264	0.435
Δ Origination v. Δ R	0.518	0.107
Δ Raw extinction v. Δ R	0.782*	0.007
Δ Extinction v. Δ R	0.762*	0.006
Δ β_{sim} v. Δ R	0.455	0.140

Spearman's rank-order correlation coefficients between raw and sampling standardized diversity parameters on the one hand (sampling standardized unless stated), and sampling probability (R) on the other. An asterisk indicates correlations that remain significant at $P < 0.05$ after correction for the effects of multiple comparisons.

Table 2. *Correlations between diversity parameters and taxic rates*

	r_s	P
Δ Origination v. Δ extinction	0.162	0.549
Δ OW diversity v. Δ origination	0.791*	0.006
Δ OW diversity v. Δ extinction	−0.196	0.543
Δ OW diversity v. Δ β_{sim}	0.343	0.276
Δ β_{sim} v. Δ origination	0.691	0.023
Δ β_{sim} v. Δ extinction	0.483	0.115

Spearman's rank-order correlation coefficients between sampling standardized diversity, spatial turnover of faunas (β_{sim}) and taxic rates. An asterisk indicates correlations that remain significant at $P < 0.05$ after correction for the effects of multiple comparisons.

sensitive to variation in time bin duration which may, therefore, confound any apparent relationship with origination rate (see above). This possibility has been tested in two ways. First, bin duration is not identified as a significant second predictor in a multiple regression including β_{sim} and origination rate (results not shown here). Secondly, however, the partial correlation coefficient between origination and β_{sim}, based on residuals from linear regressions of both against time bin duration, is low and non-significant ($r_s = 0.378$, $P = 0.227$). We therefore conclude that the apparent positive association between β_{sim} and origination rate in our data is statistically non-significant and may be an artefact of time bin duration, in part at least, although this question should be re-examined with longer time series and larger datasets.

Relationships between spatial structuring of the environment and other parameters

Correlations between taxic rates and spatial turnover of faunas, on the one hand, and spatial structuring in the environment on the other, are all relatively weak and non-significant (Table 3). Sampling probability (R) is, however, significantly and negatively correlated with spatial dispersion of unconformities, d_u – species, as one would expect, seem to have a smaller chance of being sampled when unconformities are more widely dispersed. This correlation is robust to the grid size used in computation of d_u (results not shown). It is also supported by regression analysis ($R = 0.081 - 0.82 d_u$, $se_{slope} = 0.21$, $t_{slope} = -3.86$, $p_{slope} = 0.004$, model $r^2 = 0.62$), a result that does not change substantively when other factors are included as additive or interactive terms.

Table 3. *Correlations between spatial structuring of the environment and other parameters*

	r_s	P
Δ Origination v. Δ d_u	−0.379	0.165
Δ Extinction v. Δ d_u	−0.468	0.070
Δ β_{sim} v. Δ d_u	−0.500	0.121
Δ R v. Δ d_u	−0.791*	0.006
Δ Origination v. Δ d_s	−0.229	0.411
Δ Extinction v. Δ d_s	−0.124	0.649
Δ β_{sim} v. Δ d_s	0.091	0.797
Δ R v. Δ d_s	−0.182	0.595

Spearman's rank-order correlation coefficients between measures of spatial structuring of the environment (d_u and d_s) on the one hand, and taxic rates, spatial turnover of faunas (β_{sim}) and sampling probability (R) on the other. An asterisk indicates correlations that remain significant at $P < 0.05$ after correction for the effects of multiple comparisons.

Table 4. *Correlations between stratigraphic rates, sampling probability and taxic rates*

	r_s	P
Δ p_{strat} v. Δ R	0.188	0.608
Δ q_{strat} v. Δ R	0.418	0.232
Δ Raw origination v. Δ p_{strat}	0.130	0.660
Δ Origination v. Δ p_{strat}	−0.059	0.844
Δ Raw extinction v. Δ q_{strat}	0.424	0.132
Δ Extinction v. Δ q_{strat}	0.479	0.073

Spearman's rank-order correlation coefficients between taxic rates, both raw and sampling standardized, and stratigraphic initiation and truncation rates (p_{strat} and q_{strat}, respectively).

Relationships between macrostratigraphic rates, sampling probability and taxic rates

We find no significant correlations between rates of macrostratigraphic initiation and truncation and sampling probability (Table 4). Likewise, we see no association between macrostratigraphic rates and taxic rate of origination, either raw or sampling standardized (Table 4). Although the relationship between extinction and macrostratigraphic truncation is not statistically significant, the two are positively correlated, in agreement with previous observations (Peters 2006a). Given the influence of time bin duration on macrostratigraphic initiation rate (see above), we tested these correlations using residuals from linear regressions of all quantities against time bin duration. The resultant partial correlations coefficients are little different from the raw coefficients shown in Table 4 (results not shown here).

Discussion

Key findings from our analyses are summarized in Figure 6. In essence, our results suggest that there might be at least two rather different dynamics affecting the real and apparent diversity history of New Zealand Cenozoic molluscs. First, sampling standardized origination rate is positively related to sampling standardized diversity. The causal direction in this association is unknown: increased origination rate might simply result in increased diversity, or increased diversity might promote origination via positive ecological feedbacks relating to, say, enhanced niche diversity.

Secondly, we find good evidence to suggest that extinction rate, both apparent and sampling standardized, and *apparent* diversity fluctuations vary in concert with sampling probability (R). The causal pathway between extinction and sampling must be indirect: if a species is extinct, then it cannot be

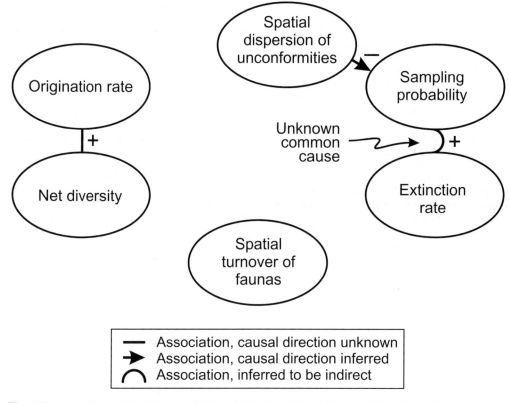

Fig. 6. Summary of key relationships identified here. ' + ' and ' − ' signs indicate positive and negative correlations, respectively. These relationships pertain to sampling standardized measures and short-term patterns of variation. Compare with Figure 1.

sampled, and if it is extant, then sampling must be controlled by some other factor. Sampling probability is itself related to spatial structuring of the environment, as measured using dispersion of unconformities (d_u), and the direction of causation is inferred to be from environmental structuring → sampling probability (Fig. 6). Specifically, for intervals when non-deposition or erosion was localized in discrete areas of uplift, sea-floor scour or sediment bypass – that is, by inference, the environment was highly structured – then sampling probability is high. Conversely, when non-deposition or erosion affected broad swaths of the shelf, then (unsurprisingly) sampling probability is low.

Unexpectedly, we find no convincing evidence for a link between spatial turnover of faunas (β_{sim}) and other diversity metrics or spatial structuring of the environment. We acknowledge, however, that such effects may have operated at spatial and temporal scales much finer than we can detect here.

From these results, therefore, and to answer the question posed in the Introduction, we would argue that there may be a common-cause effect

linking spatial structuring of the environment, biases in the fossil record and diversity dynamics. This common cause, however, is manifest most strongly in the control of extinction rate and is not linked significantly to our proxies for spatial structuring of faunas, origination rate, or short-term changes in standing diversity. We explain and explore these findings in more detail below.

The strong positive correlation that exists between sampling probability and raw diversity in our data is, perhaps unsurprisingly, removed entirely by sampling standardization. In contrast, the positive correlation between extinction and sampling probability remains unchanged following sampling standardization. A similar association between times of high sampling standardized extinction and times of high sampling probability has been observed in an analysis of global marine diversity (Foote 2003). These observations are the basis for our claim that any common-cause effect on sampling and the diversity dynamic is likely to have operated through controls on extinction rate and not on net diversity directly (see further discussion below). (It should be stressed that we rule out a simple and direct sampling effect on our sampling standardized extinction rate, which was constrained during computation to honour per-time bin sampling probability estimated from the FRED dataset. The lack of significant correlation between sampling standardized origination rate and sampling supports our inference that direct sampling effects do not drive our sampling standardized rates estimates.)

That said, we find that short-term changes in sampling standardized net diversity were in fact controlled primarily by changes in origination rate and *not* extinction rate. This result is consistent with observations for global, post-Palaeozoic marine genera (Foote 2000). Similarly, low correlations between coeval origination and extinction rates and between coeval extinction rate and diversity, as observed in our data, are consistent with patterns seen in global marine invertebrate genera (Alroy 2008). In contrast, however, we do not find significant correlations between origination and lagged extinction, or extinction and lagged diversity that have been observed in the global fauna (cf. Alroy 2008). This may suggest that, over the past 40 Ma, diversity-dependent dynamics either exerted relatively weak control of New Zealand molluscan diversity, or were manifest on a finer time-scale than we can resolve (i.e. less than the mean time bin duration of c. 3.5 Ma). Exploration of this question is beyond the scope of the present paper.

We find no statistically significant relationship between rates of initiation and truncation of stratigraphic packages and either taxic rates or sampling probability, even after correcting for time bin duration effects on macrostratigraphic initiation. This result contrasts with the findings of Peters (2005) for global Phanerozoic marine genera and the North American rock record, where there are significant associations between these measures. We note, however, that as in Peters' study, the strongest apparent relationship in the New Zealand molluscan data is between extinction and macrostratigraphic truncation; lack of significance here may reflect, in part, low statistical power due to the short time series available to us.

What then could be the common cause that links extinction rate and sampling probability in the New Zealand record? As noted by others, there is an expectation and supporting evidence to suggest that changes in relative sea-level have played a key role in driving both sampling probability (Smith 2001; Peters 2005, 2006a) and true extinction rate (Newell 1952; Sepkoski 1976; Hallam 1989, but see Hallam & Wignall 1999). Crampton *et al.* (2006a) inferred that there is indeed an association between second-order cycles of relative sea-level change, tied to tectonic activity, and the quality of the Cenozoic marine fossil record in

New Zealand. Specifically, they posited that enhanced preservation of the fossil record occurred in mid-cycle positions and destruction of the record and/or low preservation potential occurred close to sequence boundaries and base-level falls. Such effects have been predicted by modelling experiments (Holland 2000), observed in the geological record of Europe (Smith & McGowan 2007), and are supported here by the association between dispersion of unconformities and sampling probability.

At the same time, we expect that tectonic cycles of uplift and erosion will have affected habitat area, partitioning, variety, and interconnectedness, and also ocean circulation, water mass properties, and nutrient availability. All of these factors could have affected local extinction rate. Importantly, analyses of faunal diversification, both for global Phanerozoic marine faunas and New Zealand Cenozoic molluscs, support models of pulsed turnover and, in particular, pulsed extinction (Foote 2005; Crampton *et al.* 2006b). The pulsed model indicates that turnover was focused, on average, at discrete points in time, rather than being distributed in a stochastically uniform way. Pulses of high extinction, it seems, may have been separated by periods of very low or negligible extinction (Foote 2007), although this pattern is not evident at the resolution of the time bins used here. These observations are consistent with the inference that episodic abiotic environmental change may, to a substantial extent, have paced extinction through the Phanerozoic. (This hypothesis does not, however, preclude a role for density-dependent ecological interactions in regulating diversity (e.g. Alroy 2008; Finnegan *et al.* 2008)).

At the resolution of our data, we note that the two times of lowest extinction rate, at c. 20 and c. 14 Ma (Fig. 4), both lie close to second-order sequence boundaries (King *et al.* 1999, see also Crampton *et al.* 2006a). These two sequence boundaries correspond to very major tectonic events and overall base-level falls in the New Zealand region, related to reorganizations of the plate boundary geometry. First, starting at c. 25 Ma and culminating at c. 20 Ma, the earliest significant manifestation of the modern plate boundary coincided with onset of subduction, widespread uplift, allochthon emplacement and, in the north, extensive volcanism (Rait *et al.* 1991; Herzer 1995; Wood & Stagpoole 2007; Uruski 2010). Secondly, between c. 15 and c. 12 Ma there was a major crustal closure event and increase in deformation rate, extensive uplift, local subsidence, and reorientation of the volcanic arc (Herzer 1995; Wood & Stagpoole 2007). Lastly, although less conspicuous than these two events, a similar decrease in extinction rate at c. 5 Ma coincided with the most recent second-order sequence boundary and what was a major acceleration of uplift, erosion, and sedimentation (King *et al.* 1999; Wood & Stagpoole 2007).

The key point is that the intervals of low extinction rate apparently coincide with base-level falls; in other words, extinction rate is *low* when relative sea-level is low. This pattern is the opposite of that expected if diversity dynamics were being driven primarily by species-area affects, such that continental flooding increased habitat area, relaxed extinction pressures, and caused diversification (e.g. Newell 1967; Johnson 1974; Sepkoski 1976; Smith 2001; Wall *et al.* 2009). Although we acknowledge that the relationship between sea-level and habitable area is in fact likely to be highly complex and non-monotonic (Holland 2010*b*), our findings are consistent with the lack of a species-area effect observed both in this fauna (Crampton *et al.* 2006*b*) and in temperate marine bivalves (Harnik *et al.* 2010), and with the negative correlation between origination rate and sea-level observed in global Phanerozoic marine faunas (Cárdenas & Harries 2010).

More generally, the lack of significant associations between spatial turnover of faunas, spatial structuring of the environment, and any of our other diversity metrics supports the inference that simple habitat area and partitioning effects have not clearly been dominant and direct controls of large-scale Cenozoic molluscan diversity dynamics in New Zealand. This conclusion is unexpected. Again, however, we caution that these results may, to some extent, reflect low statistical power.

In any case, taken at face value, these findings suggest that other factors – such as nutrient availability or oceanographic effects (e.g. Cárdenas & Harries 2010) – have controlled extinction rate and, thereby, the relationship between environment, sampling, and diversity. We cannot, as yet, identify key drivers of extinction rate (or origination rate, for that matter), but we note that relationships between environmental variables and biotic consequences may be complex and are likely to be magnitude and scale-dependent. For example, a recent study (Algeo & Twitchett 2010) argued that greatly increased sedimentation rates during the end-Permian mass extinction may have contributed to the biotic crisis because of the harmful effects of siltation and eutrophication. In contrast, our data suggest that increased sediment fluxes during tectonic pulses coincided with times of lowered extinction rate.

The decoupling of sampling biases and environmental structuring from short-term variations in origination rate and diversity that is inferred here mirrors, in some ways, the finding of Wall *et al.* (2009) that outcrop area (i.e. sampling) operates largely independently of original habitat area to shape our view of apparent diversity history. Unlike Wall *et al.*, however, we conclude that

there *is* a common-cause that simultaneously affected diversity, through an influence on extinction rate, and preservation of the fossil record.

Conclusions

Short-term variations in the diversity of New Zealand shelf molluscs over the past 40 Ma, and our perceptions of diversity change, apparently have been controlled by two rather different dynamics:

- Variations in sampling standardized diversity were related primarily and positively to changes in origination rate and *not* to changes in extinction rate.
- A common cause that affected simultaneously the quality of the fossil record and diversity dynamics operated primarily through its effect on extinction rate; times of increasing extinction rate tend to correspond to times of increasing sampling probability.

The quality of the fossil record is correlated inversely with spatial dispersion of unconformities that is, in turn, likely to have been controlled (at the temporal scale studied here) by tectonic processes manifest as second-order sequence stratigraphic cycles. Likewise, there is evidence to suggest that the extinction record is pulsed and has been paced by episodic environmental change, including second-order cycles of base-level change. The exact nature of the common cause linking sampling probability and extinction rate, however, remains unknown, although it is *unlikely* to have been related strongly or directly to gross scale environmental partitioning *per se*.

Contrary to our expectations, spatial structuring of the environment is not significantly correlated with spatial turnover (structuring) of faunas and does not seem to have been a strong or direct driver of changes in origination rate, extinction rate, or net diversity. Whereas these findings could reflect, to some extent, low statistical power resulting from the short time series available for analysis, the key point is that these effects, significant or not, are relatively small.

To answer the question posed at the start of this paper – to what extent did spatial partitioning of the environment influence molluscan diversity and the preservation of that diversity – we suggest that partitioning *did* influence the quality of the fossil record, but did *not* exert strong effect on sampling standardized diversity dynamics at the spatial scale of New Zealand and the temporal scale of stages.

For advice and discussion on the nature of the tectonic and stratigraphic records in New Zealand, and on statistical methods, we thank K. Bland, G. Browne, D. Harte, P. King, T. Mitchell, D. Rhoades, V. Stagpoole,

R. Sutherland and R. Wood. Thanks also to A. Smith and W. Kiessling for constructive and insightful comments that improved the manuscript. We acknowledge use of information contained in the New Zealand Fossil Record File (http://www.fred.org.nz/). Early stages of this work were supported by the Marsden Fund (contract GNS0404). JSC thanks organisers of the Lyell Meeting 2010 'Comparing the geological and fossil records: implications for biodiversity studies' for both the invite to participate in, and support to attend, the meeting.

Appendix A

Details of methods

Estimation of diversity

First, we employed the by-lists unweighted standardization (LUW, Smith *et al.* 1985). In this method, for each time bin, a fixed quota of 39 collection lists was drawn at random, without replacement, and the number of unique species was tabulated. This process was repeated 500 times and the average number of taxa in each time bin was calculated (Fig. A1). The standard error was computed as the standard deviation over all replicates. The quota was determined by the number of collections present in the most poorly sampled time bin. Secondly, we used the by-lists occurrences weighted standardization (OW, Alroy 1996). In this method, collection lists were drawn at random, without replacement, until a fixed quota of 109 species occurrences was accumulated, the number of taxa was tabulated, and the process was repeated 500 times (Fig. A1). Lastly, we used the recently developed shareholder quorum standardization (SQ, Alroy 2010). Although rather more complex than the other methods, in essence this approach samples to a quota based on 'coverage' – the summed frequencies of taxa that have

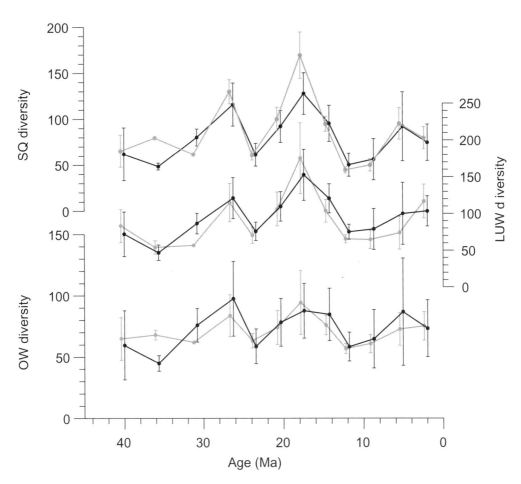

Fig. A1. Time series of sampling standardized diversity metrics examined here. Black curves employ the three-timer sampling probability correction; grey curves do not (see text for explanation). All error bars are ±1 standard error.

been sampled at least once, where each frequency is the proportion of all occurrences accounted for by the species in question. The first time a taxon is encountered during random sampling, its full frequency is counted. In this way the method gives due recognition to uncommon taxa. Importantly, it does not sample to a fixed number of taxa, proportion of taxa, number of collections, or proportion of collections. During computation, the coverage quota q was adjusted separately for each time bin such that $q_{adjusted} = q/u$, where

$$u = 1 - \frac{s_1}{N - n_d},$$

s_1 = number of singleton species, n_d = the number of occurrences of the dominant species, and N is the total number of occurrences in a given time bin (see Alroy 2010 for a full explanation). As for the other sampling standardizations, results given here are based on 500 resampling trials (Fig. A1).

For all three sampling standardization methods and following Alroy et al. (2008), we adjusted the standardized diversity counts using the three-timer sampling probability in order to minimize otherwise uncorrected short-term variation in sampling. The three-timer sampling probability is

$$R_3 = \frac{3T}{(3T + PT)},$$

where $3T$ = number of species sampled immediately before, during and after a time bin, and PT = number of species sampled immediately before and after the time bin, but not during. Diversity counts were multiplied by R_{3tot}/R_{3i}, where R_{3tot} is the aggregate three-timer sampling probability calculated over all time bins for a given sampling trial, and R_{3i} is the three-timer probability calculated for a given time bin and sampling trial. Use of the three-timer correction, although desirable, substantially inflates error bars on diversity estimates. Temporal patterns of estimates with and without the three-timer correction are, however, very similar (Fig. A1). Given this, and consistency between the three different sampling standardization methods (Table A1), we have confidence in the major patterns of diversity change inferred.

Table A1. *Correlations between sampling standardized measures of diversity*

	r_s	P
Δ SQ diversity v. Δ OW diversity	0.895	≪0.001
Δ SQ diversity v. Δ LUW diversity	0.867	<0.001
Δ OW diversity v. Δ LUW diversity	0.874	<0.001

Spearman's rank-order correlation coefficients between the different measures of diversity calculated here (all three adjusted using the three-timer correction, see Appendix A text for details).

Estimation of taxic rates

Inverse survivorship modelling of taxic rates uses numerical maximization of a likelihood function to find origination, extinction and sampling rates that yield the best fit between predicted forward and backward survivorship probabilities and the corresponding observed values (Foote 2003). Uncertainties in the parameter estimates, due to variance in the data together with imprecision in the optimization, were estimated by bootstrap resampling of biostratigraphic ranges (with replacement) 500 times. The model employed here assumes pulsed turnover of faunas (Foote 2005) – that is, that turnover occurred mainly at time bin boundaries – an assumption that is supported for the New Zealand mollusc data (Crampton et al. 2006b). Although inverse survivorship modelling allows for joint estimation of origination, extinction and sampling rates, here we constrained per-time bin sampling probability using values estimated directly from the FRED dataset using standard gap analysis (e.g. Paul 1982, as modified by Foote & Raup 1996; see main text). Constraining sampling rates in this way reduced uncertainties in the estimation of taxic rates.

Measures of spatial turnover of faunas

The first measure of spatial turnover of faunas tested here is based on β_w (Whittaker 1960), the simplest and most widely used beta diversity index (Koleff et al. 2003):

$$\beta_w = \frac{\gamma}{\bar{\alpha}},$$

where γ = total regional diversity and $\bar{\alpha}$ = average community diversity. β_w is a 'broad sense' measure *sensu* Koleff et al. (2003) that confounds effects of changes in species richness and species composition. For each time bin, we calculated average alpha diversity as the geometric mean of individual collection diversities, treating palaeontological collections as approximations of community assemblages. Use of the geometric mean is desirable given the strongly right-skewed distributions of collection diversities that vary over two orders of magnitude (although results do not change significantly if the median is used instead). Because the value of β_w is proportional to total diversity, it is highly sensitive to sample size. Hence, we used the OW sampling standardization protocol described above. Values of β_w used in our analyses are means based on a quota of 100 occurrences and 500 resampling trials; the standard error was computed as the standard deviation over all replicates. Because the resampling quota was determined by the most poorly sampled time bin and is comparatively low, error bars are correspondingly large. Increasing the quota to 275 occurrences means that error bars are substantially reduced but two time bins are excluded from the analysis. The time series based on quotas of 100 and 275 are, however, almost identical ($r_s = 0.997$, $P < 0.000$) and we therefore regard the major structure in the full time series, based on a quota of 100 occurrences, as reliable despite the large error bars (Fig. A2).

The second and preferred measure of spatial turnover used here is based on the β_{sim} formulation of Simpson (1943), as re-expressed by (Koleff *et al.* 2003). It is a 'narrow sense' measure that focuses on compositional differences rather than variations in richness (cf. β_w, see above). Of the many formulations evaluated by Koleff *et al.* (2003), it is probably the one that performs best overall according to their criteria. For two 'quadrats', one the focal and one the neighbour:

$$\beta_{sim} = \frac{\min(b, c)}{\min(b, c) + a},$$

where the quadrats are the units of sampling, a = number of species shared between the quadrats, b = number of species present in the neighbouring but not the focal quadrat, and c = number of species present in the focal but not the neighbouring quadrat. We experimented with sampling quadrats based on individual collections and on 1:50 000 map sheets. Time series derived from these two approaches are very similar and, for consistency with our collection-based standardizations of diversity and β_w, all results reported here are based on collections as quadrat units. For each time bin, we computed β_{sim} for all pairwise combinations of focal and neighbouring collection and calculated the average. Trials using the complete dataset and the OW subsampling protocol, with a range of quota sizes, confirmed that the estimation of β_{sim} is essentially insensitive to sample size. For this reason, the final results used for interpretation (Figs 4 & A2) are based on the entire dataset and standard errors were estimated using bootstrap resampling, with replacement. Because calculation of β_{sim}

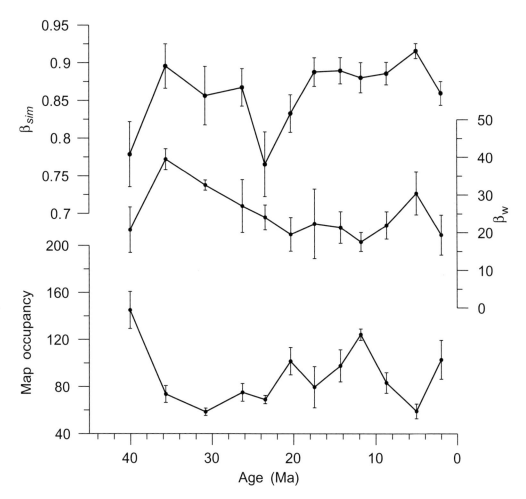

Fig. A2. Time series of faunal spatial turnover and map occupancy metrics examined here. Note that map occupancy is expressed as percentage deviation from the mean (see text for explanation). All error bars are ± 1 standard error.

is time consuming for bins with many collections, the boot-strap resampling was limited to 100 trials and a maximum of 100 collections per time bin.

Lastly, we calculated average map sheet occupancy as an independent proxy for spatial structuring of faunas. For a given species, map sheet occupancy is the number of 1:50 000 topographical map sheets occupied as a proportion of the total number of sheets represented in a given time bin. Logically, map sheet occupancy and spatial turnover are expected to be inversely correlated, and this is the pattern we observe (Table A2, Fig. A2). Again, the measure is somewhat sensitive to sample size and therefore results were sampling standardized using the OW protocol and 500 resampling trials; the standard error was computed as the standard deviation over all replicates. As in the case of β_w, error bars are large for a resampling quota of 100 occurrences, but two time bins are excluded if the quota is increased. Given that patterns of temporal variation are almost identical if one uses quotas of 100 or 275 occurrences ($r_s = 0.967$, $P < 0.001$), for interpretation we simply expressed the two time series on a common scale by dividing by their respective means, combined them, and used values for the larger quota except for the two time bins that failed to meet this quota (Fig. A2).

We tested the impact of both 'under-' and 'over-sampled' collections on the measurement of spatial turnover and map occupancy. Undersampled collections are those containing just a small number of species, and we might assume that these are either incompletely reported or taphonomically highly culled. Oversampled collections, on the other hand, are those containing relatively very high diversities that can be interpreted as atypical 'lagerstätte'. Both these types of collections might bias estimates of spatial turnover of faunas, although they are not expected to affect measures of map occupancy. In fact, experiments in which we removed collections with between 1, 2 or 3 species, and the upper 5%, 1% or 0.5% of most diverse collections, show that major patterns of temporal variation in our measures of spatial turnover are not unduly biased by these potentially problematic samples (e.g. $r_s = 0.874$, $P < 0.000$, for β_{sim} with and without single species collections and 0.5% of the most species-rich collections). As

predicted, map occupancy is essentially insensitive to the presence of under- or oversampled collections ($r_s = 0.923$, $P < 0.000$, for a quota of 100 occurrences, with and without single species collections and 0.5% of the most speciose collections). Results reported throughout the text, therefore, are for analyses in which all collections have been retained (Fig. A2).

Measures of spatial structuring of the environment

Our indices of spatial dispersion of shelfal deposition (d_s) and non-deposition/erosion (d_u) were calculated as follows. Shelfal marine units are those inferred to have been deposited, wholly or in part, at shelf depths in open marine environments (excluding units restricted to estuarine environments). New Zealand was divided into an arbitrary grid of equal-area cells following projection of data using a Lambert cylindrical equal-area projection; we experimented with grids of 20 cells by 20 cells (each cell 3531 km²), 30 cells by 30 cells (each cell 1569 km²), and 40 cells by 40 cells (each cell 883 km²); reported results are based on the 20 × 20 grid and we note instances where interpretations are sensitive to the grid used.

For each of the 403 stratigraphic sections, we considered only the interval lying between the base of the oldest stratigraphic unit and the top of the youngest stratigraphic unit and, for each time bin, we counted three quantities. First, we counted the number of sections, s_s, containing shelfal units within the time bin and the number of cells, c_s occupied by these sections. For this measure, different lithologies and facies were not discriminated (see below). Secondly, we counted the number of sections and cells containing non-shelfal units within the time bin (these quantities are not used further here). Lastly, we counted the number of sections, s_u, containing intervals of unconformity and/or non-deposition and the number of cells, c_u, occupied by these sections, according to the following protocol. Unconformities were counted only if they lie entirely within the time bin in question or at one of its boundaries, and if they are bounded above and/or below by shelfal units. These protocols were designed to ensure that counted unconformities, if erosional, are localized in time to reflect environmental events that occurred during the span of time bin, and to ensure that they reflect environmental changes that affected shelfal environments.

We then picked at random s_s sections from all those present within the time bin (sampled without replacement from the set of all sections, including those with non-shelf deposition or unconformities) and counted the number of cells occupied by this sample, c_{sr}. For each of j such randomizations, we calculated the proportional difference between the number of cells containing shelfal deposits and the corresponding number for the random array:

$$d_{s,j} = \ln(c_s/c_{sr,j}).$$

Table A2. *Correlations between different measures of spatial turnover of faunas*

	r_s	P
$\Delta\,\beta_{sim}$ v. $\Delta\,\beta_w$	0.608	0.040
Δ Map occupancy v. $\Delta\,\beta_{sim}$	−0.504	0.099
Δ Map occupancy v. $\Delta\,\beta_w$	−0.825*	0.002

Spearman's rank-order correlation coefficients between the different measures of spatial turnover of the fauna calculated here (β_{sim} and β_w) and map occupancy (see Appendix A text for details). An asterisk indicates correlations that remain significant at $P < 0.05$ after correction for the effects of multiple comparisons.

The random array was always drawn from the sections actually present in the time bin; therefore, spatial structure in the distribution of sections was accounted for. The reported values of d_s for each time bin are the averages of 100 such randomizations and the standard error was computed as the standard deviation over all replicates. If d_s is equal to zero, the distribution of shelfal deposits is indistinguishable from random. Positive and negative values of d_s correspond to shelfal deposits that are more dispersed and more clustered, respectively, than expected by chance. Most of our average measures are negative (see Fig. 4), but we refer to them in a relative sense as more dispersed if they are less negative. Values of d_u were calculated in the same way, using s_u, c_u and $c_{ur.j}$.

We also calculated an analogous set of measures of spatial dispersion of shelf-only lithofacies, distinguishing coarse (conglomerate and breccia), medium (sandstone) and fine (siltstone and mudstone) siliciclastics, carbonate, volcanic rocks, and unconformities. The resulting temporal patterns of dispersion of shelfal lithofacies show uniformly low correlations with our measures of diversity and spatial turnover of faunas (results not shown here). This is surprising but, as noted in the main text, we suggest that structuring of lithofacies may be expressed, and may influence biodiversity dynamics, at finer spatial and temporal scales than resolved here. This question will be examined elsewhere.

References

ALGEO, T. J. & TWITCHETT, R. J. 2010. Anomalous Early Triassic sediment fluxes due to elevated weathering rates and their biological consequences. *Geology*, **38**, 1023–1026.

ALROY, J. 1996. Constant extinction, constrained diversification, and uncoordinated stasis in North American mammals. *Palaeogeography, Palaeoclimatology, Palaeoecology*, **127**, 285–311.

ALROY, J. 2000. New methods for quantifying macroevolutionary patterns and processes. *Paleobiology*, **26**, 707–733.

ALROY, J. 2008. Dynamics of origination and extinction in the marine fossil record. *Proceedings of the National Academy of Sciences of the USA*, **105**, 11 536–11 542.

ALROY, J. 2010. The shifting balance of diversity among major marine animal groups. *Science*, **329**, 1191–1194.

ALROY, J., MARSHALL, C. R. ET AL. 2001. Effects of sampling standardization on estimates of Phanerozoic marine diversification. *Proceedings of the National Academy of Sciences of the USA*, **98**, 6261–6266.

ALROY, J., ABERHAN, M. ET AL. 2008. Phanerozoic trends in the global diversity of marine invertebrates. *Science*, **321**, 97–100.

BUSH, A. M., MARKEY, M. J. & MARSHALL, C. R. 2004. Removing bias from diversity curves: the effects of spatially organized biodiversity on sampling-standardization. *Paleobiology*, **30**, 666–686.

CÁRDENAS, A. L. & HARRIES, P. J. 2010. Effect of nutrient availability on marine origination rates throughout the Phanerozoic eon. *Nature Geoscience*, **3**, 430–434.

CONNOLLY, S. R. & MILLER, A. I. 2001. Joint estimation of sampling and turnover rates from fossil databases: capture-mark-recapture methods revisited. *Paleobiology*, **27**, 751–767.

COOPER, R. A. (ed.) 2004. *The New Zealand geological timescale*. Institute of Geological and Nuclear Sciences, Monograph, **22**, 1–284. Lower Hutt, New Zealand.

COOPER, R. A., MAXWELL, P. A., CRAMPTON, J. S., BEU, A. G., JONES, C. M. & MARSHALL, B. A. 2006. Completeness of the fossil record: estimating losses due to small body size. *Geology*, **34**, 241–244.

CRAMPTON, J. S., BEU, A. G., COOPER, R. A., JONES, C. M., MARSHALL, B. & MAXWELL, P. A. 2003. Estimating the rock volume bias in paleobiodiversity studies. *Science*, **301**, 358–360.

CRAMPTON, J. S., FOOTE, M. ET AL. 2006a. Second-order sequence stratigraphic controls on the quality of the fossil record at an active margin: New Zealand Eocene to Recent shelf molluscs. *Palaios*, **21**, 86–105.

CRAMPTON, J. S., FOOTE, M. ET AL. 2006b. The ark was full! Constant to declining Cenozoic shallow marine biodiversity on an isolated mid-latitude continent. *Paleobiology*, **32**, 509–532.

CURRAN-EVERETT, D. 2000. Multiple comparisons: philosophies and illustrations. *American Journal of Physiology: Regulatory, Integrative and Comparative Physiology*, **279**, R1–R8.

ELLISON, A. M. 2010. Partitioning diversity. *Ecology*, **91**, 1962–1963.

FINNEGAN, S., PAYNE, J. L. & WANG, S. C. 2008. The Red Queen revisited: reevaluating the age selectivity of Phanerozoic marine genus extinctions. *Paleobiology*, **34**, 318–341.

FOOTE, M. 2000. Origination and extinction components of taxonomic diversity: Paleozoic and post-Paleozoic dynamics. *Paleobiology*, **26**, 578–605.

FOOTE, M. 2001. Inferring temporal patterns of preservation, origination, and extinction from taxonomic survivorship analysis. *Paleobiology*, **27**, 602–630.

FOOTE, M. 2003. Origination and extinction through the Phanerozoic: a new approach. *The Journal of Geology*, **111**, 125–148.

FOOTE, M. 2005. Pulsed origination and extinction in the marine realm. *Paleobiology*, **31**, 6–20.

FOOTE, M. 2007. Extinction and quiescence in marine animal genera. *Paleobiology*, **33**, 261–272.

FOOTE, M. & RAUP, D. M. 1996. Fossil preservation and the stratigraphic ranges of taxa. *Paleobiology*, **22**, 121–140.

FOOTE, M., CRAMPTON, J. S., BEU, A. G., MARSHALL, B. A., COOPER, R. A., MAXWELL, P. A. & MATCHAM, I. 2007. Rise and fall of species occupancy in Cenozoic fossil molluscs. *Science*, **318**, 1131–1134.

GREEN, R. H. 1966. Measurement of non-randomness in spatial distributions. *Researches on Population Ecology*, **8**, 1–7.

HALLAM, A. 1989. The case for sea-level change as a dominant causal factor in mass extinction of marine invertebrates. *Philosophical Transactions of the Royal Society of London, B*, **325**, 437–455.

HALLAM, A. & WIGNALL, P. B. 1999. Mass extinctions and sea-level changes. *Earth-Science Reviews*, **48**, 217–250.

HARNIK, P. G., JABLONSKI, D., KRUG, A. Z. & VALENTINE, J. W. 2010. Genus age, provincial area and the taxonomic structure of marine faunas. *Proceedings of the Royal Society, B*, **277**, 3427–3435.

HENDY, A. J. W. 2009. The influence of lithification on Cenozoic marine biodiversity trends. *Paleobiology*, **35**, 51–62.

HERZER, R. H. 1995. Seismic stratigraphy of a buried volcanic arc, Northland, New Zealand and implications for Neogene subduction. *Marine and Petroleum Geology*, **12**, 511–531.

HOLLAND, S. M. 2000. The quality of the fossil record: a sequence stratigraphic perspective. *Paleobiology; Supplement*, **26**, 148–168.

HOLLAND, S. M. 2010*a*. Additive diversity partitioning in palaeobiology: revisiting Sepkoski's question. *Palaeontology*, **53**, 1237–1254.

HOLLAND, S. M. 2010*b*. The common-cause hypothesis: initial predictions of habitable area from simple geometric models. *Geological Society of America Abstracts with Programs*, **42**, 138. Geological Society of America, Boulder, Colorado, USA.

HOLLIS, C. J., BEU, A. G. *ET AL.* 2010. Calibration of the New Zealand Cretaceous–Cenozoic timescale to GTS2004. *GNS Science Report*, **2010/43**, 1–20.

JOHNSON, J. G. 1974. Extinction of perched faunas. *Geology*, **1**, 479–482.

KING, P. R., NAISH, T. R., BROWNE, G. H., FIELD, B. D. & EDBROOKE, S. W. 1999. Cretaceous to Recent sedimentary patterns in New Zealand. *Institute of Geological and Nuclear Sciences, Folio Series*, **1**, 1–35. Lower Hutt, New Zealand.

KOLEFF, P., GASTON, K. J. & LENNON, J. J. 2003. Measuring beta diversity for presence–absence data. *Journal of Animal Ecology*, **72**, 367–382.

LAYOU, K. M. 2007. A quantitative null model of additive diversity partitioning: examining the response of beta diversity to extinction. *Paleobiology*, **33**, 116–124.

MAURER, B. A. & NOTT, M. P. 1998. Geographic range fragmentation and the evolution of biological diversity. *In*: McKINNEY, M. L. & DRAKE, J. A. (eds) *Biodiversity Dynamics; Turnover of Populations, Taxa, and Communities*. Columbia University Press, New York, 31–50.

McKINNEY, M. L. 1990. Classifying and analysing evolutionary trends. *In*: McNAMARA, K. J. (ed.) *Evolutionary Trends*. Belhaven Press, London, 28–58.

MILLER, A. I. & FOOTE, M. 1996. Calibrating the Ordovician Radiation of marine life: implications for Phanerozoic diversity trends. *Paleobiology*, **22**, 304–309.

NEWELL, N. D. 1952. Periodicity in invertebrate evolution. *Journal of Paleontology*, **26**, 371–385.

NEWELL, N. D. 1967. Revolutions in the history of life. *Special Papers of the Geological Society of America*, **89**, 63–91.

PAUL, C. R. C. 1982. The adequacy of the fossil record. *In*: JOYSEY, K. A. & FRIDAY, A. E. (eds) *Problems of Phylogenetic Reconstruction*. Academic Press, London, 75–117.

PETERS, S. E. 2005. Geologic constraints on the macroevolutionary history of marine animals. *Proceedings of the National Academy of Sciences of the USA*, **102**, 12326–12331.

PETERS, S. E. 2006*a*. Genus extinction, origination, and the durations of sedimentary hiatuses. *Paleobiology*, **32**, 387–407.

PETERS, S. E. 2006*b*. Macrostratigraphy of North America. *The Journal of Geology*, **114**, 391–412.

PETERS, S. E. & FOOTE, M. 2001. Biodiversity in the Phanerozoic: a reinterpretation. *Paleobiology*, **27**, 583–601.

PETERS, S. E. & FOOTE, M. 2002. Determinants of extinction in the fossil record. *Nature*, **416**, 420–424.

R DEVELOPMENT CORE TEAM. 2010. *R: A Language and Environment for Statistical Computing*. R Foundation for Statistical Computing, Vienna, Austria.

RAIT, G., CHANIER, F. & WATERS, D. W. 1991. Landward- and seaward-directed thrusting accompanying the onset of subduction beneath New Zealand. *Geology*, **19**, 230–233.

RAUP, D. M. 1975. Taxonomic diversity estimation using rarefaction. *Paleobiology*, **1**, 333–342.

RAUP, D. M. 1976. Species diversity in the Phanerozoic: an interpretation. *Paleobiology*, **2**, 289–297.

ROSENZWEIG, M. L. 1995. *Species Diversity in Space and Time*. Cambridge University Press, Cambridge.

SEPKOSKI, J. J. 1976. Species diversity in the Phanerozoic: species-area effects. *Paleobiology*, **2**, 298–303.

SESSA, J. A., PATZKOWSKY, M. E. & BRALOWER, T. J. 2009. The impact of lithification on the diversity, size distribution, and recovery dynamics of marine invertebrate assemblages. *Geology*, **37**, 115–118.

SIMPSON, G. G. 1943. Mammals and the nature of continents. *American Journal of Science*, **241**, 1–31.

SMITH, A. B. 2001. Large-scale heterogeneity of the fossil record: implications for Phanerozoic biodiversity studies. *Philosophical Transactions of the Royal Society of London, B*, **356**, 351–367.

SMITH, A. B. & McGOWAN, A. J. 2007. The shape of the Phanerozoic marine palaeodiversity curve: how much can be predicted from the sedimentary rock record of western Europe? *Palaeontology* **50**, 765–774.

SMITH, E. P., STEWART, P. M. & CAIRNS, J. J. 1985. Similarities between rarefaction methods. *Hydrobiologia*, **120**, 167–170.

URUSKI, C. 2010. New Zealand's deepwater frontier. *Marine and Petroleum Geology*, **27**, 2005–2026.

WALL, P. D., IVANY, L. C. & WILKINSON, B. H. 2009. Revisiting Raup: exploring the influence of outcrop area on diversity in light of modern sample-standardization techniques. *Paleobiology*, **35**, 146–167.

WHITTAKER, R. H. 1960. Vegetation of the Siskiyou Mountains, Oregon and California. *Ecological Monographs*, **30**, 279–338.

WHITTAKER, R. H. 1972. Evolution and measurement of species diversity. *Taxon*, **21**, 213–251.

WOOD, R. A. & STAGPOOLE, V. M. 2007. Validation of tectonic reconstructions by crustal volume balance: New Zealand through the Cenozoic. *Geological Society of America Bulletin*, **119**, 933–943.

Disentangling palaeodiversity signals from a biased sedimentary record: an example from the Early to Middle Miocene of Central Paratethys Sea

MARTIN ZUSCHIN[1]*, MATHIAS HARZHAUSER[2] & OLEG MANDIC[2]

[1]*University of Vienna, Department of Palaeontology, Althanstrasse 14, A-1090 Vienna, Austria*

[2]*Natural History Museum Vienna, Burgring 7, A-1010 Vienna, Austria*

*Corresponding author (e-mail: martin.zuschin@univie.ac.at)

Abstract: Changes in molluscan diversity across the 3rd order sequence boundary from the Lower to the Middle Miocene of the Paratethys were evaluated in the context of environmental bias. Taken at face value, quantitative data from nearshore and sublittoral shell beds suggest a transition from low-diversity Karpatian (Upper Burdigalian) to highly diverse Badenian (Langhian and Lower Serravallian) assemblages, but environmental affiliation of samples reveals a strong facies shift across the sequence boundary. Ordination methods show that benthic assemblages of the two stages, including 4 biozones and four 3rd order depositional sequences over less than four million years, are developed along the same depth-related environmental gradient. Almost all samples are from highstand systems tracts, but Karpatian faunas are mostly from nearshore settings, and Badenian faunas are strongly dominated by sublittoral assemblages. This study emphasizes the importance of highly resolved stratigraphic and palaeoenvironmental frameworks for deciphering palaeodiversity patterns at regional scales and highlights the effort required to reach the asymptote of the collector's curve. Abundance data facilitate the recognition of ecological changes in regional biota and it is suggested that in second and higher order sequences the facies covered within systems tracts will drive observed diversity patterns.

The quality of the fossil record of biodiversity is strongly influenced by the rock record (Holland 2000; Smith 2007). The amount of sedimentary rock preserved has strongly fluctuated over time and is very similar to corresponding diversity patterns, suggesting that a major bias exists (Raup 1976; Miller & Foote 1996; Smith 2001; Peters & Foote 2001; Smith & McGowan 2007; Barrett *et al.* 2009). Alternatively, it suggests that both the rock record and diversity are driven by a common underlying factor, such as sea-level change (Peters 2005, 2006), a signal that can be regionally obscured at tectonically active margins (Crampton *et al.* 2003). The change in the proportion of onshore to offshore sediments preserved in the record, however, is probably as important as changes in the volume of rock preserved (Smith *et al.* 2001; Crampton *et al.* 2003, 2006). Although global datasets are comparatively robust to such biases (e.g. Marx & Uhen 2010), sequence stratigraphical architecture undoubtedly controls patterns of faunal change on a local and regional scale (Bulot 1993; Brett 1995, 1998; Patzkowsky & Holland 1999; Smith *et al.* 2001; Smith 2001). Specifically, most changes in first and last occurrences of species, and widespread changes in species abundance and biofacies, occur at sequence boundaries and at major

transgressive surfaces (Holland 1995, 1999, 2000). It is therefore important to evaluate stage-level changes in taxonomic diversity, at the temporal scale of millions of years, in the context of rock volume- and environmental bias to ensure that these changes are not simply driven by sequence architecture (Smith 2001).

The present study focuses on diversities of two regional Miocene stages of the Paratethys, an epicontinental sea whose history is closely linked to the Alpine orogeny and that covered vast parts of Central and Eastern Europe (Rögl 1998, 1999) (Fig. 1). Standing diversity of the Central Paratethys indicates a strong increase in species richness at the boundary from the Karpatian (Upper Burdigalian) to the Badenian (Langhian and Lower Serravallian), which is interpreted as a major faunal turnover associated with the Langhian transgression (Harzhauser *et al.* 2003; Harzhauser & Piller 2007). Based on a comprehensive echinoderm dataset, however, it has been suggested that the low diversity of the Karpatian was rather caused by non-preservation of suitable habitats (Kroh 2007). In this study we use a species abundance dataset of benthic molluscs to evaluate the influence of environmental bias on the faunal change. Previous molluscan species lists from the area are not useful for

From: McGOWAN, A. J. & SMITH, A. B. (eds) *Comparing the Geological and Fossil Records: Implications for Biodiversity Studies.* Geological Society, London, Special Publications, **358**, 123–139.
DOI: 10.1144/SP358.9   0305-8719/11/$15.00 © The Geological Society of London 2011.

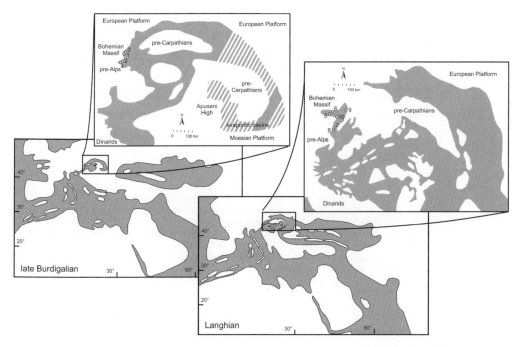

Fig. 1. Karpatian and Badenian palaeogeography of the Central Paratethys (modified after Rögl (1998) and Kovác *et al.* (2004*a*, 2007)) with approximate positions of studied localities. Karpatian (i.e. Upper Burdigalian) localities include: 1 = Laa a.d. Thaya; 2 = Kleinebersdorf; 3 = Neudorf bei Staatz; 4 = Korneuburg SPK. Badenian (i.e. Langhian and Lower Serravalian) localities include: 5 = Grund; 6 = Immendorf; 7 = Gainfarn; 8 = St. Veit; 9 = Borský Mikuláš, 10 = Niederleis.

this purpose because they are biased in favour of larger shells and biostratigraphically and palaeogeographically useful species, but stable temporal and spatial patterns of diversity can only be deciphered using large bulk samples from extensive field work (Koch 1978; Kosnik 2005).

The Vienna Basin and adjacent basins have now been systematically studied for almost two centuries for molluscs and other invertebrates. Based on the publication of a visiting French geologist (Prévost 1820), these basins were key areas for the foundation of the concept of the Tertiary in the early 19th century (Rudwick 2005, pp. 546–549; Vávra 2010). The stratigraphy of the Central Paratethys is comparatively well studied (for review see Piller *et al.* 2007) and the taxonomic composition of the Central Paratethys molluscan fauna very well known (e.g. Schultz 2001, 2003, 2005). Compilations on the standing diversity of Neogene stages were published recently (Harzhauser *et al.* 2003, Harzhauser & Piller 2007). With respect to species-abundance patterns, it has been shown that at the scale of outcrops, shell beds, and samples most species are rare and diversity is patchy (Zuschin *et al.* 2004*a*, 2006), a pattern that is also evident for the total assemblage studied here. Diversity is influenced by taphonomic processes, for example

by size sorting during tempestitic transport (Zuschin *et al.* 2005). Finally, it has been suggested that diversities of the marine Paratethys are lower than those of contemporary adjacent basins because diversity curves have rather gentle slopes when compared with such curves from the Miocene Boreal bioprovince (Kowalewski *et al.* 2002).

So far, however, studies dealing with potential biases of the raw diversities, including sampling efficiency, stage duration, fossil preservation or rock record bias, are scarce for the Paratethys (Kroh 2007). Studies on the quantitative composition of fossil molluscan lagerstätten have only been performed during the last few years (see references in Table 1). The present contribution is the first attempt to link this information to better understand one of the strongest diversity turnovers in the Central Paratethys, the transition from the Karpatian to the Badenian (Harzhauser *et al.* 2003; Harzhauser & Piller 2007).

Geological setting

The Paratethys was an epicontinental sea ranging from the French/Swiss border region in the west to the Transcaspian area (east of Lake Aral in

Table 1. *Basic data of the studied assemblages*

Locality	Section	Stage international	Stage regional	Biozone benthic foraminifers	Formation	Sequence stratigraphy (3rd order)	Systems tract	Age	Geographical position		No. of shell beds	No. of samples	References
									Latitude	Longitude			
Laa a.d. Thaya	Wienerberger AG	Burdigalian	Karpatian	Uvigerina graciliformis	Novy Prerov Fm	Tb.2.2	Late HST	16.5	48°43'07"	16°24'57"	1	4	Unpublished data
Kleinebersdorf	Kleinebersdorf Sandpit Lehner	Burdigalian	Karpatian	Uvigerina graciliformis	Korneuburg Fm	Tb.2.2	Late HST	16.5	48°29'37"	16°23'44"	1	3	Zuschin et al. 2004a
	Kleinebersdorf Sandpit Wohlmuth	Burdigalian	Karpatian	Uvigerina graciliformis	Korneuburg Fm	Tb.2.2	Late HST	16.5	48°29'42"	16°23'48"	1	3	Zuschin et al. 2004a
Korneuburg	Korneuburg SPK	Burdigalian	Karpatian	Uvigerina graciliformis	Korneuburg Fm	Tb.2.2	Late HST	16.5	48°21'28"	16°23'14"	96	110	Unpublished data
Neudorf bei Staatz		Burdigalian	Karpatian	Uvigerina graciliformis	Novy Prerov Fm	Tb.2.2	Late HST	16.5	48°43'07"	16°30'14"	1	6	Unpublished data
Grund		Langhian	Badenian	Lower Lagenidae Zone	Grund Fm	Tb.2.3	Early HST	15	48°38'18"	16°03'48"	5	5	Zuschin et al. 2004b, 2005
Immendorf		Langhian	Badenian	Lower Lagenidae Zone	Grund Fm	Tb.2.3	Early HST	15	48°39'00"	16°07'49"	5	25	Zuschin et al. 2006
Niederleis	Niederleis Buschberg	Langhian	Badenian	Lower Lagenidae Zone	Lanžhot Fm	Tb.2.3	Early HST	15	48°33'48"	16°24'17"	4	4	Mandic et al. 2002
	Niederleis Bahnhof	Langhian	Badenian	Lower Lagenidae Zone	Lanžhot Fm	Tb.2.3	Early HST	15	48°32'25"	16°24'39"	5	5	Mandic et al. 2002
St. Veit a.d Triesting		Langhian	Badenian	Upper Lagenidae Zone	Lanžhot Fm	Tb.2.3	Late HST	14.5	47°55'55"	16°08'53"	9	9	Unpublished data
Gainfarm	Gainfarm 1	Langhian	Badenian	Upper Lagenidae Zone	Lanžhot Fm	Tb.2.3	Late HST	14.5	47°56'45"	16°10'59"	7	8	Zuschin et al. 2007
	Gainfarm 2	Langhian	Badenian	Upper Lagenidae Zone	Jakubov Fm	Tb.2.4	TST	14	47°56'40"	16°10'57"	14	14	Zuschin et al. 2007
Borsky Mikulas		Serravallian	Badenian	Bolivina/Bulimina Zone	Studienka Fm	Tb.2.5	Early HST	13	48°36'20"	17°11'57"	3	17	Švagrovský 1981; Unpublished data

Kazakhstan) in the east. Its development started during the Late Eocene to Oligocene and was strongly linked to the alpine orogeny. It was separated from the Mediterranean by the newly formed land masses of the Alps, Dinarides, Hellenides, and the Anatolian Massif. Afterwards, it experienced a complex history of connection and disconnection with the Mediterranean Sea (Rögl 1998, 1999; Popov *et al.* 2004). The present study focuses on shell beds of the Vienna Basin and the North Alpine Foreland Basin; in terms of palaeogeography, they were part of the Central Paratethys, which ranged from southern Germany in the west to the Carpathian Foredeep, Ukraine in the east, and from Bulgaria in the south to Poland in the north (Fig. 1). Due to the complex geodynamic history, a regional chronostratigraphic stage system (Fig. 2) is used in the Central Paratethys. The two stages of interest here are the Karpatian and the Badenian. The Karpatian stage is characterized by a strong tectonic reorganization in the Central Paratethys area, leading to a change from west–east trending basins towards rift and intra-mountain basins (Rögl & Steininger 1983; Rögl 1998; Kovác *et al.* 2004*b*). Associated with this geodynamic impact is the abrupt, discordant progradation of upper Karpatian fossiliferous estuarine to shallow

marine deposits over macrofossil-poor lower Karpatian offshore clays in the North Alpine Foreland Basin and in the Carpathian Foredeep (Adámek *et al.* 2003). The climate was subtropical with warm and wet summers and rather dry winters (Harzhauser *et al.* 2002; Kern *et al.* 2010). The early Middle Miocene is marked by a widespread marine transgression following a major drop in sea-level at the Burdigalian/Langhian transition (Haq *et al.* 1988; Hardenbol *et al.* 1998). The regression was intensified by regional tectonic movements, the so-called Styrian phase (Stille 1924; Rögl *et al.* 2006). Sediments of the Langhian transgression are commonly eroded or reduced in thickness at the basin borders, with continuous sedimentation occurring only in bathyal settings of the basin centres (Hohenegger *et al.* 2009). In shallow-marine environments of the Vienna Basin, erosion of up to 400 m took place between the youngest preserved Karpatian and the oldest preserved Badenian sediments (Strauss *et al.* 2006). Due to the tectonic reorganization, however, a broad connection opened between the Mediterranean Sea and the Paratethys during the Langhian transgression, through which free faunal exchange occurred (Rögl 1998; Studencka *et al.* 1998; Harzhauser *et al.* 2002; Harzhauser & Piller 2007). The rising

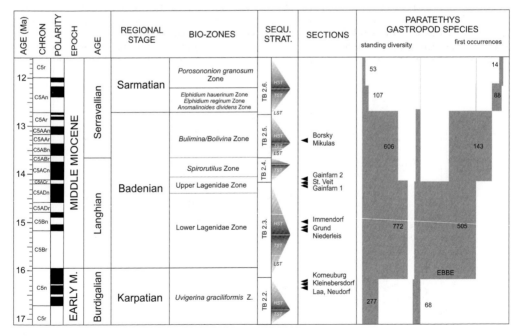

Fig. 2. Stratigraphic details for the studied sections and standing diversity of Karpatian and Badenian gastropods compiled from regional species lists and monographs (after Harzhauser & Piller 2007). The sections belong to six formations and four 3rd order sequence stratigraphic cycles and are all, except Gainfarn 2, from early or late HSTs (cf. Table 1). EBBE = Early Badenian Build-up Event.

sea-level and the Middle Miocene climatic optimum potentially strongly influenced marine life in the Central Paratethys (Harzhauser *et al.* 2003). In contrast to the Karpatian, the Badenian stage is characterized by highly fossiliferous offshore sands and pelites, and by carbonate platforms (corallinacean limestones and coral carpets). Several fossil groups increase strongly in diversity at the onset of the Badenian (Fig. 2). This event has been explicitly worked out for gastropods, with 505 taxa having their first occurrences (FOs), and for foraminifers, with FOs of 82 taxa (Harzhauser & Piller 2007). These authors dubbed this event as 'Early Badenian Build-up Event' (EBBE).

Material and methods

We studied benthic molluscs from 10 localities from the Karpatian (Upper Burdigalian) to the Badenian (Langhian and Lower Serravallian), covering all available fossil lagerstätten in the Vienna Basin and the North Alpine Foreland Basin that were amenable to bulk sampling (Fig. 3, Table 1). All samples are from siliciclastic pelitic, sandy and gravelly sediments, are characterized by aragonite and calcite preservation and were sieved through a 1 mm mesh. Detailed palaeoecological and taphonomical studies have been published on some of the sections (see references in Table 1). The shell beds of the respective localities were deposited between 16.5 and 12.7 Ma and belong to six formations, four 3rd order sequence stratigraphic cycles (Tb.2.2 to Tb.2.5 of Hardenbol *et al.* 1998),

and are mostly part of highstand systems tracts (HST); only one section belongs to a transgressive systems tract (TST) (Strauss *et al.* 2006). All fossiliferous Karpatian assemblages belong to a single regional benthic foraminifera biozone, and the studied Badenian assemblages to three such biozones (Table 1) (*Uvigerina graciliformis* zone, Lower and Upper Lagenidae zones and *Bolivina/Bulimina* zone; Grill 1943; Steininger *et al.* 1978). The faunal transition from the Karpatian to the Badenian is studied at the level of stages and biozones. For the purpose of this study, samples are environmentally assigned to the intertidal to very shallow sublittoral (<1 m water depth), termed as nearshore for the rest of the paper, and to the deeper sublittoral (few metres to several tens of metres of water depth). Palaeoenvironmental designations of samples were based on palaeogeographical positions of localities and actualistic environmental requirements of dominant molluscan taxa. Independent data from foraminifera confirm our assignments and suggest a total range of depositional water depths from intertidal to several tens of metres (pers. comm. Holger Gebhart, Patrick Grunert, Johann Hohenegger & Fred Rögl, 2009). Logarithmic scale rank abundance plots of family level data were used to compare community organization between stages and the data were fit to geometric series, log-series, broken stick and log-normal abundance models using the program PAST (Hammer *et al.* 2001). Species accumulation curves were computed to compare species richness between stages, biozones and environments using the program Estimates with 50 sample order

Fig. 3. Map of sample localities in Austria and Slovakia. Karpatian (i.e. Upper Burdigalian) localities include: 1 = Laa a.d. Thaya; 2 = Kleinebersdorf; 3 = Neudorf bei Staatz; 4 = Korneuburg SPK. Badenian (i.e. Langhian and Lower Serravalian) localities include: 5 = Grund; 6 = Immendorf; 7 = Gainfarn; 8 = St. Veit; 9 = Borský Mikuláš, 10 = Niederleis (modified after Sawyer & Zuschin 2011).

128   M. ZUSCHIN *ET AL.*

randomizations without replacement. (Colwell 2009). Diversity was measured as species richness and as evenness, which is based on the proportional abundance of species (for a review see Magurran 2004). To compensate for sampling effects in species richness we used Margalef's diversity index. The Simpson index, which is affected by the 2–3 most abundant species, and the Shannon index, which is more strongly affected by species in the middle of the rank sequence of species, were used as measures of evenness (see Gray 2000 for discussion). All indices were calculated using the program PAST (Hammer *et al.* 2001). The Margalef index was calculated with the equation

$$D_{Mg} = (S-1)/\ln N$$

where S = the total number of species and N = the total number of individuals. The Simpson index is expressed as $1 - D$ and was calculated with the equation

$$D = \sum_{i=1}^{S} \frac{n_i(n_i-1)}{N(N-1)}$$

where S = the total number of species, n_i = the number of individuals in the *i*th species and N = the total number of individuals. The Shannon index was calculated with the equation

$$H = -\sum_{i=1}^{S} p_i \ln p_i$$

where S = the total number of species, and p_i = the proportion of individuals found in the *i*th species. Species richness, the Simpson index and the

Shannon index were chosen because they are the most commonly employed measures of diversity (Lande 1996). It should be mentioned, however, that the underlying statistical distribution of a sample will generally influence the constancy of evenness measures and that the Shannon index is particularly sensitive to sample size (Lande 1996; Magurran 2004; Buzas & Hayek 2005). Non-metric multidimensional scaling (NMDS, Kruskal 1964) was used as an ordination method to evaluate the presence of environmental gradients in the dataset and was performed using the software package PRIMER (Clarke & Warwick 1994). Surface outcrop areas and their environmental affiliation of the Karpatian and Badenian in Austria are adapted from Kroh (2007) and were calculated from digital 1:200 000 scale map series of the Geological Survey of Austria for the Burgenland (Pascher *et al.* 2000) and Lower Austria and Vienna (Schnabel 2002).

Results

Sampling intensity was very high (213 samples, yielding 494 species from >49 000 shells), but the species accumulation curve for the total assemblage does not level off (Fig. 4). The number of families, genera and species, however, is significantly higher for Badenian than for the Karpatian assemblages (Fig. 5a). While in the Karpatian sampling intensity was sufficient to cover diversity at all hierarchical levels, for the Badenian the diversity of species and genera do not show a tendency to level off (Fig. 5b).

Strong differences in the abundances of dominant families and in the shape of the rank abundance plot of family level data indicate environmental

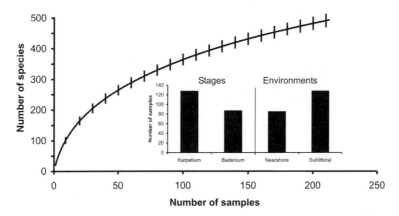

Fig. 4. Species accumulation curve of the total assemblage with 95% confidence intervals. Inset: number of samples per environment and stage. Sampling intensity was very high but the species accumulation curve does not level off.

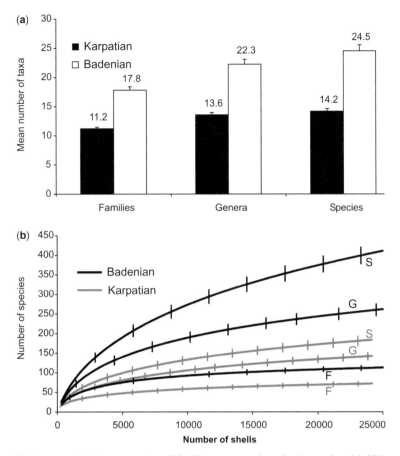

Fig. 5. Diversities in stages. (**a**) Average number of families, genera and species in samples with 95% confidence intervals. (**b**) Taxa accumulation curves for families, genera and species of the total assemblages with 95% confidence intervals. At all three taxonomic levels diversities differ significantly between the Karpatian and Badenian. For the Badenian the species- and genus accumulation curves do not show a tendency to level off, although sampling intensity was very high. F = families; G = genera; S = species.

differences between shelly assemblages of the two stages (Fig. 6). Karpatian molluscan assemblages are dominated by neritid and potamidid–batillariid gastropods, which indicate the prevalence of tidal flat deposits, whereas the Badenian molluscan assemblages are dominated by corbulid and venerid bivalves and rissoid gastropods, which all indicate the preponderance of sublittoral conditions (Fig. 6a). In accordance rank abundance plots suggest higher evenness for the total Badenian assemblage (Fig. 6b) and diversity indices are significantly higher for sublittoral than for nearshore samples in our dataset (Fig. 7). An environmental bias may therefore explain the apparent faunal turnover. In fact, in the Karpatian more samples derive from nearshore environments, whereas the Badenian is strongly dominated by sublittoral samples. This difference is even more pronounced when

considering biozones. In the Lower Lagenidae zone, samples are exclusively from the sublittoral; nearshore samples of the Badenian only occur in the Upper Lagenidae zone and in the *Bolivina/Bulimina* zone (Fig. 8).

At the level of stages and biozones the environmental affiliations of samples correlate with diversities, which are high wherever assemblages are dominated by samples from the sublittoral (Fig. 9). An exception is the *Bulimina/Bolivina* zone, but there the sampling intensity was by far the lowest (Table 1). Species accumulation curves of environments within stages and biozones are always steeper for sublittoral than for nearshore assemblages. Differences between environments within time slices are significant except for the Karpatian (i.e. the *Uvigerina gracliformis* zone) as indicated by overlapping confidence intervals (Fig. 10). Strong

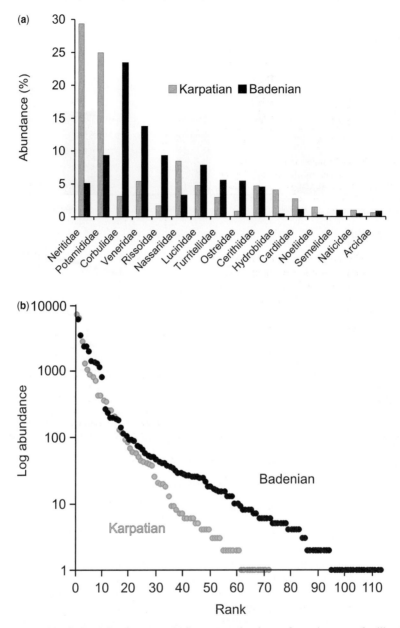

Fig. 6. Abundances of family level data for stages. (**a**) Percentage abundance of most important families. Karpatian assemblages are dominated by families indicative of tidal flat deposits and Badenian assemblage are dominated by families indicative of sublittoral conditions. (**b**) Rank abundances of families suggest higher evenness in Badenian assemblages. The log normal abundance model best describes the data in both stages (Karpatian: Chi-square $= 5.265$, $p = 0.7289$, Badenian: Chi-square $= 11.73$, $p = 0.2287$); the other three tested models have very low p-values, which implies bad fits.

diversity differences between sublittoral assemblages at the level of stages and biozones indicate habitat differences, most notably between the well-sampled *Uvigerina graciliformis* zone of the Karpatian and the Lower and Upper Lagenidae zones of the Badenian (Fig. 10b). In fact, an ordination of family-level data suggests the presence of a distinct water depth gradient (Fig. 11). Sublittoral

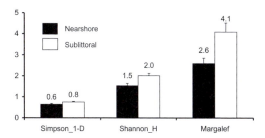

Fig. 7. Diversity indices of samples tallied based on environments. Diversity indices are significantly higher for sublittoral than for nearshore samples in our dataset and an environmental bias may therefore explain the apparent faunal turnover from the Karpatian to the Badenian.

samples from the *Uvigerina graciliformis* zone of the Karpatian represent shallower environments than those from the Lower and Upper Lagenidae zones. Differences between the latter can be explained by substrate differences. Assemblages from the Lower Lagenidae zone tend to be from

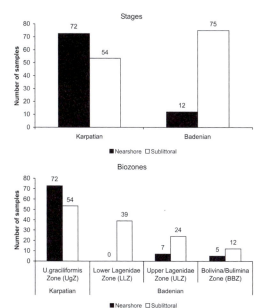

Fig. 8. Number of nearshore and sublittoral samples in stages and biozones. More Karpatian samples derive from nearshore environments, but the Badenian is strongly dominated by sublittoral samples. This environmental shift is especially pronounced at the 3rd order sequence boundary between the Karpatian *Uvigerina graciliformis* and the Badenian Lower Lagenidae zones and amplifies the impression of a diversity increase due to the Langhian transgression from a literal reading of the fossil record.

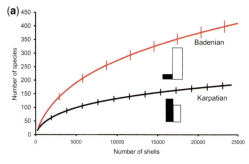

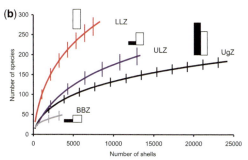

Fig. 9. Species accumulation curves with 95% confidence intervals in relation to sampled environments for stages (**a**) and biozones (**b**). Diversities are high wherever assemblages are dominated by samples from the sublittoral. An exception is BBZ, where sampling intensity was by far the lowest. LLZ = Lower Lagenidae zone; ULZ = Upper Lagenidae zone; BBZ = *Bolivina/Bulimina* zone; UgZ = *Uvigerina graciliformis* zone.

sandy environments and are therefore more diverse than those from the Upper Lagenidae zone, which are rather from pelitic environments. Environmental affiliation of Karpatian and Badenian outcrops in eastern Austria support this finding. In the Karpatian the importance of terrestrial, fluvial, fluvio-marine and limnic environments suggests that most fossiliferous marine outcrops are from nearshore environments. In the Badenian, in contrast, most outcrops preserve fully marine environments (Fig. 12) (compare also Kroh 2007).

Discussion

The importance of local and regional studies

The present study demonstrates that the quantitative evaluation of bulk samples significantly improves the understanding of regional diversity changes at temporal scales ranging from tens of thousands to a few million years and thereby confirms

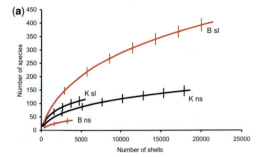

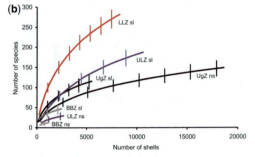

Fig. 10. Species accumulation curves with 95% confidence intervals of environments in stages (**a**) and biozones (**b**) Sublittoral environments are always more diverse than nearshore environments but for the Karpatian (i.e. the *Uvigerina gracliformis* zone) the differences are not significant as indicated by overlapping confidence intervals. Strong diversity differences between sublittoral assemblages at the level of stages and biozones are evident, most notably between the well-sampled *Uvigerina graciliformis* zone of the Karpatian and the Lower and Upper Lagenidae zones of the Badenian and point to habitat differences of the respective assemblages. K ns = Karpatian nearshore; B ns = Badenian nearshore; K sl = Karpatian sublittoral; B sl = Badenian sublittoral; LLZ ns = Lower Lagenidae zone nearshore; LLZ sl = Lower Lagenidae zone sublittoral; ULZ ns = Upper Lagenidae zone nearshore; ULZ sl = Upper Lagenidae zone sublittoral; BBZ ns = *Bolivina/Bulimina* zone nearshore; BBZ sl = *Bolivina/Bulimina* zone sublittoral; UgZ ns = *Uvigerina graciliformis* zone nearshore; UgZ sl = *Uvigerina graciliformis* zone sublittoral.

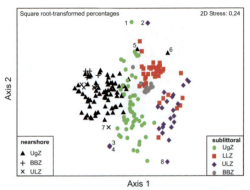

Fig. 11. Non-metric multidimensional scaling (nMDS) of family level data of the studied assemblages suggests the presence of a distinct water depth gradient along axis 1. Samples from nearshore environments of all biozones cluster at the left. Sublittoral samples from the *Uvigerina graciliformis* zone of the Karpatian represent shallower environments than those from the Lower and Upper Lagenidae zones. Differences between the latter are tentatively explained by substrate differences (samples from the Lower Lagenidae zone tend to be from sandy environments, samples Upper Lagenidae zone are rather from pelitic environments). LLZ = Lower Lagenidae zone; ULZ = Upper Lagenidae zone; BBZ = *Bolivina/Bulimina* zone; UgZ = *Uvigerina graciliformis* zone. Numbers 1–8 in the plot refer to some outliers. 1–4 are characterized by high abundances of otherwise rare taxa. In 5–7 the number of specimens is relatively low, taxonomic composition heterogeneous and environmental affiliation therefore not straight forward. 8 is a sample with very high number of shells, which are strongly dominated by one taxon.

previous authors who emphasized the importance of rigorous, extensive sampling combined within a highly resolved stratigraphic and palaeoenvironmental framework for deciphering palaeodiversity patterns (e.g. Koch 1978; Jackson *et al.* 1999; Kosnik 2005). Several lines of evidence suggest great importance of regional and local studies for the understanding of global diversity patterns. Biodiversity can be studied at a series of hierarchical scales which all contribute to an understanding of its distribution in time and space (Willis & Whittaker 2002). Diversity is, however, biologically

meaningful at local scales, where ecological processes operate and at regional scales because local communities receive species from a biogeographically delimited metacommunity (Hubbell 2001). Long-term diversity trends actually differ significantly among major regions of the world (e.g. Miller 1997; Jablonski 1998). With respect to the rock record there is a global diversity signature that relates to supercontinent cycles, but on shorter time-scales regional processes are more important and, due to heavy sampling bias, the European and North American data sets drive these patterns (McGowan & Smith 2008). Correspondingly, fossil first and last occurrences are dominated by records from these two continents (Kidwell & Holland 2002) and the Cenozoic tropics are undersampled because Europe and North America had largely moved out of the tropics by Cenozoic time (Jackson & Johnson 2001). McGowan & Smith (2008) therefore suggest focusing on the construction of regional data sets within tectonically and sedimentologically meaningful frameworks. Such regional diversity studies can typically be performed at low taxonomic

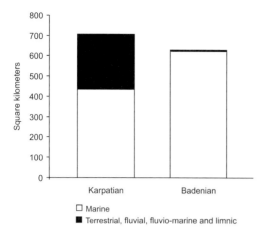

Fig. 12. Area and environments of Karpatian and Badenian outcrops in Eastern Austria (modified after Kroh 2007). Badenian outcrops mostly preserve fully marine environments. Karpatian terrestrial, fluvial, fluvio-marine and limnic environments are very prominent, suggesting that most fossiliferous marine outcrops in this stage are from nearshore environments.

levels with highly resolved stratigraphic control (e.g. Johnson & Curry 2001; Hendy *et al.* 2009). Knowledge of local abundances of organisms enables determination of sampling completeness (Koch 1987) and to recognize ecological reorganization of regional biota, which can be independent from standing diversity (Jackson *et al.* 1999; Todd *et al.* 2002). In line with these evidences, this paper highlights the sheer sampling effort that is required to reach the asymptote of the collector's curve (Figs 4, 5, 9 & 10), a feature that is well known from modern and fossil molluscan assemblages (e.g. Jackson *et al.* 1999; Bouchet *et al.* 2002; Zuschin & Oliver 2005) and which suggests that most species are rare (Gaston 1994; Harnik 2009). The use of abundance data allowed the recognition of ecological changes across a stage boundary, which drive the observed diversity increase and which can be explained by a strong environmental shift.

Environmental bias in stages and biozones

This study demonstrates strong differences in quantitative molluscan composition between two succeeding stages (Fig. 6), but it also underlines a predominance of nearshore and shallow sublittoral habitats in the studied Karpatian versus a predominance of somewhat deeper environments in the studied Badenian outcrops. Since shelf environments have a higher diversity than the physically stressed nearshore environments, the diversity increase from the Karpatian to the Badenian in our

dataset can be largely related to an environmental shift. When considering biozones, this environmental shift is especially pronounced at the 3rd order sequence boundary between the Karpatian *Uvigerina graciliformis* and the Badenian Lower Lagenidae zones. This pattern amplifies the impression of a diversity increase due to the Langhian transgression from a literal reading of the fossil record (Fig. 8). Following Jablonski (1980), it is therefore important to sample a single habitat or across a suite of habitats when evaluating diversity changes through time.

Although our data are from a relatively small subset of the Central Paratethys, they are considered as representative because a study on echinoderms from the whole Central Paratethys also showed that comparable habitats of the Karpatian and Badenian stages had very similar faunas and diversities (Kroh 2007). This author specifically stressed that the non-presence of Karpatian shallow-water carbonates in the rock record explains much of the lower echinoderm diversity compared to the Badenian. Our study adds a new aspect in demonstrating that also among siliciclastic sediments a facies shift from nearshore and shallow sublittoral habitats in the Karpatian to somewhat deeper environments in the Badenian is responsible for diversity differences.

Palaeogeography and palaeoclimate

It may be possible that for palaeogeographical reasons the non-preserved deeper shelf assemblages of the Karpatian were less diverse than their preserved Badenian counterparts. During the Karpatian a marine connection of the Central Paratethys existed only with the Mediterranean Basin, via the Slovenian 'Trans-Tethyan Trench corridor' (Bistricic & Jenko 1985). In the Badenian, open connections with the Eastern Paratethys may also have existed, although the timing of the connections is highly controversial (Rögl 1998; Studencka *et al.* 1998; Steininger & Wessely 2000; Popov *et al.* 2004). In both time slices, however, the Mediterranean Basin was at least temporarily connected to the Indo-Pacific, enabling water circulation between both oceans, although faunas differed considerably (Harzhauser *et al.* 2007). A palaeogeographical scenario for the observed diversity differences is therefore rather speculative and not supported by hard data. From a palaeoclimatological perspective the differences between the time slices are rather small. This is because the Karpatian and Lower to Middle Badenian were characterized by subtropical temperatures of the Middle Miocene climate optimum (Böhme 2003; Latal *et al.* 2006; Bruch *et al.* 2007; Kern *et al.* 2010), which enabled the presence of thermophilic molluscs at that time in the Paratethyan Basins (Harzhauser *et al.* 2003). In fact, nearshore assemblages, which are available from both time slices, do not support the scenario of higher Badenian diversities (Fig. 10a).

The sequence stratigraphic framework

In our study on 3rd order cycles from the Central Paratethys, most outcrops are from highstand systems tracts (Fig. 2, Table 1). These are internally characterized by relatively gradual biofacies replacements with major faunal turnovers occurring at sequence boundaries (Zuschin *et al.* 2007), a pattern that corresponds to sequence stratigraphic expectations (e.g. Brett 1995, 1998; Holland 2000). The dominance of HSTs corresponds well to the fact that the thickest parts of the sedimentary record were built at times of progradation and that the transgressive phases are only represented by thin levels (e.g. Jablonski 1980; Fürsich *et al.* 1991; Clifton 2006).

Among the studied sequences, however, diversities clearly depend on facies (Figs 9 & 10), which differ in a systematic way due to a biased sedimentary record. Karpatian shell beds are mostly preserved from nearshore and shallow sublittoral environments, which discordantly overlay macrofossil-poor Karpatian offshore clays, whereas from the Badenian mostly somewhat deeper shelf assemblages are recorded. This is most evident in the Lower Lagenidae zone, which completely lacks nearshore assemblages (Figs 8, 10 & 11).

But also later in the Badenian, nearshore assemblages are strikingly underrepresented when compared to the Karpatian (Fig. 8). Sequence stratigraphic models predict that nearshore sediments of the HST will be eroded during subsequent 3rd order sea-level drops. This would well explain the paucity of nearshore sediments in the three Badenian 3rd order cycles. This interpretation is supported by 3-D seismic reflection data, which reveal significant drops of relative sea-level (90–120 m) between the cycles (Strauss *et al.* 2006). The dominance of such environments and corresponding lack of somewhat deeper water shelf assemblages in the Karpatian is counterintuitive, however, and is probably related to the strong tectonic reorganization of the Central Paratethys at the Karpatian/Badenian boundary (Adámek *et al.* 2003). One explanation for the scarceness of shelf environments is the uplift of the North Alpine Foreland Basin and the subsequent retreat of the sea. Deeper marine environments became established only in the Carpathian Foredeep (Rögl 1998). In contrast, the new tectonic regime initiated rapid subsidence in small satellite basins of the Vienna Basin, where such littoral deposits escaped erosion (Wessely 1998; Kern *et al.* 2010).

Tectonics therefore affected sequence architecture in this particular setting by controlling subsidence and sedimentary input, highlighting the problem that sequence stratigraphic models were conceived for passive margin and only poorly predict sediment accumulation in tectonically active settings.

Comparison with other studies

Many studies have treated the distribution and preservation of shell beds in relation to flooding surfaces and sequence boundaries (e.g. Kidwell 1988, 1989, 1991; Banarjee & Kidwell 1991; Abbott & Carter 1997; Kondo *et al.* 1998; Fürsich & Pandey 2003). A series of others have examined palaeocommunity dynamics at local to regional scales in relation to the rock record (e.g. Patzkowsky & Holland 1999; Goldman *et al.* 1999; Olszewski & Patzkowsky 2003; Olszewski & Erwin 2004; Scarponi & Kowalewski 2004; Hendy & Kamp 2004; Dominici & Kowalke 2007; Zuschin *et al.* 2007; Tomašových & Siblík 2007). Only few studies, however, have dealt with diversity changes as related to depositional sequences. The results depend on scale, tectonic setting and environments preserved (or available to sample). Diversity seems largely to be decoupled from 1st order cycles because stage-level post-Palaeozoic marine standing diversity of western Europe increases although marine sediment outcrop area decreases (Smith 2001; see also Smith & McGowan 2007). A strong relation, however, has been proposed for 2nd order sequence stratigraphic cycles (Smith 2001). Two case studies suggest highest diversity or sampling probability at mid-cycle position at the top of transgressive systems tract intervals (Smith *et al.* 2001; Crampton *et al.* 2006), but the causes seem to differ somewhat between tectonic settings (see discussion in Crampton *et al.* 2006). At the active margin of New Zealand, for example, the best preservation of molluscan faunas is at mid-cycle position at the top of transgressive systems tracts, and poorest preservation towards the end of highstand systems tracts. This is related to continuous subsidence and creation of accommodation space (Crampton *et al.* 2006). At the passive margin of western Europe, due to minimum accommodation space, shallow-water deposition is displaced onto the cratonic interiors, where erosive loss during subsequent lowstands is most pervasive (Smith *et al.* 2001). In both areas, however, long-term diversity trends are related to distinct facies biases. In the Cenomanian/Turonian of western Europe a distinct diversity decrease can be related to an increase of offshore at the expense of onshore sedimentary facies in the course of platform drowning due to sea-level rise (Smith *et al.* 2001). In the Neogene of New Zealand an apparent decline in species diversity reflects erosion of shallow-water deposits and a relative increase of bathyal at the cost of shelf facies (Crampton *et al.* 2003). The importance of environments covered within systems tracts is finally also stressed in a study on late Quaternary 4th order sequences deposited on the Po Plain (Italy). There, transgressive systems tract samples displayed the highest, and the highstand systems tract samples the lowest diversity. At the same time, turnover across sequences is negligible and major diversity shifts across systems tracts are mostly driven by Waltherian-type environmental shifts (Scarponi & Kowalewski 2007).

Conclusions

The diversity increase between two regional stages of the Central Paratethys is largely due to an environmental shift, which is related to selective preservation and erosion of environments due to tectonics and sea-level drops. Although most samples analysed in this study stem from highstand systems tracts, diversity differences between stages and biozones are significant. Pure standing diversity estimates will reveal biogeographical relations and might capture faunal migrations aside from reflecting palaeoecological and palaeoclimatic benchmarks. They will not, however, reliably mirror biodiversity. This study therefore strongly supports the importance of environmental bias when considering faunal changes though time and suggests that in second and higher order sequences the facies covered within systems tracts will drive diversity patterns. The importance of rigorous, extensive sampling within a highly resolved stratigraphic and palaeoenvironmental framework for deciphering palaeodiversity patterns at regional scales is emphasized. The sheer sampling effort that is required to reach the asymptote of the collector's curve is highlighted and it is strongly recommended to use abundance data, which enable the recognition of ecological changes in regional biota.

We thank S. Ćorić, P. Pervesler, R. Roetzel and numerous students for help with sampling; S. Dominici, J. Hohenegger, A. Kroh, F. Rögl, J. Sawyer and A. J. Tomašových for discussions; A. Kroh and J. Reischer for providing the data for Figure 12; A. Smith and A. J. McGowan for letting us present this work at the Lyell symposium 2010; A. J. McGowan and J. Nebelsick for a stimulating reviews of the manuscript. This study was supported by project P19013-B17 of the Austrian Science Fund (FWF).

References

ABBOTT, S. T. & CARTER, R. M. 1997. Macrofossil associations from mid-Pleistocene cyclothems, Castlecliff Section, New Zealand: implications for sequence stratigraphy. *Palaios*, **12**, 188–210.

ADÁMEK, J., BRZOBOHATÝ, R., PÁLENSKÝ, P. & ŠIKULA, J. 2003. The Karpatian in the Carpathian Foredeep (Moravia). *In*: BRZOBOHATÝ, R., CICHA, I., KOVÁC, M. & RÖGL, F. (eds) *The Karpatian – a Lower Miocene Stage of the Central Paratethys*. Masaryk University, Brno, 75–92.

BANARJEE, I. & KIDWELL, S. M. 1991. Significance of molluscan shell beds in sequence stratigraphy: an example from the Lower Cretaceous Mannville Group of Canada. *Sedimentology*, **38**, 913–934.

BARRET, P. M., MCGOWAN, A. J. & PAGE, V. 2009. Dinosaur diversity and the rock record. *Proceedings of the Royal Society B*, **276**, 2667–2674.

BISTRICIC, A. & JENKO, K. 1985. Area No. 224 b1: Trans-tethyan Trench 'Corridor', YU. *In*: STEININGER, F. F., SENES, J., KLEEMANN, K. & RÖGL, F. (eds) *Neogene of the Mediterranean Tethys and Paratethys. Stratigraphic Correlation Tables and Sediment Distribution Maps*. University of Vienna, Vienna, **1**, 72–73.

BOUCHET, P., LOZOUET, P., MAESTRATI, P. & HEROS, V. 2002. Assessing the magnitude of species richness in tropical marine environments: exceptionally high numbers of molluscs at a New Caledonia site. *Biological Journal of the Linnean Society*, **75**, 421–436.

BÖHME, M. 2003. The Miocene Climatic Optimum: evidence from ectothermic vertebrates of Central Europe. *Palaeogeography, Palaeoclimatology, Palaeoecology*, **195**, 389–401.

BRETT, C. E. 1995. Sequence stratigraphy, biostratigraphy, and taphonomy in shallow marine environments. *Palaios*, **10**, 597–616.

BRETT, C. E. 1998. Sequence stratigraphy, paleoecology, and evolution: biotic clues and responses to sea-level fluctuations. *Palaios*, **13**, 241–262.

BRUCH, A. A., UHL, D. & MOSBRUGGER, V. 2007. Miocene climate in Europe – patterns and evolution: a first synthesis of NECLIME. *Palaeogeography, Palaeoclimatology, Palaeoecology*, **253**, 1–7.

BULOT, L. G. 1993. Stratigraphical implications of the relationships between ammonites and facies: examples taken from the Lower Cretaceous (Valanginian–Hauterivian) of the western Tethys. *In*: HOUSE, M. R. (ed.) *The Ammonoidea: Environment, Ecology, and Evolutionary Change*. Clarendon Press, Oxford, Systematics Association Special Volume, **47**, 243–266.

BUZAS, M. A. & HAYEK, L.-A. C. 2005. On richness and evenness within and between communities. *Paleobiology*, **31**, 199–220.

CLARKE, K. R. & WARWICK, R. M. 1994. *Changes in Marine Communities: An Approach to Statistical Analysis and Interpretation*. Plymouth Marine Laboratory, Plymouth, UK.

CLIFTON, H. E. 2006. A re-examination of facies models for clastic shorelines. *In*: POSAMENTIER, H. W. & WALKER, R. G. (eds) *Facies Models Revisited*. SEPM Special Publication, **84**, 293–337, Tulsa, OK.

COLWELL, R. K. 2009. *EstimateS: Statistical estimation of species richness and shared species from samples*. Version 8.02 Users guide and application. Published at http://viceroy.eeb.uconn.edu/EstimateS.

CRAMPTON, J. S., BEU, A. G., COOPER, R. A., JONES, C. A., MARSHALL, B. & MAXWELL, P. A. 2003. Estimating the rock volume bias in paleobiodiversity studies. *Science*, **301**, 358–360.

CRAMPTON, J. S., FOOTE, M. *ET AL.* 2006. Second-order sequence stratigraphic controls on the quality of the fossil record at an active margin: New Zealand Eocene to Recent shelf molluscs. *Palaios*, **21**, 86–105.

DOMINICI, S. & KOWALKE, T. 2007. Depositional dynamics and the record of ecosystem stability: early Eocene faunal gradients in the Pyrenean Foreland, Spain. *Palaios*, **22**, 268–283.

FÜRSICH, F. T. & PANDEY, D. K. 2003. Sequence stratigraphic significance of sedimentary cycles and shell concentrations in the Upper Jurassic–Lower Cretaceous of Kachchh, western India. *Palaeogeography, Palaeoclimatology, Palaeoecology*, **193**, 285–309.

FÜRSICH, F. T., OSCHMANN, W., JAITLY, A. K. & SINGH, I. B. 1991. Faunal response to transgressive–regressive cycles: examples from the Jurassic of Western India. *Palaeogeography, Palaeoclimatology, Palaeoecology*, **85**, 149–159.

GASTON, K. J. 1994. *Rarity*. Chapman and Hall, London.

GOLDMAN, D., MITCHELL, C. E. & JOY, M. P. 1999. The stratigraphic distribution of graptolites in the classic upper Middle Ordovician Utica Shale of New York State: an evolutionary succession or a response to relative sea-level change? *Paleobiology*, **25**, 273–294.

GRAY, J. S. 2000. The measurement of marine species diversity with an application to the benthic fauna of the Norwegian continental shelf. *Journal of Experimental Marine Biology and Ecology*, **250**, 23–49.

GRILL, R. 1943. Über mikropaläontologische Gliederungsmöglichkeiten im Miozän des Wiener Becken. *Mitteilungen der Reichsanstalt für Bodenforschung*, **6**, 33–44.

HAMMER, O., HARPER, D. A. T. & RYAN, P. D. 2001. PAST: Paleontological statistics software package for education and data analysis. *Palaeontologia Electronica*, **4**. http://palaeo-electronica.org/2001_1/past/issue1_01.htm.

HAQ, B. U., HARDENBOL, J. & VAIL, P. R. 1988. Mesozoic and Cenozoic chronostratigraphy and cycles of sea level changes. *In*: WILGUS, C. K. (ed.) *Sea-level Changes – an Integrated Approach*. Society of Economic Paleontologists and Mineralogists, Special Publications, **42**, 71–108, Tulsa, OK.

HARDENBOL, J., THIERRY, J., FARLEY, M. B. & JACQUIN, T. 1998. Mesozoic and Cenozoic sequence chronostratigraphic framework of European basins. *In*: DE GRACIANSKY, P.-C., HARDENBOL, J., JACQUIN, T. & VAIL, P. R. (eds) *Mesozoic and Cenozoic Sequence Stratigraphy of European Basins*. SEPM, Special Publications, **60**, 3–14, Tulsa, OK.

HARNIK, P. G. 2009. Unveiling rare diversity by integrating museum, literature, and field data. *Paleobiology*, **35**, 190–208.

HARZHAUSER, M. & PILLER, W. E. 2007. Benchmark data of a changing sea – Palaeogeography, Palaeobiogeography and events in the Central Paratethys during the Miocene. *Palaeogeography, Palaeoclimatology, Palaeoecology*, **253**, 8–31.

HARZHAUSER, M., PILLER, W. E. & STEININGER, F. F. 2002. Circum-Mediterranean Oligo-Miocene biogeographic evolution – the gastropods' point of view.

Palaeogeography, Palaeoclimatology, Palaeoecology, **183**, 103–133.

HARZHAUSER, M., MANDIC, O. & ZUSCHIN, M. 2003. Changes in Parathethyan marine molluscs at the Early/Middle Miocene transition: diversity, palaeogeography and palaeoclimate. *Acta Geologica Polonica*, **53**, 323–339.

HARZHAUSER, M., KROH, A., MANDIC, O., PILLER, W. E., GÖHLICH, U., REUTER, M. & BERNING, B. 2007. Biogeographic responses to geodynamics: a key study all around the Oligo-Miocene Tethyan Seaway. *Zoologischer Anzeiger – Journal of Comparative Zoology*, **246**, 241–256.

HENDY, A. J. W. & KAMP, P. J. J. 2004. Late Miocene to early Pliocene biofacies of Wanganui and Taranaki Basins, New Zealand: applications to palaeoenvironmental and sequence stratigraphic analysis. *New Zealand Journal of Geology and Geophysics*, **47**, 769–785.

HENDY, A. J. W., KAMP, P. J. J. & VONK, A. J. 2009. Late Miocene turnover of molluscan fauna, New Zealand: taxonomic and ecological reassessment of diversity changes at multiple spatial and temporal scales. *Palaeogeography, Palaeoclimatology, Palaeoecology*, **280**, 275–290.

HOHENEGGER, J., RÖGL, F. *ET AL.* 2009. The Styrian Basin: a key to the Middle Miocene (Badenian/Langhian) Central Paratethys transgressions. *Austrian Journal of Earth Sciences*, **102**, 102–132.

HOLLAND, S. M. 1995. The stratigraphic distribution of fossils. *Paleobiology*, **21**, 92–109.

HOLLAND, S. M. 1999. The New Stratigraphy and its promise for paleobiology. *Paleobiology*, **25**, 409–416.

HOLLAND, S. M. 2000. The quality of the fossil record: a sequence stratigraphic perspective. *Paleobiology Suppl.*, **26**, 148–168.

HUBBELL, S. P. 2001. *The Unified Theory of Biodiversity and Biogeography*. Princeton University Press, Princeton.

JABLONSKI, D. 1980. Apparent versus real effects of transgressions and regressions. *Paleobiology*, **6**, 397–407.

JABLONSKI, D. 1998. Geographic variation in the molluscan recovery from the End-Cretaceous extinction. *Science*, **279**, 1327–1330.

JACKSON, J. B. C. & JOHNSON, K. G. 2001. Measuring past biodiversity. *Science*, **293**, 2401–2404.

JACKSON, J. B. C., TODD, J. A., FORTUNATO, H. & JUNG, P. 1999. Diversity and assemblages of Neogene Caribbean Mollusca of Lower Central America. *Bulletins of American Paleontology*, **357**, 193–230.

JOHNSON, K. G. & CURRY, G. B. 2001. Regional biotic turnover dynamics in the Plio-Pleistocene molluscan fauna in the Wanganui Basin, New Zealand. *Palaeogeography, Palaeoclimatology, Palaeoecology*, **172**, 39–51.

KERN, A., HARZHAUSER, M., MANDIC, O., ROETZEL, R., ĆORIĆ, S., BRUCH, A. A. & ZUSCHIN, M. 2010. Millenial-scale vegetation dynamics in an estuary at the onset of the Miocene Climatic Optimum. *In*: MOSBRUGGER, V. & UTESCHER, T. (eds) *2nd NECLIME Synthesis Volume – The Neogene of Eurasia: Spatial Gradients and Temporal Trends. Palaeogeography, Palaeoclimatology, Palaeoecology*: 10.1016/j.palaeo.2010.07.014.

KIDWELL, S. M. 1988. Taphonomic comparison of passive and active continental margins: Neogene shell beds of the Atlantic coastal plain and northern Gulf of California. *Palaeogeography, Palaeoclimatology, Palaeoecology*, **63**, 201–223.

KIDWELL, S. M. 1989. Stratigraphic condensation of marine transgressive records: origin of major shell deposits in the Miocene of Maryland. *Journal of Geology*, **97**, 1–24.

KIDWELL, S. M. 1991. Condensed deposits in siliciclastic sequences: expected and observed features. *In*: EINSELE, G., RICKEN, W. & SEILACHER, A. (eds) *Cycles and Events in Stratigraphy*. Springer-Verlag, Berlin, 682–695.

KIDWELL, S. M. & HOLLAND, S. M. 2002. The quality of the fossil record. *Annual Review of Ecology and Systematics*, **33**, 561–588.

KOCH, C. F. 1978. Bias in the published fossil record. *Paleobiology*, **4**, 367–372.

KOCH, C. F. 1987. Prediction of sample size effects on the measured temporal and geographic distribution patterns of species. *Paleobiology*, **13**, 100–107.

KONDO, Y., ABBOTT, S. T., KITAMURA, A., KAMP, P. J. J., NAISH, T. R., KAMATAKI, T. & SAUL, G. S. 1998. The relationship between shellbed type and sequence architecture: examples from Japan and New Zealand. *Sedimentary Geology*, **122**, 109–127.

KOSNIK, M. A. 2005. Changes in Late Cretaceous–early Tertiary benthic marine assemblages: analyses from the North American coastal plain shallow shelf. *Paleobiology*, **31**, 459–479.

KOVÁC, M., ANDREYEVA-GRIGOROVICH, A. S. ET AL. 2004a. Karpatian Paleogeography, Tectonics and Eustatic Changes. *In*: BRZOBOHATÝ, R., CICHA, I., KOVÁC, M. & RÖGL, F. (eds) *The Karpatian – a Lower Miocene Stage of the Central Paratethys*. Masaryk University, Brno, 49–72.

KOVÁC, M., BARÁTH, I., HARZHAUSER, M., HLAVATÝ, I. & HUDÁCKOVÁ, N. 2004b. Miocene depositional systems and sequence stratigraphy of the Vienna Basin. *Courier Forschungsinstitut Senckenberg*, **246**, 187–212.

KOVÁC, M., ANDREYEVA-GRIGOROVICH, A. ET AL. 2007. Badenian evolution of the Central Paratethys Sea: paleogeography, climate and eustatic sea-level changes. *Geologica Carpathica*, **58**, 579–606.

KOWALEWSKI, M., NEBELSICK, J. H., OSCHMANN, W., PILLER, W. E. & HOFFMEISTER, A. P. 2002. Multivariate hierarchical analyses of Miocene mollusk assemblages of Europe: Paleogeographic, paleoecological, and biostratigraphic implications. *Bulletin of the Geological Society of America*, **114**, 239–256.

KROH, A. 2007. Climate changes in the Early to Middle Miocene of the Central Paratethys and the origin of its echinoderm fauna. *Palaeogeography, Palaeoclimatology, Palaeoecology*, **253**, 169–207.

KRUSKAL, J. B. 1964. Multidimensional scaling by optimizing goodness of fit to a nonmetric hypothesis. *Psychometrika*, **29**, 1–27.

LANDE, R. 1996. Statistics and partitioning of species diversity, and similarity among multiple communities. *Oikos*, **76**, 5–13.

LATAL, C., PILLER, W. E. & HARZHAUSER, M. 2006. Shifts in oxygen and carbon isotope signals in marine molluscs from the Central Paratethys (Europe) around the Lower/Middle Miocene transition. *Palaeogeography, Palaeoclimatology, Palaeoecology*, **231**, 347–360.

MAGURRAN, A. E. 2004. *Measuring Biological Diversity*. Blackwell, Oxford.

MANDIC, O., HARZHAUSER, M., SPEZZAFERRI, S. & ZUSCHIN, M. 2002. The paleoenvironment of an early Middle Miocene Paratethys sequence in NE Austria with special emphasis on paleoecology of mollusks and foraminifera. *Geobios Mémoire Special*, **24**, 193–206.

MARX, F. G. & UHEN, M. D. 2010. Climate, critters, and cetaceans: Cenozoic drivers of the evolution of modern whales. *Science*, **327**, 993–996.

McGOWAN, A. J. & SMITH, A. B. 2008. Are global Phanerozoic marine diversity curves truly global? A study of the relationship between regional rock records and global Phanerozoic marine diversity. *Paleobiology*, **34**, 80–103.

MILLER, A. I. 1997. Comparative diversification dynamics among palaeocontinents during the Ordovician Radiation. *Geobios Mémoire Spécial*, **20**, 397–406.

MILLER, A. I. & FOOTE, M. 1996. Calibrating the Ordovician radiation of marine life: implications for Phanerozoic diversity trends. *Paleobiology*, **22**, 304–309.

OLSZEWSKI, T. & ERWIN, D. H. 2004. Dynamic response of Permian brachiopod communities to long-term environmental change. *Nature*, **428**, 738–741.

OLSZEWSKI, T. D. & PATZKOWSKY, M. E. 2003. From cyclothems to sequences: the record of eustasy and climate on an icehouse epeiric platform (Pennsylvanian–Permian, North American Midcontinent). *Journal of Sedimentary Research*, **73**, 15–30.

PASCHER, G. A., HERRMANN, P., MAND, A. L., MATURA, A., PAHR, A. & SCHNABEL, W. 2000. *Geologische Karte des Burgenlandes 1:200.000*. Geologische Bundesanstalt, Wien.

PATZKOWSKY, M. E. & HOLLAND, S. M. 1999. Biofacies replacement in a sequence stratigraphic framework: middle and Upper Ordovician of the Nashville Dome, Tennessee, USA. *Palaios*, **14**, 301–323.

PETERS, S. E. 2005. Geological constraints on the macroevolutionary history of marine animals. *Proceedings of the National Academy of Sciences of the USA*, **102**, 12326–12331.

PETERS, S. E. 2006. Genus extinction, origination, and the durations of sedimentary hiatuses. *Paleobiology*, **32**, 387–407.

PETERS, S. E. & FOOTE, M. 2001. Biodiversity in the Phanerozoic: a reinterpretation. *Paleobiology*, **27**, 583–601.

PILLER, W. E., HARZHAUSER, M. & MANDIC, O. 2007. Miocene Central Paratethys stratigraphy – current status and future directions. *Stratigraphy*, **4**, 151–168.

POPOV, S. V., RÖGL, F., ROZANOV, A. Y., STEININGER, F. F., SHCHERBA, I. G. & KOVÁC, M. (eds) 2004. Lithological paleogeographic maps of Paratethys. 10 Maps Late Eocene to Pliocene. *Courier Forschungsinstitut Senckenberg*, **250**, 1–46.

PRÉVOST, C. 1820. Essai sur la constitution physique et géognostique du bassin à l'ouverture duquel est située la ville de Vienne en Autriche. *Journal de*

Physique, de chimie, et d'histoire naturelle, **91**, 347–364. and 460–473.

RAUP, D. M. 1976. Species diversity in the Phanerozoic; an interpretation. *Paleobiology*, **2**, 289–297.

RÖGL, F. 1998. Palaeogeographic considerations for Mediterranean and Paratethys seaways (Oligocene to Miocene). *Annalen des Naturhistorischen Museums in Wien*, **99A**, 279–310.

RÖGL, F. 1999. Mediterranean and Paratethys. Facts and hypotheses of an Oligocene to Miocene Paleogeography (short overview). *Geologica Carpathica*, **59**, 339–349.

RÖGL, F., ĆORIĆ, S. *ET AL*. 2006. The Styrian tectonic phase – A series of events at the Early/Middle Miocene boundary revised and stratified (Styrian Basin, Central Paratethys). *Joannea-Geologie und Paläontologie*, **9**, 89–91.

RUDWICK, M. 2005. *Bursting the Limits of Time: The Reconstruction of Geohistory in the Age of Revolution.* The Univiersity of Chicago Press, Chicago.

SAWYER, J. A. & ZUSCHIN, M. 2011. Drilling predation in molluscs from the Early and Middle Miocene of the Central Paratethys. *Palaios*, **26**, 284–297.

SCARPONI, D. & KOWALEWSKI, M. 2004. Stratigraphic paleoecology: bathymetric signature and sequence overprint of mollusk associations from upper Quaternary sequences of the Po Plain, Italy. *Geology*, **32**, 989–992.

SCARPONI, D. & KOWALEWSKI, M. 2007. Sequence stratigraphic anatomy of diversity patterns: late Quaternary benthic mollusks of the Po Plain, Italy. *Palaios*, **22**, 296–305.

SCHNABEL, W. (ed.). 2002. *Geologische Karte von Niederösterreich 1:200.000.* Mit einer Legende und kurzen Erläuterung von. Geologische Bundesanstalt, Wien.

SCHULTZ, O. 2001. Bivalvia neogenica (Nuculacea–Unionacea). *In*: PILLER, W. E. (ed.) *Catalogus Fossilium Austriae.* Verlag der Österreichischen Akademie der Wissenschaften, Wien, **1/1**, 1–379.

SCHULTZ, O. 2003. Bivalvia neogenica (Lucinoidea–Matroidea). *In*: PILLER, W. E. (ed.) *Catalogus Fossilium Austriae.* Verlag der Österreichischen Akademie der Wissenschaften, Wien, **1/2**, 381–690.

SCHULTZ, O. 2005. Bivalvia neogenica (Solenoidea–Clavagelloidea). *In*: PILLER, W. E. (ed.) *Catalogus Fossilium Austriae.* Verlag der Österreichischen Akademie der Wissenschaften, Wien, **1/3**, 691–1212.

SMITH, A. B. 2001. Large-scale heterogeneity of the fossil record: implications for Phanerozoic biodiversity studies. *Philosophical Transactions of the Royal Society, London, Series B*, **356**, 351–367.

SMITH, A. B. 2007. Marine diversity through the Phanerozoic: problems and prospects. *Journal of the Geological Society, London*, **164**, 1–15.

SMITH, A. B. & McGOWAN, A. J. 2007. The shape of the Phanerozoic marine palaeodiversity curve: how much can be explained from the sedimentary record of western Europe. *Palaeontology*, **50**, 765–774.

SMITH, A. B., GALE, A. S. & MONKS, N. E. A. 2001. Sea-level change and rock-record bias in the Cretaceous: a problem for extinction and biodiversity studies. *Paleobiology*, **27**, 241–253.

STEININGER, F., RÖGL, F. & MÜLLER, C. 1978. Geodynamik und paleogeographische Entwicklung des Badenien. *In*: PAPP, A., CICHA, I., SENES, J. & STEININGER, F. F. (eds) *Chronostratigraphie und Neostratotypen: Miozän der Zentralen Paratethys 6.* Slowakische Akademie der Wissenschaften, Bratislava, 110–116.

STEININGER, F. F. & WESSELY, G. 2000. From the Tethyan Ocean to the Paratethys Sea: Oligocene to Neogene stratigraphy, paleogeography and palaeobiogeography of the circum-Mediterranean region and the Oligocene to Neogene Basin evolution in Austria. *Mitteilungen der Österreichischen Geologischen Gesellschaft*, **92**, 95–116.

STILLE, H. 1924. *Grundfragen der vergleichenden Tektonik.* Gebrüder Bornträger, Berlin.

STRAUSS, P., HARZHAUSER, M., HINSCH, R. & WAGREICH, M. 2006. Sequence stratigraphy in a classic pull-apart basin (Neogene, Vienna Basin). A 3D seismic based integrated approach. *Geologica Carpathica*, **57**, 185–197.

STUDENCKA, B., GONTSHAROVA, I. A. & POPOV, S. V. 1998. The bivalve faunas as a basis for reconstruction of the Middle Miocene history of the Paratethys. *Acta Geologica Polonica*, **48**, 285–342.

ŠVAGROVSKÝ, J. 1981. Lithofazielle Entwicklung und Molluskenfauna des oberen Badeniens (Miozän M4d) in dem Gebiet Bratislava Devinska Nová Ves. *Západné Karpaty, Seria Paleontológia*, **7**, 1–203.

TODD, J. A., JACKSON, J. B. C., JOHNSON, K. G., FORTUNATO, H. M., HEITZ, A., ALVAREZ, M. & JUNG, P. 2002. The ecology of extinction: molluscan feeding and faunal turnover in the Caribbean Neogene. *Proceedings of the Royal Society, B*, **269**, 571–577.

TOMAŠOVÝCH, A. & SIBLÍK, M. 2007. Evaluating compositional turnover of brachiopod communities during the end-Triassic mass extinction (Northern Calcareous Alps): removal of dominant groups, recovery and community reassembly. *Palaeogeography, Palaeoclimatology, Palaeoecology*, **244**, 170–200.

VÁVRA, N. 2010. Constant Prévost und seine geognostischen Studien im Wiener Becken (1820). *In*: HUBMAN, B., SCHÜBL, E. & SEIDL, J. (eds) *Die Anfänge geologischer Forschung in Österreich.* Scripta geo-historica – Grazer Schriften zur Geschichte der Erdwissenschaften. Leykamverlag, Graz, **4**, 59–78.

WESSELY, G. 1998. Geologie des Korneuburger Beckens. *Beiträge zur Paläontologie*, **23**, 9–23.

WILLIS, K. J. & WHITTAKER, R. J. 2002. Species diversity-scale matters. *Science*, **295**, 1245–1248.

ZUSCHIN, M. & OLIVER, P. G. 2005. Diversity patterns of bivalves in a coral dominated shallow-water bay in the northern Red Sea – high species richness on a local scale. *Marine Biology Research*, **1**, 396–410.

ZUSCHIN, M., HARZHAUSER, M. & MANDIC, O. 2004a. Spatial variability within a single parautochthonous Paratethyan tidal flat deposit (Karpatian, Lower Miocene Kleinebersdorf, Lower Austria). *Courier Forschungsinstitut Senckenberg*, **246**, 153–168.

ZUSCHIN, M., HARZHAUSER, M. & MANDIC, O. 2004b. Taphonomy and Palaeoecology of the Lower Badenian (Middle Miocene) molluscan assemblages at Grund (Lower Austria). *Geologica Carpathica*, **55**, 117–128.

ZUSCHIN, M., HARZHAUSER, M. & MANDIC, O. 2005. Influence of size-sorting on diversity estimates from tempestitic shell beds in the middle Miocene of Austria. *Palaios*, **20**, 142–158.

ZUSCHIN, M., HARZHAUSER, M. & MANDIC, O. 2007. The stratigraphic and sedimentologic framework of fine-scale faunal replacements in the middle Miocene of the Vienna Basin (Austria). *Palaios*, **22**, 284–295.

ZUSCHIN, M., HARZHAUSER, M. & SAUERMOSER, K. 2006. Patchiness of local species richness and its implication for large-scale diversity patterns: an example from the middle Miocene of the Paratethys. *Lethaia*, **39**, 65–78.

The deep-sea microfossil record of macroevolutionary change in plankton and its study

DAVID B. LAZARUS

Museum für Naturkunde, Invalidenstrasse 43, 10115 Berlin
(e-mail: david.lazarus@mfn-berlin.de)

Abstract: The deep-sea planktonic microfossil record (foraminifera, coccolithophores, diatoms, radiolaria and dinoflagellates) provides a unique resource for palaeobiology. Despite some geographical gaps due to poor regional preservation, and intermittant time intervals lost to erosion, most time periods for each Cenozoic planktonic biogeographical province are preserved. Vast numbers of specimens and numerous deep-sea cores provide abundant material and the opportunity to tightly integrate macroevolutionary and palaeoenvironmental data. Current documentation of this record is mixed. Catalogues for foraminifera and coccolithophores offer nearly complete species-level clade histories, but taxonomy for siliceous microfossils is incomplete. Published occurrence data is primarily stratigraphic and covers only a fraction of the total preserved diversity. Age models for some sections are excellent (accuracy *c.* 100 kya) but for many other sections are still poor. Taxonomic errors, age model errors and reworking displace fossil occurrences in time, complicating palaeobiological analysis. With additional taxonomic work, careful collection of whole fauna/floral assemblage occurrence data, improved age models, and the development of better data filtering and analysis tools to deal with data outliers the deep-sea microfossil record can deliver its promise of providing the most complete, detailed record of macroevolutionary change available to science.

Supplementary material: Supplementary Appendix is available at http://www.geolsoc.org.uk/ SUP18485

Palaeontology has benefited in the last decades from studies of larger-scale patterns of fossil occurrence over geological time. Beginning largely with the development of a global Phanerozoic database of family ranges by Sepkoski (Sepkoski 1978; Sepkoski *et al.* 1981), some of the major patterns in life's evolution have been quantitatively documented and evaluated. More recently research has shifted to using large fossil occurrence databases (e.g. the Paleobiology Database, PaleoDB, Alroy *et al.* 2001). Such databases record the occurrence of many taxa in numerous geological samples, which have been placed into a palaeontologically meaningful framework of geological time and palaeoenvironmental setting. Occurrence databases, unlike earlier databases such as Sepkoski's that recorded only first and last occurrences of taxa, allow many advanced types of data analysis, including resampling to compensate for uneven data density. Despite these advances in data quality and analytical methods, many uncertainties remain, largely centred on the major imperfections of the fossil record, including loss of primary diversity during preservation, systematic biases in the completeness of the preserved record over time, and incomplete recovery of the preserved record both temporally and geographically (Smith 2007).

The deep-sea marine microfossil record has long been recognized as a potential source of highly complete, species-level data for palaeobiological research, and has been used for an array of studies in the 1970s to 1990s of microevolution (e.g. Hays 1970; Malmgren & Kennett 1981; Lazarus 1986; Lazarus *et al.* 1995a, b; review in Norris 2000), as well as some studies of macroevolution (e.g. Harper & Knoll 1975; Corfield & Shackleton 1988; Stanley *et al.* 1988; Pearson 1996). Based on this early work, palaeobiology has repeatedly been recommended as a research theme for development within the deep-sea drilling programs (planning documents for ODP and IODP, e.g. COSOD II (1988) and COMPLEX (Pisias & Delaney 1999)). Despite this, micropalaeontology has played a relatively minor role in palaeobiological research, and palaeobiology a minor role in deep-sea drilling research. Various reasons for this exist, including allocation of very limited micropalaeontologist effort to higher priority palaeoclimate research, and relatively limited knowledge of the biology of fossilizing forms. The lack of suitable synthesis tools for macroevolution research, such as a marine microfossil equivalent to PaleoDB, has also played a role. Recently however, macroevolutionary studies based on fossil occurrence datasets

From: McGowan, A. J. & Smith, A. B. (eds) *Comparing the Geological and Fossil Records: Implications for Biodiversity Studies.* Geological Society, London, Special Publications, **358**, 141–166.
DOI: 10.1144/SP358.10   0305-8719/11/$15.00 © The Geological Society of London 2011.

have been extended to the marine micropalaeontologic record. In contrast to earlier studies
based on first and last occurrence databases (e.g.
Stanley *et al.* 1988; Pearson 1996), these newer
studies have been based on detailed occurrence
data (Spencer-Cervato 1999; Finkel *et al.* 2005,
2007; Kucera & Schönfeld 2007; Liow & Stenseth
2007, 2010; Foote *et al.* 2008; Rabosky & Sorhannus 2009), primarily by exploiting the recent development of large databases of deep-sea microfossil
occurrences (e.g. Neptune: Lazarus 1994; Spencer-
Cervato 1999). Neptune is a database of deep-sea
marine microfossil occurrence data, based on the
reports of the Deep-Sea Drilling Project and the
Ocean Drilling Program. It contains *c.* 500 000
records of individual species' occurrences in
c. 300 deep-sea sections. Each occurrence record
is linked to the section sample from which it was
observed, and the samples are assigned numeric
ages in turn as each section in the database is
linked to a numeric age model. Taxa concepts in
Neptune have been reviewed by taxonomic experts
and many synonyms and other variant uses resolved.
Neptune is currently available as an online resource
at the Chronos portal (www.chronos.org) and is
linked to the data analysis system of the Paleobiology Database (www.pbdb.org). A new version of
Neptune is being developed at the Museum für
Naturkunde in Berlin.

Many of these new palaeobiological studies
are being carried out not by micropalaeontologists
but also palaeontologists from other subdisciplines
or by evolutionary biologists. Although interdisciplinary research is a very positive development, it increases the need for scientists to
understand data collected from material, and with
methods, outside their own area of direct expertise.
This paper has two goals. First, it is meant to summarize, from a marine micropalaeontologist's perspective, some of the salient characteristics of the
deep-sea planktonic microfossil record relevant to
palaeobiological studies, particularly those using
DSDP-ODP data such as in Neptune. It is essential
to recognize several unique characteristics of the
living biology and preservation of ocean plankton,
and understand the current state of material recovery, data collection and analyses. Second, practical
suggestions are offered on how to improve both
data collection and analyses, particularly to deal
with the distinct types of errors that characterize
current deep-sea microfossil data. Reviews of a
broad nature are naturally affected by the experi-
ences of the author. Although the observations and
comments of this paper are meant to be broadly
applicable, readers should note that the author's
own experience is primarily with Neogene radiolarians, and some comments in the text below may
reflect a resultant bias.

The deep-sea microfossil record

The pelagic realm

The deep-sea planktonic microfossil record derives
from the pelagic ocean biota, an extremely
diverse, complex, and poorly understood environment. Although multicellular nekton also plays a
role, the majority of organisms, measured either
by diversity or biomass, are planktonic, unicellular,
and include protists, bacteria and viruses. There is a
great diversity of higher level taxa within these
primary categories. Adl *et al.* (2005) for example
recognize 26 'first rank' categories of unicellular
eucaryotes below the *c.* kingdom level of super-
group, not including Metazoa or single genus
categories. The large majority of these groups are
probably present in the plankton, although comprehensive surveys are still largely lacking. For
example, of the 10 first-rank groups with photo-
synthetic plastids, 7 are currently known from
the marine phytoplankton (Simon *et al.* 2009).
Although comprising but a single high-rank taxon,
the Metazoa include extremely important pelagic
groups such as krill. Currently, there are *c.* several
thousand metazoan species known from 11 phyla
(Census of Marine Zooplankton, www.cmarz.org,
24.6.2010, excluding protists). Bacteria are also
extremely diverse (tens of thousands of strains
even in local water samples (Sogin *et al.* 2006)
and are much more directly coupled in food-web
ecology to protist plankton, particularly as prey,
than is true for metazoan ecosystems. This high-
level diversity is thus similar to the *c.* 30 marine
benthic metazoan phyla from which the traditional
invertebrate palaeontological record is formed.

The biogeographical distribution of species in
the marine plankton is quite different than on land
or in shallow benthos. Individual species are found
throughout one or more of a small number (*c.* one
dozen) of very large (thousands of km scale) biogeographical provinces, created by the geophysically
driven surface ocean circulation, which have
remained stable over long periods of geological
time (McGowan 1974, 1986; Lazarus 1983,
Fig. 1). As these circulation systems are mixed by
current motion on scales of decades or less,
fine-scale biogeographical patchiness is associated
with highly dynamic features of ocean circulation
such as gyres and filaments, with longevities of
only days to months (Knauss 2005). In addition to
these short-term structures, major, more or less per-
manent environmental gradients exist within the
primary biogeographical regions, including horizontal variation in primary water properties, depth
stratification, shelf-pelagic differences, and seasonal variation. These variations collectively allow
the ocean environment to be subdivided further

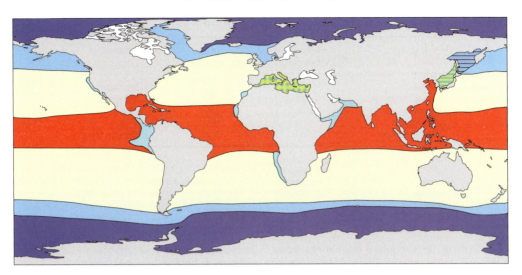

Fig. 1. Generalized global marine plankton provinces for microfossil producing groups. Based on synthesis of early plankton literature by Goll (1976), modified by author's own experience. Map is largely conformable with regions defined solely on water characteristics by Longhurst (1998), although distinctiveness of equatorial upwelling as top-level division not prominent in latter publication. Polar provinces often have highly divergent endemic biotas, although significant bipolarity is also known (Stepanjants *et al.* 2006). Low latitude biotas by contrast are highly similar across ocean basins. Marginal ocean basins (Mediterranean, Japan Sea, Sea of Okhotsk) have somewhat mixed or modified biotas derived from adjacent open-ocean regions.

into *c.* 100 regions (Longhurst 1998), although species are normally distributed across entire provinces, crossing these largely arbitrary boundaries subdividing a continuous, graded environment.

The total populations of species that inhabit these environments are very large. For most, small individual size (1–100 μm) and the closely related large numbers of individuals/volume water increases the population size even further, so that for many species effective population sizes (in an evolutionary sense) are many orders of magnitude larger than for typical shelf benthic or terrestrial metazoan species (Lazarus 1983) although in such large regions partial isolation by distance can occur, so that species populations are not fully panmictic (Casteleyn *et al.* 2010).

Preservation of pelagic marine plankton in the deep-sea fossil record

Although it is often stated that the deep-sea microfossil record is unusually taxonomically, stratigraphically and geographically complete (Prothero & Lazarus 1980; Lazarus 1983), this has to be qualified in certain important respects. First, from the 26 'first rank' level of higher protist taxa cited above, representatives of only five are known at all in the deep-sea microfossil record (Foraminifera, Radiolaria, Haptophytes, Stramenopiles and Alveolates; Table 1), while no deep-sea fossil record is

preserved of the bacteria. Fossilizable metazoan plankton are restricted to the holoplanktonic ostracodes (but only rarely, as most species lack calcified shells: Angel 2010) and pteropods, although these latter, being made of highly soluable aragonite, are only preserved in relatively shallow-water sediments (e.g. <1 km water depth, which represents only a few percent of the samples in deep-sea microfossil databases such as Neptune) (Lalli & Gilmer 1989). By contrast, of the *c.* 30 marine metazoan phyla, *c.* 10 have reasonably good (shelf benthos) fossil records, while a further *c.* 10 are preserved occasionally in 'Lagerstätten' (Valentine 1995; M. Aberhan, pers. comm. June 2010). Thus, the deep-sea microfossil record is in fact concentrated in the records of only a rather small number of high level groups, and is less representative of the total high level ecosytem diversity than the traditional invertebrate palaeontological record. Fortunately, both phytoplankton and zooplankton are represented by deep-sea marine microfossils, so that basic elements of ecosystem structure are available for study.

Preservation at the species level

Before progressing further, it is appropriate to briefly define what is meant by microfossil species. Although palaeontologists occasionally closely concern themselves with species and

Table 1. *Major groups of marine plankton which yield a significant fossil record*

Fossil group	Living diversity	% Species preservation	Mineralogy
Foraminifera	50	100	Carbonate
Coccolithophores	200+	<50	Carbonate
Diatoms	1500	≪50	Silica
Radiolarians	400	90	Silica
Dinoflagellates	1500	*c.* 15	Organic carbon

Diversity and preservation values are approximate. Information sources: Foraminifera: Arnold & Parker 1999; de Vargas *et al.* 2004; Coccolithophores: de Vargas *et al.* 2004; Young *et al.* 2005; Diatoms: Sournia *et al.* 1991; Radiolarians: Lazarus *et al.* in press at www.radiolaria.org; www. marinespecies.org (Protozoa > Sarcomastigophora > Radiolaria > Polycystinea); Dinoflagellates: Sournia *et al.* 1991; Head 1996.

speciation (e.g. the punctuated equilibrium debate of the 1970s–80s, Gould & Eldredge 1993) for the most part, species are secondary to genera and higher level taxa in palaeontological thinking and analyses. Thus even recent textbooks largely present a simplistic, classical, animal-centric view of species and speciation (e.g. Mayr 1963; Foote & Miller 2006; Benton & Harper 2009), as the complexities of species and speciation are rather peripheral to most practical research. For deep-sea microfossil records species concepts are by contrast essential for research as species are the core operational units. In this paper, the term refers to the classically defined morphospecies. Although cryptic or semi-cryptic genetically defined species are widely reported within classically defined morphospecies of living planktonic foraminifera, coccolithophores and diatoms, the coherence of microfossil morphospecies over time, including their consistent timing of origination and extinction (which forms the backbone of Cenozoic global stratigraphy), consistent relation to specific environments (which has provided the basis for much palaeoclimate research including CLIMAP), suggests that the traditional morphospecies is a valid biological entity, even if we do not yet understand well what it is. It has been noted that the species-structure characteristics of plants, protists and marine invertebrate organisms in general have many similarities, and thus all may tend to share, at least to some degree, a similar pattern of species structure (Lazarus 1983; Palumbi 1992). In plants this includes the common occurrence of species complexes, hybridization, and other mechanisms that produce a somewhat reticulate phylogenetic pattern (Grant 1971; Arnold 1997). Similar patterns are increasingly being found also in marine organisms, including protists and invertebrates (Arnold & Fogarty 2009), and lateral gene transfer between protists is also becoming better known (Boto 2010). Marine protist plankton

morphospecies may thus represent clusters of intermittently hybridizing species within species complexes (as indeed are known among groups of protists, e.g. diatoms (Arnold & Fogarty 2009; Casteleyn *et al.* 2009) and primarily asexual forms such as *Leishmania* (Waki *et al.* 2007)), or they may be clusters of parallel evolving species constrained to a similar niche and form by adaptive selection (super-species, *sensu* de Vargas *et al.* 2004).

Preservation of species level diversity in these fossilizing groups is variable, from well under 50% to nearly 100% (Table 1). In particular, planktonic foraminifera and polycystine radiolaria have nearly complete levels of preservation at the species level in modern deep-sea sediments. In addition, because the processes that control preservation are well known and based on general features of ocean circulation and chemistry, past degrees of preservation completeness can be estimated from geological knowledge.

Preservation of calcitic carbonate microfossils (foraminifera, coccolithophores, and minor groups such as *Bolboforma*) is largely controlled by the robustness of the shells themselves (observable in fossil specimens) and the depth in the ocean at which calcite carbonate begins (lysocline) and completes dissolution (carbonate compensation depth, or CCD). Carbonate microfossils in modern sediments are mostly well preserved, except in the deep basins of the subtropical Pacific, in polar regions, and beneath upwelling zones, where corrosive bottom waters (and, in the latter, high levels of sedimentary organic carbon) dissolve most shells (Fig. 2). The history of ocean carbonate chemistry, particularly the CCD, is reasonably well known, despite lack of detailed data in the earlier Cenozoic on very short time-scales (Kennett 1982; Broecker 2008; Lyle *et al.* 2009), and sufficient areas of ocean basin have been available for good preservation (at the scale of global species presence/ absence) of calcitic microfossils throughout most

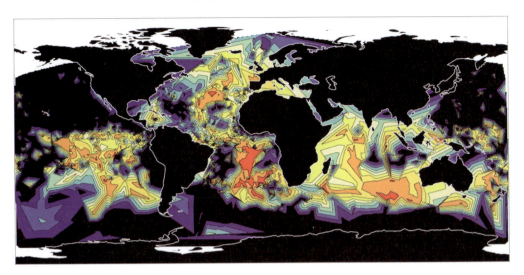

Fig. 2. Distribution of calcium carbonate (primarily composed of planktonic microfossils) in deep-sea sediments. Replotted from data given in Archer (1996), downloaded May 2011 from www.pangaea.de. Contours in 10% increments from >90% (red) to <10% (black).

of the last 100 Ma. Carbonate microfossils tend to undergo diagenesis with time and burial depth: the original microcrystalline shell material gradually recrystallizes into larger crystals, and eventually the shell form is lost. This process significantly affects many carbonate microfossil assemblages. It has been extensively studied however (due to its effect on geochemical signals often extracted from fossil shells) and is not normally extensive enough to obliterate morphospecies key taxonomic characters, at least in the unconsolidated to weakly lithified deep-sea sediments that predominate in the last c. 100 Ma.

Deep-sea microfossils made from opaline silica (primarily the high diversity groups diatoms and radiolarians, though low diversity groups such as silicoflagellates are also known) have a different, and more complex, pattern of preservation, due to the different behaviour of the marine silica system (Ragueneau et al. 2000). Silica is highly undersaturated in all ocean waters and thus dissolution of siliceous shells occurs to some degree almost everywhere on the sea floor. However, if the flux of siliceous material is sufficiently high, and the sediments are not too strongly diluted by non siliceous material, a small fraction of the shells will survive dissolution and become part of the microfossil record. The balance of shell production and dissolution today allows for nearly complete preservation, at a global scale, of the relatively robust radiolarian species record, but the loss of a large fraction of the diatom species record. Preservation varies strongly with geographical location, from excellent preservation underneath areas of high export productivity

to regions barren of siliceous microfossils under oligotrophic central gyes (Fig. 3). These patterns are controlled by the primary patterns of wind-driven circulation (upwelling) and deep-ocean circulation, which provides nutrients to upwelling intermediate water layers. Despite several shifts over time, regions of good preservation are sufficient to provide, again at a global scale, nearly complete species preservation over the last 100 Ma for the radiolarians, and proportionally less for the diatoms. There is also evidence that the current patterns of silica dissolution are relatively extreme, with significantly less dissolution of biogenic silica by low silica ocean water in earlier periods of time (Muttoni & Kent 2007; Lazarus et al. 2009). The extremely low concentration of silica in ocean waters today is thought to reflect the high abundance and productivity of marine diatoms. As this group's evolutionary radiation occurred through the Cenozoic, in earlier Cenozoic and late Cretaceous times their ability to remove silica, and thus create (for siliceous shells) highly corrosive ocean waters, was also less. Earlier Cenozoic radiolarian and diatom shells are more thickly silicified, probably reflecting higher levels of ocean water silica availability (Lazarus et al. 2009). These thicker shells are also more dissolution resistant, so the completeness of the siliceous fossil record should, in principle, be even more complete in earlier time intervals than today. Siliceous microfossils however are much more strongly affected by diagenesis than carbonate microfossils. At sediment burial depths of more than a few hundred metres, the primary opaline silica changes, via

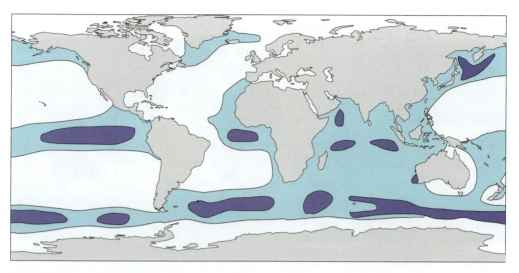

Fig. 3. Distribution of biogenic silica (diatoms and radiolaria) in deep-sea sediments. Based primarily on data and models in Leinen *et al.* (1986) and Heinze *et al.* (2003). Opal concentration in surface sediments, carbonate free basis. Dark blue: >50%; light blue: >10%; white: 0–10%.

intermediate mineral phases, into microcrystalline quartz. Much of the diatom silica in particular is mobilized and redeposited as chert (Calvert 1974; Muttoni & Kent 2007; Moore 2008). At typical (compacted) siliceous sediment accumulation rates of *c.* 1 cm/ka, this implies a major loss of siliceous (primarily diatom) microfossil record within and below the early Palaeogene, and this is indeed observed (Fenner 1985). Radiolarian shells, being much more robust, are not as much affected by this process, but even in late Cretaceous sediments, they typically do lose a significant proportion of the original diversity, particularly in more diagenetically altered land sections (Empson-Morin 1984). Much higher quality preservation of radiolarians in older time intervals is restricted to unusual sediment phases, such as concretions (De Wever *et al.* 2001), which are normally found only from land sections.

Organic walled microfossils, in deep-sea sediments primarily dinoflagellates, are preserved only when the organic substances comprising the skeleton are able to resist the degradation of organic matter that occurs in most marine sediments. Most dinoflagellate cell walls decompose rather quickly and are rarely preserved in sediments (Head 1996). Instead, the fossil dinoflagellate record is composed of the more resistant, sporopollenin-like cell walls of cysts – a special phase of the life cycle that only a fraction of living species create. Estimates vary but probably only 10–20% of living species form cysts during their life cycle, and these are concentrated in only a few families (Head 1996). Thus the fossil record of dinoflagellates is necessarily a rather biased subset of the

original diversity, although trends through time can still be documented and interpreted (Fensome *et al.* 1996, 1997).

Preservation, being for most deep-sea microfossils essentially a dissolution regulated process, can also be more directly determined from the preservation state of morphological characters of the shells themselves. The presence–absence of delicate structures, relative frequency of complete v. broken shells and other characteristics that are general for the fossil group can be used to approximately judge preservation state, and, when compared to patterns of modern preservation, provide an independent, if approximate, estimate of how complete species-level preservation is in fossil assemblages. Preservation estimates of this sort are routinely used to estimate potential bias in biostratigraphic and palaeoceanographic studies.

From the above it cannot be absolutely excluded that significant numbers of species were not preserved even for planktonic foraminifera and radiolarians, even in 'well preserved' past sediments. Very weakly silicified radiolarians or extremely thin-shelled planktonic foramifera taxa could have evolved and subsequently become extinct without leaving any trace of a record. However, such caveats are equally true for other types of fossils. Based on available evidence, the preservation of these two groups seems highly complete, and still quite good at the species level for a few phytoplankton groups as well.

Although relatively complete at the level of species and biogeographically on global scales, the preserved record is altered in many ways from the original plankton. The first is the loss of most geographical pattern on scales less than a few hundred kilometres, due to lateral transport of material during the transition from plankton to sediment, which occurs over a period of a few weeks to months (Honjo *et al.* 1982). Lateral transport in deep-sea microfossils is thus greater than is typical for most other fossil groups (Kowalewski & Bambach 2008) although substantial transport of plants and vertebrate fossils in riverine settings can sometimes occur. Equally important is the blurring of time resolution by bioturbation. Individuals from any one preparation are normally a mixture of individuals from different times – on scales of thousands of years (Berger & Heath 1968). Lost also are any direct representations of depth or seasonal partitioning of the environment in the water column. There are exceptions to this rule, in laminated sediments from enclosed basins with high productivity (Pike & Kemp 1999) or where bottom water anoxia eliminates benthic life (Kennett & Ingram 1995), but such sediments are relatively rare. Bioturbation created limitations of time averaging are in principle not any different than those for other fossil groups, where time averaging on scales typically in thousands of years occur (Kidwell 1998; Kowalewski & Bambach 2008). What is distinctive is that deep-sea microfossil assemblages in typical preparations are complete on time-scales of years or longer (a typical microfossil preparation of a few thousand individuals will typically have one or more specimens per year within it (Lazarus 1989)). Most other fossil records have gaps in deposition (diastems) that range from hundreds to thousands of years between short (annual or less) intervals of deposition, although these deposits may contain significant amounts of material from diastem intervals as well (Kowalewski & Bambach 2008).

Once preserved as fossils, deep-sea sediments do not undergo the same degree of erosion that is characteristic of the shallow benthic or land record. Local slumping and regional-scale erosion by bottom current activity is well known, but relatively limited in impact (Moore & Heath 1977; Moore *et al.* 1978). Most sediment records remain physically intact until they are permanently lost to subduction of the oceanic plate. The usable record is limited to approximately the last 100 Ma, by which *c.* 80% of the plate area has been lost to subduction (calculated from curves given in Parsons 1982). Despite this great advantage, there are some problems specific to deep-sea sedimentation. Detecting the existence of hiatuses for example is more difficult than in shallow water deposits, or those exposed on land. The uniformity of lithology in pelagic oozes, the lack of visible sedimentary structures (particularly in the few cm of 'exposure'

in cored material) means that hiatuses must primarily be detected by biostratigraphy or other geochronological means. Although distinguishing hiatuses in land sections is also not easy, facies shifts, distinct erosional or even angular unconformities at least help to identify suspect intervals. Erosion also has a significant side-effect that degrades the quality of the deep-sea microfossil record – reworking of older material into younger sediments. This is discussed in a separate section below.

From the above, it can be stated that, despite some significant gaps (e.g. siliceous microfossils from low productivity central gyre regions, or globally in the early Paleocene; carbonate microfossils in deep Pacific basins or in late Neogene polar and upwelling regions; diatoms in older Palaeogene or Cretaceous sediments), most of the biogeographical provinces during the last *c.* 100 Ma have preserved records of these fossil groups in deep-sea sediments (Prothero & Lazarus 1980; Lazarus 1983; COSOD II 1988; Pisias & Delaney 1999). Thus, high levels of completeness (locally and globally, at least for the Cenozoic) make these groups particularly attractive for palaeobiological (macroevolution) research, as in principle nearly complete histories of species-level evolution for entire clades are preserved in the fossil record. Such records dramatically reduce the need to devise complex statistical strategies to compensate for highly incomplete primary data, a need that currently consumes much effort in other areas of palaeontological research, and whose success is still quite uncertain. Our current ability to use this uniquely complete deep-sea microfossil record in palaeobiological research is however influenced by the current state of sediment recovery and documentation, as well as methods used in determining and removing the influence of hiatuses and reworking. These are discussed below.

Recovery of deep-sea fossil material

Deep-sea sediments have been sampled for scientific study in several ways. The most common type of sampling is of the uppermost layers only, primarily using piston corers. Most oceanographic institutions hold hundreds, if not thousands of such cores in their core repositories, and tens of thousands are available globally (Fig. 4). The relatively short length of most cores (*c.* 10 m) means however that at typical deep-sea sedimentation rates of a couple cm/ka only the last few hundred thousand years are recovered, and their primary use to date has been in studies of late Pleistocene–Recent climate change. A few cores have been taken from areas where older sediments outcrop, so that occasionally early Cenozoic or even Mesozoic

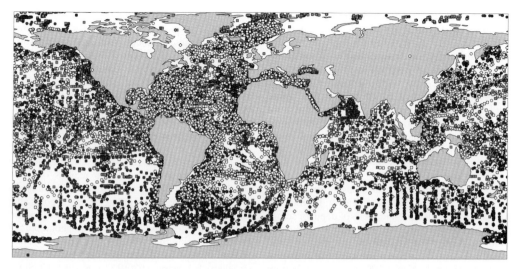

Fig. 4. Location map of deep-sea sediment (mostly piston core) sampling from selected marine institutions (www.ngdc. noaa.gov). Colours/symbols refer to core length: solid blue circles, <1 m; green squares, 1–10 m; yellow diamonds, 10–50 m; orange triangles, >50 m; open circles, no length information.

material is recovered. Most evolutionary processes take place over time-scales of tens of thousands to many millions of years, and most piston cores are thus of somewhat limited use for evolutionary research. By contrast, they provide an extraordinary global record of biological dynamics (ecological scale behaviour) over the last few glacial–interglacial cycles. Piston core information is currently highly scattered and incomplete, but as publication data increasingly becomes routinely digitally archived, it is gradually becoming possible to access and build global datasets of basic core information. However the majority of cores do not have any detailed information on age within section, or on fossil content.

Most of the deep -sea fossil record currently used for palaeobiological research comes from deep-sea drilling. The ocean margins have been extensively drilled by the oil industry but coring is discontinuous, sediments are hemipelagic and fossil preservation variable, and very little material (or data) is available for academic research. Deep-sea drilling by the Deep Sea Drilling Program and its successor programs Ocean Drilling Program and Integrated Ocean Drilling Program provide the vast majority of material older than c. 500 ka. Legs and Sites drilled have been consecutively numbered and, as of this writing, have reached 327 and 1360, respectively (Fig. 5). Early drilling (e.g. the first few hundred Sites) yielded incompletely recovered sediment columns and considerable physical disturbance of cored material, but better coring technology and changed policies have provided mostly complete, relatively undisturbed cored sections

since. Science goals have also changed over time. While the majority of Legs up to c. Leg 150 were focused on recovery of Cenozoic sedimentary sections, and only a minority on other science goals such as basement basalt, geochemistry or geophysics, increasingly scientific priorities have shifted to these latter, while sedimentary targets, if present, are often restricted to relatively short key time intervals (late Neogene, Paleocene–Eocene boundary, etc.). Thus the rate at which the global deep-sea record of microfossil evolution over the last 100 Ma is recovered has decreased substantially, although normally each year yields a few new high quality sections to the accumulating archive. It should be noted as well that the combined effects of extremely small size and extremely high fossil density (often effectively 100% in pelagic oozes) means extremely high numbers of specimens. At a very rough estimate there are c. 10^{15} individual microfossils already recovered and available for study in the current deep-sea drilling core archives. This is at least a million times more fossils than are available from the combined global repositories of the world's natural history museums (cumulatively almost certainly $<10^9$ specimens).

Deep sea drilling materials are normally studied by a large interdisciplinary team of scientists assigned to each Leg, so that (despite increasing gaps in coverage in recent years due to shortages of trained taxonomists) for each fossil group recovered by each Leg, there is usually a specialist report giving occurrence and taxonomic information. The primary goals of these reports are

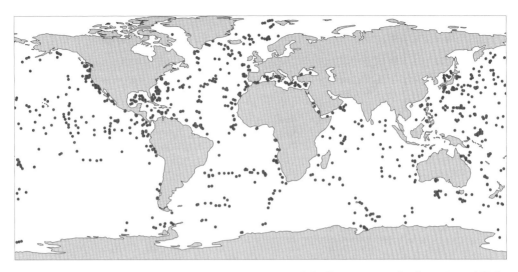

Fig. 5. Location of DSDP and ODP drill sites. Not all sites recovered significant amounts of sediment (some drilled basalt), and completeness of recovery varies.

biostratigraphic and palaeoceanographic, although significant amounts of new primary descriptive taxonomy are also included. Deep-sea drilling materials, having been studied by numerous other specialists, normally have a wealth of detailed palaeoenvironmental data available from the same sections where the fossil occurrences have been recorded. This is a major advantage for palaeobiological research as precise temporal integration of evolutionary and environmental history is relatively easy and information-rich.

Below approximately 100 Ma, most of the deep sea record has been lost to subduction, and thus the pelagic record must increasingly be examined from pelagic sedimentary sections now exposed on land. These exist in large numbers, particularly for the Mesozoic, but they are both stratigraphically and geographically fragmentary in coverage, and have frequently been disturbed to some degree by the tectonic processes that have transferred and exposed them on land (e.g. for radiolarians: Sanfilippo & Riedel 1985; Sanfilippo *et al.* 1985; De Wever *et al.* 2001). There are no global archives of information on their extent, and their scientific study is equally fragmented. No databases exist as yet that summarize more than a small fraction of the microfossil occurrence data that has been reported from land sections. Most of what has been entered is for radiolaria and exists within the PaleoDB. Land sections, if adequately synthesized, do however have the potential to extend the usable deep-sea microfossil record to much older time intervals, and can also occasionally provide extraordinarily well preserved assemblages of microfossils

(Bown *et al.* 2007), including taxa rarely, if ever, preserved in deep-sea sections.

How well scientific sampling of the deep-sea sediment record has recovered the history of marine plankton can be most easily seen by examples of actual recovery for different fossil groups on a global scale from different time intervals – either from summaries of sample distributional data from the Neptune database, or from independent sampling efforts such as the Micropaleontologic Reference Center (MRC) collections. The latter is a global network of identical collections of prepared deep-sea microfossil reference material obtained from selected DSDP and ODP drill sites (Lazarus 2006). The MRC collections contain several thousand samples for each of the four most important microfossil groups: planktonic foraminifera, calcareous nannofossils (coccolithophores), radiolarians and diatoms. While for some sections (those also incorporated in Neptune) numerical age models are available, most samples are dated only to approximately stage level interval ages. Much more material than this exists of course in the primary core repositories, but the MRC sampling provides a good example of the type of coverage that can in practice be achieved for palaeobiological research.

Recovery of deep-sea sediment v. time is shown in Figure 6, based on the MRC database (Lazarus 2006). The recovery of material by fossil group mostly closely matches the overall pattern of recovery of sediments by age, as determined by preliminary analyses of age data in the ODP Janus databases. The curves show a strong decline with

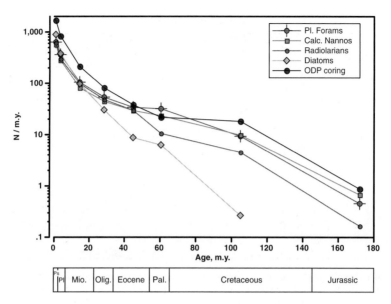

Fig. 6. Recovery of deep-sea microfossils by age interval, based on samples in the MRC database. From Lazarus (2006).

increasing age, approximating a power-law relationship. This is probably less due to the erosional loss of material with increasing age as the 'top down' recovery by drilling – virtually all drilled sections recover some late Neogene, but only a few penetrate as far as the early Palaeogene. Recovery patterns v. age are also biased by scientific drilling priorities, which, particularly in recent years, have favored numerous sections from the late Neogene for palaeoclimate work, and relatively few for older Cenozoic or Mesozoic time intervals. The decline in the relative recovery of siliceous microfossils, particularly diatoms in older time intervals reflects a true decrease in the geographical spread of siliceous sediments in much of the Palaeogene (despite an Eocene interval of widespread silica), loss of fossils, primarily diatoms, to diagenesis in deeply buried sections, and again, policy decisions on drilling targets, which favor carbonate microfossil rich sections that can be used in geochemical studies of palaeoceanography. Despite the general decline in availability with time, for most microfossil groups, over most of the Cenozoic, the MRC collections alone provide a sample density of *c.* 100 samples/ Ma which in principle, is more than adequate to document species level evolution on a global scale.

Maps showing the actual availability of MRC materials for any given microfossil group for different time intervals illustrate the high extent of coverage relative to the biogeographical provinces in the plankton of Figure 1 (Fig. 7; full set at http:// iodp.tamu.edu/curation/mrc/maps.html). Late

Cretaceous recovery of the marine microfossil plankton record is less complete, particularly for siliceous microfossils, but still generally very good, particularly when supplemented by additional, albeit somewhat fragmentary, pelagic sections which have been exposed by tectonic activity on land.

In summary, the deep-sea microfossil record of the last *c.* 100 Ma, at least for some microfossil groups, is nearly complete at the species level, and enough material has been recovered by deep-sea drilling to provide a nearly complete coverage for study of evolutionary patterns, at least for the Cenozoic. However, this does not mean that existing data is always a reasonably complete and accurate representation of this record. In reality, our documentation of the record is rather mixed. For some groups, at least at the level of first and last occurrences, our data are already quite good, and offer much better histories of evolution at the species level than virtually any other type of fossil data. For other groups, and in particular for more detailed occurrence data, there are a variety of problems, albeit temporary and solvable, that currently affect their use in palaeobiological research.

IRAT: imperfections in the existing dataset

Despite, and ironically in part because of this unusually complete, detailed fossil record, palaeobiological research using this record also faces

Late Oligocene Forams

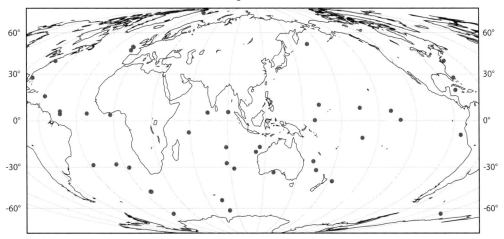

Fig. 7. Geographical coverage of plankton provinces in typical sample sets. Example here is a MRC samples map for the mid-Cenozoic Late Oligocene. Each point represents one or typically several stratigraphically distributed samples from a section within the given time interval. Full sets of maps for both calcareous and siliceous microfossil groups are available at the MRC website (http://iodp.tamu.edu/curation/mrc.html).

some unique problems in data content. Some of these are inherent in the fossil record itself, some are due to the way in which the deep-sea fossil record has been studied. These problems can be summarized as an acronym: IRAT, which stands for incomplete data, reworking, age model errors, and taxonomic error.

Incomplete data

Our ability to extrapolate from observations of species occurrences in samples to the original biodiversity for these samples depends on several factors, among them the relationship between recorded species diversity and true total diversity of the original assemblage. It has often been assumed, particularly in recent years, that the diversity of palaeontological species lists from samples are proportional, not only to palaeontological effort, but also to the diversity of the sample itself. The model (for discussion purposes called 'A') essentially is that a palaeontologist collects a certain number of specimens, limited by resources available, and identifies all the species found in his sample. In this model, the palaeontologist's processed material is a random subsample of the total diversity, and the higher the true diversity is in a sample, the higher also will be the sampled diversity. Much micropalaeontological data however are collected using quite different models, and in these, the correlation between reported diversity and actual diversity is weak, or even absent.

In addition to model 'A' – the random sampling of diversity model just described, there are two other common styles of data collection ('B' and 'C'). Both of these are based on the idea of a data matrix of species by samples, and the desire to uniformly record the abundance (or at least presence absence) of each taxon in each sample. Uniformity is expected as in both biostratigraphic and palaeoenvironmental analyses, the absence of specimens is used, not only the presence. Dynamically altering the list during the data collection phase may require re-examination of samples already processed to check for the presence/abundance of species newly added to the list, and the additional work this imposes means it is largely not done. In model B, the palaeontologist wishes to collect data for biostratigraphic or palaeoenvironmental analysis. A list of taxa – biostratigraphic markers, or palaeoenvironmental 'counting groups' is chosen based on prior knowledge, and the occurrence data for these taxa – and only these taxa – are recorded for the sample set. There is always a trade off between improved biostratigraphic or palaeoenvironmental information (with more taxa being recorded) on the one hand, and more temporally precise documentation (more densely spaced samples) on the other. Generally the number of taxa is not only fixed but kept to the minimum judged necessary to yield an acceptable answer, so that time remains to examine enough samples to provide the best temporal resolution possible. Thus data of this type tends to report only a rather small subset of the total diversity in a sample, and the numbers of taxa reported is fixed – it is not related to the diversity in a sample at all. The only exception is for samples where true diversity is below the size of the list, due for example to poor preservation. Lastly, there is data collected via model 'C'. This is something of a hybrid between models A and B. As in model B, data in any given study is collected according to a pre-set matrix of taxa names, so that, within the data set, there is no correlation between changing true diversity and the numbers of species recorded. However, unlike A, in model C the list of taxa recorded is less rigidly fixed. While pre-defined taxa, for example stratigraphic indicators, may still make up much of the list, the author also chooses to report at least some of the other taxa found in the assemblage, based on a variety of largely personal critera, such as their interest to him/her for a taxonomic revision. As each author collecting data this way tends to choose a different subset of taxa to report, there is a correlation between total reported diversity and number of studies published. Indeed, because the 'free' part of the list can be seen as random sampling with replacement, as the total number of studies increases, duplication of taxa becomes more frequent and the correlation between total diversity and total study effort shows a flattening curve similar to that for diversity within samples when sampled via model A. However, the average reported diversity per sample/study simply reflects the average practical size of a taxonomic list, and does not have the necessary relationship to actual real sample diversity.

In practice, all three types of data are commonly collected by deep-sea marine micropalaeontologists, although the authors of the studies frequently do not explicitly indicate which model they used. However, the bulk of the data probably is collected under model C, so that the correlation of diversity as reported to true diversity is problematic. Data collected under model C will generally show a good correlation between sample availability and total diversity, but this is due to the strong correlation, at least in deep-sea drilling material, between taxonomic effort, in numbers of studies done (and thus, via the cumulative effect of the 'open' part of each investigator's taxonomic list) total reported diversity) and sample availability. This correlation of studies to samples derives from long-standing deep-sea drilling program policy to insure that recovered core materials are more or less uniformly examined by assigned micropalaeontological specialists, who are expected to produce a set of data matrices ('range charts') as the core component of their report. Given the c. power law relationship between core recovery and geological age (Fig. 4), and the approximately constant amounts of core

recovery per Leg, this also means that younger time intervals will have many more published lists and thus higher reported total diversity.

Figure 8 illustrates the extent of underreporting of species diversity in typical deep-sea microfossil datasets. For both carbonate and siliceous fossils, whether regionally or globally, the large majority of species are reported in deep-sea drilling literature only from a rather small number of samples and sedimentary sections, although some species are reported from a much larger number of samples and sections (Fig. 8a). This reflects in part the rarity of many species, but primarily the effects of using pre-defined, largely stratigraphically oriented lists in recording species occurrences: stratigraphic marker species are much more common in the available data than are other forms. Similarly, the diversity of species in samples (Fig. 8b) are typically much smaller than the diversity expected if observations of each species were fully recorded: most non-stratigraphic marker species are simply not recorded in most samples. Although this latter problem is certainly more severe for the highly diverse, taxonomically less well studied siliceous microfossil groups, it is a significant limitation even for calcareous groups such as planktonic foraminifera, where many workers do try to record diversity more completely. This can be indirectly inferred from analyses such as Figure 8b and the average diversity for the Cenozoic of 51. This diversity value is derived from Stewart & Pearson (2000). This, together with the recently released update for macroperforate forms (Aze et al. 2011) is the best currently published database for Cenozoic planktonic foraminifera ranges. Although the latter number is a global value and thus will be higher than any local one, in modern foraminifera, diversity in low-mid latitude regions is near the maximum diversity of the group (Rutherford et al. 1999), and the majority (nearly 2/3) of Neptune foraminifera samples come from these regions. Thus most Neptune samples should have diversities near the global value, for example near 50 and not, as shown in Figure 8, a median value of only 10 taxa.

It should be noted here that the degree to which other fossil data, such as the shallow marine invertebrate data that comprises the bulk of databases such as PaleoDB, is actually recorded according to the requirements of diversity subsampling (e.g. model 'A' above) is open to debate. While some workers undoubtedly have recorded all taxa encountered, others presumably have only searched for stratigraphic markers (e.g. ammonites) or only taxa of personal interest. If enough workers have studied a time interval, then the multiplicity of personal biases should average out to an effectively random sampling of the time interval. If not, hard to detect

(a)

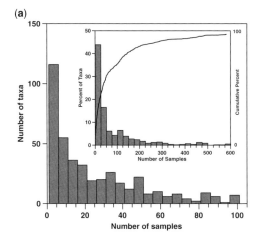

(b)

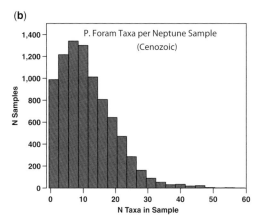

Fig. 8. Incompleteness of existing occurrence data for microfossil species, calculated from the DSDP-ODP published occurrence data held in Neptune. (**a**) Number of samples in which a radiolarian taxon occurs. Inset plot shows the full dataset and a cumulative curve, both by percent, while the main plot shows absolute counts, but only for a subset of the data in the interval holding the bulk of the values, for example <100 samples; (**b**) planktonic foraminiferal species diversity in samples. In both cases, if species diversity were completely recorded the mean values would be several times higher than actually observed. For radiolarian occurrences (a), given mean longevities of species of nearly 10 Ma, typical species geographical ranges (of about 1/3 the total global sampling distribution, for example tropical, polar or temperate), and sampling densities in Neptune (120 samples/Ma, Cenozoic average, from Neptune database, 8.2010) the average value in 'a' should be: 120 samples/Ma × 10 Ma/3, for example *c.* 400 hundred samples, v. actual mean of near 100 samples. For (b) within-sample diversity of planktonic foraminifera, the average Cenozoic diversity, calculated with the more conservative data of Stewart & Pearson (2000 – see also Fig. 10) is 51, while the actual mean value for Neptune data is 12.

biases may remain. At least with the existing marine microfossil data, the biases – for example biostratigraphic/palaeoenvironmental primary goals, are usually clearly stated, even if the precise degree to which this has affected data recording is often not precisely defined.

Reworking

In most other types of fossil record, fossilization is so rare that the chance of an older fossil being reworked (well beyond the range of primary temporal averaging of normal assemblages) and then preserved a second time ('remanié') is small, even if precise values are unknown (Kowalewski & Bambach 2008). In deep-sea sediments by contrast fossil abundances are enormous, and even if a small percentage of eroded/resuspended material is re-preserved, there will be a persistent trace amount of older material scattered throughout younger sediments. Reworked material of this sort (which differs substantially in its effect on the fossil record from the much smaller scale reworking associated with bioturbation within the mixed layer) is often found in sediments deposited just above hiatuses, compounding the problem of hiatus determination and demarcation. The extent of reworking is at the moment hard to objectively estimate, as it has rarely been studied in any detail (Johnson 1972), and no general quantitative estimates are available on its extent. Also, some marine micropalaeontologists do not even enter occurrences into their data tables if they feel the specimens are reworked, only noting the presence of reworking in a general way in the text of their reports. However documented, reworking, albeit normally minor, is mentioned in many, perhaps most of the papers on microfossil occurrences published in the reports of the DSDP and ODP. Reworking's effect on existing data is most problematic for taxa (typically the majority of the assemblage diversity) which are *not* stratigraphic markers, and whose stratigraphic range is thus not all that well known. Reworked occurrences of such species cannot be *a-priori* defined and filtered out of the data, so that the ranges of less well known species are more likely to have range extension artifacts due to reworking. Even for well known species, *a-priori* removal can be problematic, as it assumes that existing published ages for species first and last occurrences are correct. In fact, it is increasingly clear that many, if not most species first and last occurrences are at least to some degree diachronous (Johnson & Nigrini 1985; Moore *et al.* 1993; Spencer-Cervato *et al.* 1993) and published calibrations from one geographical region may not fully capture the stratigraphic range of species in other regions of the ocean. Rejecting observational data based on published expected ranges may thus lead to a circularity and errors of incomplete, incorrectly truncated ranges in observational data.

Age model problems

Deep-sea sediment geochronology is highly advanced and in general much more precise than for most other types of sedimentary rock record. Many more recently drilled deep-sea sections, extending over tens of millions of years, and with abundant microfossil content, now have age models with mean errors of c. 100 ka, and some have been dated to mean accuracies nearer 10 ka (see below). These sections thus form a globally distributed, if still incomplete, framework for extremely high chronological resolution palaeobiological research. However significant errors can still exist in the age models of many other, often older sections (Lazarus 1994; Spencer-Cervato et al. 1994; Lazarus et al. 1995a, b; Spencer-Cervato 1999), and these errors can become significant when data is compiled across large numbers of sections in large-scale syntheses.

Biostratigraphy using marine microfossils is still the primary method of determining geological age in deep-sea marine sediments, but this method has known limits of precision. Errors of as much as 1 Ma are not unusual, particularly in higher latitude sections where diachroniety of microfossil biostratigraphic datums is common (Spencer-Cervato et al. 1994; Cervato & Burckle 2003) and regional hiatuses complicate interpretation of palaeomagnetic polarity data. Although quite precise when judged by the general standards of palaeontology, a 1 Ma error within the typically short, <10 Ma time-span of individual deep-sea microfossil species (Stanley et al. 1988; Lazarus 2002) is still significant. More important are larger geochronological errors that occasionally occur in age models for deep-sea sediment sections at depth intervals where sedimentation rates are low, or where a time-interval is missing due to a hiatus. As noted above, the uniformity of pelagic sediment fabrics and bioturbation make physical identification of hiatuses difficult, and scatter of biostratigraphic events at these levels on age-depth plots (probably due to sediment reworking at these levels) make age-model development for these time intervals difficult and uncertain. At such levels age errors can be quite large, sometimes exceeding 10 Ma or more (Spencer-Cervato 1999). Examples of typical well constrained and poorly constrained age models based on biostratigraphy are shown in Figure 9.

Various methods have been developed in recent years to deal with the uncertainties of developing age models purely on the basis of biostratigraphy. These include palaeomagnetic stratigraphy, geochemical age proxies such as Sr isotope ratios, and orbital tuning of age models using cyclically varying sedimentary or geochemical proxies sensitive to climate referenced to orbital-scale climate cyclicity curves derived from astronomical principles (Hilgen 2008). Such methods can frequently resolve uncertainties in age-models and add substantial precision, and are often developed to examine key time intervals in the past at high temporal resolution (late Neogene ice-ages, Paleocene–Eocene hyperthermal event, the Cretaceous–Tertiary boundary extinction event). There are also many methods by which biostratigraphy itself can be improved, including regional instead of global calibrations, use of larger numbers of taxa, and advanced quantitative methods of analysis (Gradstein 1985; Sadler 2010). As noted above, enough sections have been studied with these methods to form the nucleus of a global, high resolution framework for palaeobiological (and palaeoceanographic) research. Nonetheless, due to manpower constraints and a priority to study palaeoenvironmental change at key time intervals, the large majority of sites and time intervals from which micropalaeontologic data has been generated have yet to be re-examined using these new tools. Thus there is a significant degree of age error in much of the existing microfossil occurrence dataset. It should be emphasized that, in comparison to much other fossil data, where time bins are at best 10 Ma, and age errors are very frequently 10 Ma or more, the deep-sea microfossil data is already quite good. But it could be even better.

Taxonomy

The unusually complete species-level preservation, and the relatively simple morphologies of shells of these primarily protist groups, mean that many morphologically similar species are preserved in the record, creating problems of morphological overlap between individuals of closely related forms and relatively common uncertainties or inconsistencies in the identification of specimens to species. Convergence and iterative evolution are well known phenomena in marine microfossil plankton (Cifelli 1969; Coxall et al. 2007), and can also create erroneous species ranges. Even when correctly identified, marine microfossil taxonomic practice also has a significant effect on diversity and derived metrics (speciation and extinction rates), in that many species are named morphological ranges in evolving phyletic lineages. Thus for any moment in time, more than one morphospecies may be reported for a single evolving phyletic lineage. The resultant inflation of apparent diversity may or may not be important, depending on the type of evolutionary analysis being done. Taxonomic errors in species identification, of whatever sort,

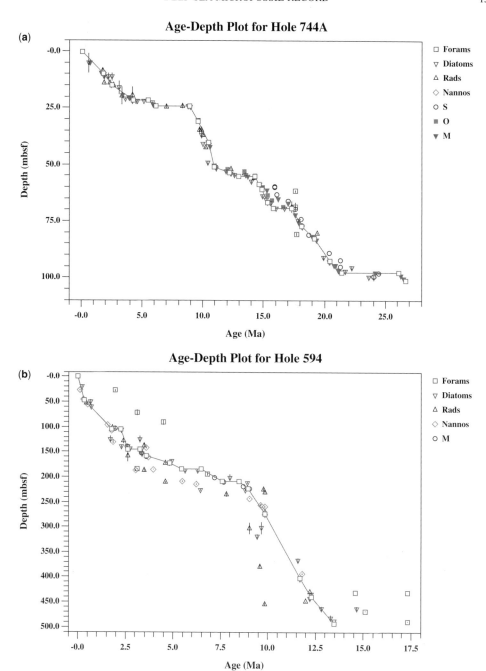

Fig. 9. Examples of age models for deep-sea drilled sections, from Lazarus *et al.* (1995*b*) and used in Neptune. (**a**) ODP Site 744 recovered a Neogene section from the Kerguelen Plateau, Antarctic which, despite incomplete recovery, can be accurately dated over most of the interval with mutually consistent siliceous and (in the lower Miocene) calcareous biostratigraphic data, and a largely conformable interpretation of the palaeomagnetic stratigraphy (M, Harwood *et al.* 1992). (**b**) DSDP Site 594 recovered a high sedimentation rate mid-late Neogene section from near the Polar Front, south of New Zealand. Biostratigraphic data is inconsistent, both within and between microfossil groups. Even in the 0–10 Ma interval (*c.* 0–250 mbsf) the age model is uncertain by as much as 2 Ma, and is even worse below this level. Various reasons may be responsible, including reworking of material by strong regional bottom currents and diachronous species ranges near biogeographical boundaries associated with the front. S, strontium isotope age estimates; O, oxygen isotope events.

are temporally neutral, but can significantly displace the reported occurrences of species into both younger and older time intervals.

The ranges of higher level taxa, particularly genera, are subject to substantial error when the defining characteristics are based on inappropriate characters. Although largely eliminated in the taxonomy of calcareous microfossil groups, these problems are still present, for example in the generic definitions of Cenozoic (though not older) radiolarians. These are still based, in part, on the highly artificial classification of Haeckel (1887). Indeed, for many years in the early part of the twentieth century, the uncritical application of Haeckelian generic concepts yielded radiolarian generic ranges from Palaeozoic to Recent, and the (quite incorrect) belief that radiolarians were not useful for biostratigraphy (Moore et al. 1952).

Despite these problems, the taxonomic units of deep-sea micropalaeontology – morphospecies – are more closely based on natural biological categories than the genera and families primarily used in other areas of palaeobiology, and in principle should reduce the distortions of primary pattern that occur when using higher ranked taxa as proxies for species level diversity (Lane & Benton 2003; Bertrand et al. 2006).

Reworking, age model errors and macroevolutionary metrics

Before going further, it is useful to know, at least approximately, the overall scale of the problem. Lazarus et al. (in press, and summarized below; see also Kucera & Schönfeld 2007) have made a preliminary comparison of ranges of selected tropical biostratigraphic marker species in raw radiolarian data from Neptune with the reported calibrated age for these species first and last occurrences. Their comparison suggests that the raw occurrence data for radiolarians in Neptune includes approximately 5% displaced occurrences for last occurrence events, and c. 3% displaced data for first occurrence events. Thus, the absolute values of c. 8% (total) provide an estimate of the degree to which existing deep-sea drilling data has been affected by various sorts of error. Of the variety of reasons why species occurrences may be displaced, most are symmetric with respect to age, but reworking is asymmetric, affecting only last occurrences. Thus the difference (2% in their preliminary estimate) is a reasonable order of magnitude approximation of the extent to which occurrences of tropical Cenozoic radiolarians, and possibly more generally marine microfossils, have been affected by reworking, while the remainder (c. 6%) is a rough estimate of the combined effect of all other errors (e.g. age models and taxonomy).

The consequence of the combined effect of reworking, age model errors and taxonomic errors for databases such as Neptune, even if they are only a few percent of the data, is that species ranges as recorded in compilations of occurrences in multiple sections are typically extended, often well beyond their true range. Although the vast bulk of the occurrence data is from the known range of a species, there are typically at least a few records of a species' occurrence from time intervals well outside the accepted range for the species (Spencer-Cervato 1999; Kucera & Schönfeld 2007; Lazarus et al. in press). This is very much in contrast to the situation in most other areas of palaeontology, where the highly incomplete nature of the fossil record means that the ranges of most species, both in local sections and in compilations of data, are thought to be for the most part partial, truncated ranges of a longer true original stratigraphic range (Shaw 1964). This in turn has significant consequences for the use of the deep-sea microssil record for evolutionary studies. Because most taxa in almost any fauna or flora (including most deep-sea microfossil assemblages) are rare, even with more complete data (as proposed below), the occurrences of many species in individual samples will be sporadic. This sporadic occurrence effect will be enhanced for those taxa which are both rare and relatively susceptible to dissolution. To effectively estimate the range of most species it is therefore desirable to make use of 'range-through' methods, that is making use of occurrences of taxa in younger and older time intervals to infer their existence as well in intermediate intervals where the taxon has not (yet) been recorded. However, range-through methods (in addition to possible biases due to uneven sampling, which can be corrected by appropriate data collection methods, see below) are by their nature very sensitive to the existence of data errors that have artificially extended the range of taxa, such as those described above for the deep-sea microfossil record. Calculation of the diversity history of marine plankton groups from the reported deep-sea microfossil record using simple range-through methods thus can produce severely inaccurate results.

This can be seen when a simple total diversity curve calculated from 'raw' Neptune data is compared to the known diversity history for one of the better studied groups- planktonic foraminifera (Fig. 10). In this figure not only is the total reported diversity from Neptune significantly higher than most literature estimates for any given time interval (Stewart & Pearson 2000, which is in part based on Kennett & Srinivasan 1983) the dynamics of the diversity curve do not reflect what is believed to be the overall pattern of planktonic foraminiferal diversity change over the Cenozoic (Fig. 10).

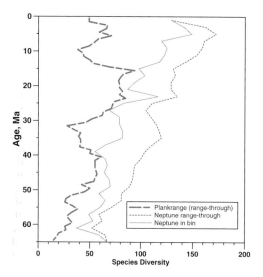

Fig. 10. Diversity of Cenozoic plantkonic foraminifera using two different data sources: the Stewart & Pearson (2000) Plankrange list of species FADS and LADS; and directly computed from species occurrence data in the Neptune database. The Neptune data allows computing both the range-through and within-bin diversity, the former being most similar to the FAD-LAD data of the Plankrange list.

Insight into the source of this discrepancy can be obtained by individually checking the taxa occurrences that make up the reported diversity in the raw Neptune curve at a level where the discrepancy between the Neptune value and the accepted diversity value from the literature is large. This has been done for the taxa reported for the 5–6 Ma (Miocene/Pliocene boundary) time interval (details in Supplementary Appendix).

The total diversity of early Pliocene planktonic foraminifera is not all that precisely fixed in a survey of the published literature, although data in the Stewart and Pearson database yield a value of 71. This is somewhat higher than recent diversity (c. 50, according to taxonomy used between 44 and 64, of which only one third are reasonably common (Arnold & Parker 1999; Stewart & Pearson 2000). Additional difficulties arise in determining the taxonomic status of names – not all of these have been fully resolved in a global compilation such as Neptune, and the taxonomy used by the foraminiferal specialists for Neptune and by Stewart & Pearson (2000) is frequently significantly different. Out of a total calculated range-through diversity of c. 140 valid planktonic foraminiferal taxa in the Neptune database, at least 38 taxa are not present in this bin but are inferred to exist by range-through interpolation. Of these, only 8 are in conformance with catalogue ranges for example

Stewart & Pearson (2000). A full 30 species are 'stratigraphically displaced' taxa, whose inferred presence at 5 Ma are noticeably in conflict with published age range estimates for the taxa. Several of these 'erroneous' occurrences were examined to determine the primary source of error. Although a quantitative evaluation of all these 'displaced' taxa is beyond the scope of this paper, some of the 'stratigraphically displaced' taxa apparently are based on reliable occurrence data, and thus are not displaced but instead reflect outdated values for ranges in the catalogue (Supplementary Appendix). The majority of these 'displaced' taxa however appear to be due to taxonomic or age model errors, or reworking in the original source data used as input to the database. For example, the latest Cretaceous/ basal Palaeogene species *Acarinina intermedia*, with over 80 Neptune occurrence records in this time interval, also has rare (single) occurrences in late Pliocene–Pleistocene sediments (Supplementary Appendix). Of the remaining c. 100 presumed valid species (based on Neptune's taxonomy) recorded within the 6–5 Ma bin, 57 can be clearly matched to species that occur in this time interval as well in Stewart & Pearson (2000), a further 23 species can be matched to Stewart and Pearson taxa but the ages given by them are not conformable with the Neptune data, and the remaining c. 20 species could not (by an admittedly non-planktonic foraminiferal expert) be matched to taxa in the Stewart and Pearson catalogue at all. These latter unmatched species may represent additional diversity not covered by Stewart and Pearson, unidentified taxonomic synonyms, or both. Some of the 23 species with conflicting age estimates are probably true range extensions compared to the given catalogue values, as these are sometimes based on older literature. Many however are also likely to be due to reworking, taxonomic error or age model problems. *Globorotalia truncatulinoides* for example, has been studied in detail and its evolutionary origin from *G. crassaformis* in the mid Pliocene documented (Lazarus *et al.* 1995*b*), so that basal Pliocene occurrences must be in error.

The total Neptune range through diversity calculated from species validated against the Stewart and Pearson catalogue is 65, a value that is very close to the value given by Stewart & Pearson (2000). This suggests that, although the catalogue compilations may also contain some omissions or incomplete range estimates, a very significant fraction of the total calculated diversity in this bin is erroneous: between 40 and 50 species, v. 65 valid entries, among the taxa which could be resolved in this simple comparison. Errors of this magnitude, while perhaps not any worse than probable errors (of under, rather than over-reporting of diversity) in other types of palaeontological data (e.g. marine invertebrates and other macro-organismal data in PaleoDB) are still much larger than need be, and significantly degrade the quality of results that are obtained from macroevolutionary analysis of the deep-sea microfossil record. Furthermore, as the effects of reworking are cumulative over time, there is a built-in non-uniform bias – a 'ratchet' effect, by which total diversity increases in younger time-intervals due to the accumulation of reworked occurrences from an increasingly long prior historical record.

Solutions

Incomplete data

Trite though it may sound at first, the solution to highly incomplete occurrence data on species from any given time interval, and in particular for any given sample, is simply to collect complete data. This is actually a remarkable aspect of the deep-sea microfossil record, as in reality there are few, if any, other parts of the palaeontological record where incomplete data can be transformed into complete data by any realistic amount of additional data collection. The fossil record of dinosaurs, hominids, or even Palaeozoic brachiopods will not become complete by a few additional field seasons of collecting. Yet, given the virtually unlimited numbers of specimens and samples available for study (and their inherent completeness of preservation at the species level), a carefully chosen program of re-examination of the recovered record should be able to provide nearly complete species level data for all time intervals and biogeographical regions.

The sampling requirements for such a dataset have already been outlined in past deep-sea drilling planning documents (COSOD II 1988; COMPLEX report: Pisias & Delaney 1999), and, even without explicit recognition as a goal by the drilling programs, are gradually reaching maturity as parallel palaeoceanographic targets are sampled by drilling campaigns. Essentially, what is needed is at least one complete, accurately dated, composite section covering the entire last 100 Ma (or as close to this as possible) for each existing or inferred past biogeographical province. Sampling of each composite section at c. 100–200 ka spacing would be sufficient to provide material at all evolutionary scales above the largely ecological, Milankvitch frequency behaviour of the biotic systems. Although not explicitly targeting this goal, the existing MRC collections already provide a good first approximation to such a sample set. Further, many, albeit partial, high chronological resolution sections are now available from time intervals and many biogeographical provinces. Targeted study of these sections, and extension of microfossil study and high-resolution

age-modelling to other suitable sections with well preserved material can provide the foundation for a much more robust, global occurrence-level dataset. Collection of new data also will allow better control of sampling effects and the effective use of range-through methods.

The numbers of specimens that need to be examined in each sample to insure reasonably complete documentation of preserved biodiversity are also manageable. Micropalaeontologists already often examine samples of several thousand specimens, in order to determine the presence or absence of (unfortunately frequently) rare biostratigraphical marker taxa. The primary difference is that, instead of skimming past the large majority of specimens and other taxa without documenting them, a careful record needs to be made of the entire assemblage. While adding significantly to the time needed to collect data for a sample, such a documentation effort is a realistic goal for future work. The precise numbers needed will vary by fossil group and geographical region, and largely parallel the sample diversity.

Prior approaches to dealing with 'RAT' imperfections of deep sea microfossil data

Several methods have been used to deal with the problems of reworking, age model error and taxonomic misidentification that result in species occurrences in sediments younger or older than the original stratigraphic range. Displaced occurrences can be detected in principle by their different distributional characteristics within large, multiple section/author sample sets in comparison to in situ data. In situ occurrences are normally relatively uniformly distributed within biogeographical provinces, and usually are fairly consistent in relative abundance over time. Displaced occurrence data by contrast is usually more sporadic and clustered, in association with the more local sources that cause it: for example individual reworked or misdated samples, authors of discrepant taxa concepts etc.

The most widely used method (as noted above), practiced to some degree by most marine micropalaeontologists, is, when reporting data for an individual section, to identify some occurrences of species in samples as anomalous, based on prior knowledge of the known range of the taxon and the consistency of distributional behaviour. These occurrences may be marked in the reported occurrence data table (aka range chart) in various ways (r-reworked;? questionable occurrence; etc.), or in many cases simply omitted from the reported data completely, and only mentioned briefly in the body of the published paper. Although this approach has helped to maintain consistency in biostratigraphic practice, it is often very subjective (although sometimes differences in preservation can provide an objective basis for identifying reworked specimens) and can lead to circular reasoning and the inability to recognize true differences in species ranges from new data when these are in conflict with *a-priori* assumptions. It is also only applicable for those taxa with relatively well known stratigraphic ranges for example biostratigraphic marker taxa, or common taxa with frequently reported occurrences within the true biogeographical range, but not to less well known or rare taxa, which can in most microfossil groups comprise the majority of species in the assemblages.

Analytical methods

In working with compilations of data across several sections, a variety of different analytical methods can be employed, some of which provide a more robust, general, and objective basis for dealing with stratigraphic outliers. In her synthesis of data from Neptune, Spencer-Cervato (1999) noted the relative prevalence of chronologically displaced taxa in sections from levels near hiatuses in the section's age model, and excluded all data from intervals near these levels from her analyses. This method certainly helps remove a significant amount of erroneous data but does not address other sources of error such as reworking in otherwise continuous sections, taxonomic mis-identification etc. Using within-bin diversity estimates, or sampling methods that effectively smooth occurrence information across nearby bins (e.g. boundary crossers, PaleoDB) avoid at least part of the range extension problem, but at the cost of missing substantial numbers of rare taxa. Resampling methods such as those used (though for other reasons) by the Paleobiology Database can also help, at least when the sample number is set to relatively small values. In this approach, the subsample is likely to miss the relatively rare displaced occurrence records for a species and only select from the much more common records within the species' true range. Resampling as a method to deal with outliers though also has several limitations. Resampling at a level sufficient to eliminate significant numbers of displaced occurrence records is also likely to miss sampling the full range of, or even completely miss rare taxa, which, as noted above, usually make up the majority of the reported diversity. Although perhaps necessary to compensate for unevenly sampled compiled pre-existing data, a better approach is, as noted above, to collect more complete, uniformly sampled data for analysis. Resampling itself is a subjective procedure in that the reported total diversity depends significantly on the size of the subsample chosen, and thus only relative changes in calculated diversity can be recovered by this method. Resampling also makes the assumption that the basic shape of the

individuals-diversity (equability) curve is similar for all samples being compared. As in modern plankton (and in many other environments as well) there is a strong shift in equability with changing total diversity (less diverse assemblages are also less equable: Lipps 1979; Boltovskoy 1987) this assumption is likely to be frequently violated in real-world studies. Resampling also assumes that sample size or numbers are directly related to true diversity, which as noted above, is frequently, or even mostly not true for existing data sets. Lastly, resampling, as a way to deal with outliers, does not actually identify the sources of outlier error, that is, the samples in the database from which the problematic data come. This is a significant limitation, as ideally any analytical protocol should contain methods for identifying the source of problematic data, with the ultimate goal of allowing this data to be re-examined and corrected, not merely statistically masked.

Liow & Stenseth (2007) and Liow *et al.* (2010) recently examined the general pattern of plankton species occurrences over time, using as data the number of sections per time interval for numerous species in the Neptune database. They calculated a metric that can be seen as a proxy for species first and last occurrences from this data, by fitting a hat-shaped curve to each species occurrence distribution over time, and arbitrarily designating the 50% of maximum occurrence (inflection) points of the fitted curve as 'species rise' or 'fall', respectively. This method (hats for RATs) has the advantages of being derived from an explicit ecological model of how species occurrences are expected to behave over time, and makes use of total occurrence information for species, not just end-point data. However, this method, as a means for outlier identification, has some disadvantages as well. The rise and fall points in the fitted curve chosen reflect properties such as maximum abundance as much as they do the biologically, and temporally different phenomena of origin and extinction; the percent cutoff levels are arbitrary; nor is it clear that a hat-shaped model is in fact appropriate for many species' abundance histories. Lastly, the curve-fitting procedure is complex and computationally intensive.

Detection of data outliers is in fact a widespread problem in data analysis and many different approaches have been developed (e.g. Hodge & Austin 2004). Although too numerous to review here, it should be noted that many are based on iterative removal of successive outlier data points using selection criteria, and many, if not most, use some sort of theoretical distribution model to develop the selection rule (e.g. Rosner 1983, which uses the Student t distribution).

Lazarus (2008; Lazarus *et al.* in press) suggest a multi-step procedure to improve the accuracy of species ranges in deep-sea microfossil data

(Fig. 11). As in Liow *et al.*'s (2007, 2010) hat method, species occurrence records compiled across multiple sections are used, but instead of fitting a theoretical curve, a simple percent trim of the most extreme oldest and youngest occurrence records is made. Use of a percentage lacks the sophistication of a fitted curve, but does respect concentrations of occurrence data, regardless of the overall shape of the distribution of occurrences v. time within the species range, and can be seen as non-parametric and thus applicable regardless of knowledge of model appropriateness. The percent used is in addition, not arbitrary but based on an empirical calibration – of the amount of data that needs to be trimmed from the subset of biostratigraphical marker species occurrence data to match their (presumed to be accurate) calibrated first and last occurrences in the deep-sea microfossil record. Lastly, the now identified individual species

occurrence 'outlier' records are matched to the samples they come from, and the relative abundance of outlier occurrences is calculated for each sample in each section, allowing the reliability of samples, and entire sections, to be estimated (Fig. 12). This in turn provides useful indications of the sources of error (reworking, age model errors, taxonomy) that can then be remedied by targeted new data collection and analysis.

Taxonomy

The various types of taxonomic problems described above vary substantially by fossil group, and will need an equal variety of solutions. For some clades, particularly the siliceous microfossil groups, a great deal of classic primary morphospecies descriptive work still needs to be done, and synthesized in suitable reference catalogues. For

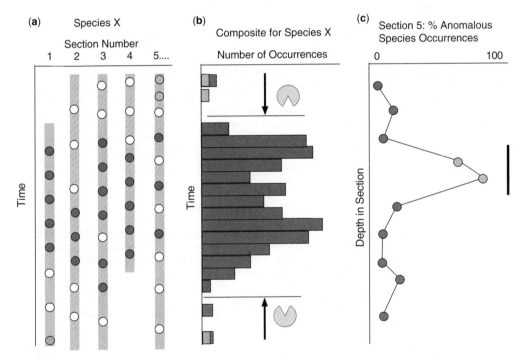

Fig. 11. Schematic illustration of the pacman method for identifying and reducing the frequency of data outliers in deep-sea microfossil occurrence databases. (**a**) Occurrences of a species (coloured dots) and samples without the species (open dots) in 5 different sections, plotted to a common time-scale. (**b**) Compilation of occurrence data for a species over time from multiple sections. A pre-defined calibrated fraction (see text) of the occurrence data from the top and bottom of the occurrence distribution is distinguished from the main data ('pacman' symbols and arrows): blue data are accepted as the valid distribution for subsequent analyses, the rest are defined as outliers (red, green, orange). (**c**) Procedure is repeated for multiple species and all occurrence data in each sample is evaluated for fraction of data in sample marked in step b as an outlier. In this example (not scale – depth in section, not age), section 5 of (a) has two samples (orange in (a) and (c), marked by black bar on side of (c)) in which the majority of occurrences are not conformable with the bulk of species occurrence data. The age models, sedimentologic integrity and taxonomic identifications of these samples should be re-examined and the existing data from these samples removed from analyses until the source of discrepancies is resolved.

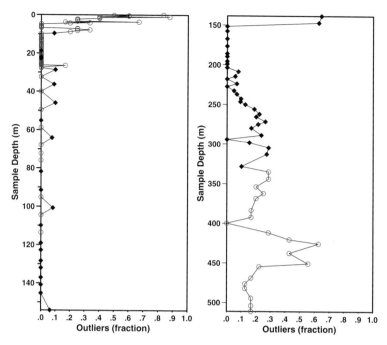

Fig. 12. Result of pacman profiling of two deep-sea drilled sections from DSDP Leg 85, equatorial Pacific, using the global Cenozoic radiolarian occurrence data in Neptune (from Lazarus *et al.* in press). The pacman trim values were set to 5% of youngest occurrences and 3% of oldest occurrences for the global dataset. The radiolarian data from Holes 572A and 573B (Nigrini 1985) were then used to compute profiles of stratigraphic outlier frequency (fraction of outliers in occurrences per sample) for each section. Samples with fewer than 10 species are plotted with open circles. Section 572A (0–150 mbsf, late Miocene–Recent) shows generally very low levels of outlier frequency except in the uppermost few meters of the section. This site was drilled and nearly continuously recovered using a hydraulic-piston corer, and has a generally well constrained, coherent age-model. Section 573B (140–510 mbsf, late Miocene–late Eocene) was drilled with a rotary corer – a method more likely to disturb stratigraphic integrity – and has a less well resolved age model. Section recovery was also poor below *c.* 250 mbsf.

radiolarians and perhaps other groups, many more phylogenetic analyses are needed to establish more natural generic and higher level taxonomic concepts. In many cases, studies of taxonomy and systematics using molecular methods on living material are badly needed to provide a baseline for taxonomic work on extinct forms. More work is also needed to determine to what extent genetic species can actually be identified using refined morphological criteria within the current more broadly defined morphospecies. Initial investigations suggest that for planktonic foraminifera differences are very subtle, and this may not be generally feasible (Huber *et al.* 1997; Morard *et al.* 2009, though see also de Vargas *et al.* 2004), while for coccolithophores, size and subtle character differences suggest that a more refined, genetic species level taxonomy may at least be partially possible (de Vargas *et al.* 2004; Young *et al.* 2005). Distinguishing anagenetic v. cladogenetic species boundaries is in principle not difficult as the distinction is usually made in

the original species descriptions. This information does however need to be collated and integrated in the taxonomic references used in analysing primary occurrence datasets.

Discussion, general recommendations and conclusions

Palaeontology offers a deep-time perspective on the evolution of life, and how biological evolution has been influenced by, and influences, the evolution of the non-living Earth system. Despite many successes, palaeontology is still dogged by the incompleteness and biases of the fossil record. Current efforts to understand the history of biodiversity and how this history relates to current issues of interest such as climate change are challenged by these gaps and biases, and success in overcoming them is still very uncertain. Deep-sea plankton microfossils offer a unique opportunity to sidestep most of the problems of incompleteness and bias, providing near-ideal material for scientific study of these questions. Despite many caveats, there are at least some clades of zooplankton, and to a lesser degree also phytoplankton, where preservation at the species level is extremely good, global recovery of primary biogeography is possible, specimen numbers are essentially unlimited, and detailed, closely linked parallel data on environmental change is widely available. Such material is ideally suited to studying both microevolution and macroevolution, and offers as well the opportunity to understand how these two types of evolution are related, as the individual microevolutionary patterns that underlie the larger, macroevolutionary one are also preserved and available for analysis. The completeness of the deep-sea microfossil record also offers an unusual opportunity to combine molecular approaches based on living representatives with nearly complete, detailed clade histories based on the fossil record.

So far, the impact of this extraordinary fossil record on palaeobiological research has been fairly modest, despite multiple relevant studies on a variety of palaeobiological subjects on scales ranging from microevolution to macroevolution. In part this may just represent taxonomic parochialism – evolutionary processes in protist plankton may well not be a very good analogy, at least in detail, for processes operating on metazoans, and thus the results of such research are seen as irrelevant to the study of the 'main' record of life's evolution, for example metazoans. Yet most of the history of life on Earth, most of the Earth's current biomass, and the dominant biological influences on the Earth's geochemical cycles are all unicellular. It may be time to acknowledge a broader view of life and establish the evolution of unicellular organisms as one of the 'main' subjects of palaeobiological research. Furthermore, it is to be hoped that at least some mechanisms of evolution are universal, and thus suited to study in any group of organisms. For those seeking palaeontological material for such purposes, deep-sea plankton microfossils may be a better choice than the less complete fossil record of other groups of organisms.

Other reasons for micropalaeontology's limited impact on palaeobiology remain to be addressed. Biological understanding of these fossilizing groups is still very superficial and requires much additional work. It has been pointed out in this paper that, although the documentation of the record is already significantly better at the species level than for most other types of fossil data, the available data still fall short of this materials' true potential. Catalogues of first and last occurrence data (suited to many but not all types of analyses) do not exist for some groups, and are not always up to date with current data on occurrences from more recent deep-sea drilling. Much of the available occurrence data has been collected for purposes other than documenting diversity, and new methods need to be developed to optimally analyse such data. Various processes have created outliers in the distribution of taxa over time and data thus needs careful data filtering to use in palaeobiological research. It has been shown how in principle these problems can be dealt with, so that obtaining data that reflects the high quality of the fossil record is entirely feasible. Palaeobiologists however must realize that the work outlined is substantial, and is unlikely to be carried out by micropalaeontologists employed to produce biostratigraphic reports or study palaeoceanography. If marine micropalaeontologic palaeobiology is to properly develop, institutions with a mandate to study palaeobiology (natural history museums, academic departments) will need to invest resources to make this happen. The opportunity is there – are we going to make use of it?

The author wishes to thank B. Kotrc for a detailed and extremely helpful review of an early draft of this manuscript; two anonymous reviewers for constructive comments and the conference organizers for the opportunity to present this review. Thanks to J. Renaudie for creating Figures 2 and 4 and base maps for Figures 1 and 3.

References

ADL, S. M., SIMPSON, A. G. B. ET AL. 2005. A new higher level classification of Eurcaryotes with emphasis on the taxonomy of protists. *Journal of Eukaryotic Microbiology*, **52**, 399–451.

ALROY, J., MARSHALL, C. R. ET AL. 2001. Effects of sampling standardization on estimates of Phanerozoic marine diversification. *Proceedings of the National Academy of Sciences*, **98**, 6261–6266.

ANGEL, M. V. 2010. *Marine planktonic ostracods.* http://species-identification.org/about

ARCHER, D. 1996. An atlas of the distribution of calcim carbonate in sediments of the deep sea. *Global Biogeochemical Cycles*, **10**, 159–174.

ARNOLD, A. J. & PARKER, W. C. 1999. The biogeography of planktonic foraminifera. *In:* SEN GUPTA, B. K. (ed.) *Modern Foraminifera.* Kluwer, London, 103–122.

ARNOLD, M. L. 1997. *Natural Hybridization and Evolution.* Oxford University Press, Oxford.

ARNOLD, M. L. & FOGARTY, N. D. 2009. Reticulate evolution and marine organisms: the final frontier? *International Journal of Molecular Sciences*, **10**, 3836–3860.

AZE, T., EZARD, T. H. G., PURVIS, A., COXALL, H., STEWART, D. R. M., WADE, B. S. & PEARSON, P. N. 2011. A phylogeny of Cenozoic macroperforate planktonic foraminifera from fossil data. *Biological Reviews*, doi: 10.1111/j.1469-185X.2011.00178.x

BENTON, M. J. & HARPER, D. A. T. 2009. *Introduction to Paleobiology and the Fossil Record*. John Wiley, New York.

BERGER, W. H. & HEATH, G. R. 1968. Vertical mixing in pelagic sediments. *Journal of Marine Research*, **26**, 134–143.

BERTRAND, Y., PLEIJEL, F. & ROUSE, G. W. 2006. Taxonomic surrogacy in biodiversity assessments, and the meaning of Linnean ranks. *Systematics and Biodiversity*, **4**, 149–159.

BOLTOVSKOY, D. 1987. Sedimentary record of radiolarian biogeography in the equatorial to Antarctic western Pacific Ocean. *Micropaleontology*, **33**, 267–281.

BOTO, L. 2010. Horizontal gene transfer in evolution: facts and challenges. *Proceedings of the Royal Society, B*, **277**, 819–827.

BOWN, P. R., DUNKLEY JONES, T. ET AL. 2007. A Paleogene calcareous microfossil Konzervat–Lagerstätte from the Kilwa Group of coastal Tanzania. *Geological Society of America Bulletin*, **120**, 3–12.

BROECKER, W. 2008. A need to improve reconstructions of the fluctuations in the calcite compensation depth over the course of the Cenozoic. *Paleoceanography*, **23**, 883–896.

CALVERT, S. E. 1974. Deposition and diagenesis of silica in marine sediments. *In*: HSU, K. J. & JENKYNS, H. C. (eds) *Pelagic Sediments, on Land and Under the Sea: Proceedings of a Symposium*. Wiley-Blackwell, New York, 273–299.

CASTELEYN, G., ADAMS, N. G., VANORMELINGEN, P., DEBEER, A.-E., SABBE, K. & VYVERMAN, W. 2009. Natural hybrids in the marine diatom Pseudo-nitzschia pungens (Bacillariophyceae): genetic and morphological evidence. *Protist*, **160**, 343–354.

CASTELEYN, G., LELIAERT, F. ET AL. 2010. Limits to gene flow in a cosmopolitan marine planktonic diatom. *Proceedings of the National Academy of Sciences USA*, **107**, 12952–12957.

CERVATO, C. & BURCKLE, L. H. 2003. Pattern of first and last appearance in diatoms: oceanic circulation and the position of the Polar Fronts during the Cenozoic. *Paleoceanography*, **18**, 1055.

CIFELLI, R. 1969. Radiation of Cenozoic planktonic foraminifera. *Systematic Zoology*, **18**, 154–168.

CORFIELD, R. M. & SHACKLETON, N. J. 1988. Productivity change as a control on planktonic foraminiferal evolution after the Cretaceous/Tertiary boundary. *Historical Biology*, **1**, 323–343.

COSOD II WORKING GROUP MEMBERS. 1988. *Report of the Second Conference on Scientific Ocean Drilling*. COSOD II. European Science Foundation, Strasbourg, France.

COXALL, H., WILSON, P. A., PEARSON, P. N. & SEXTON, P. F. 2007. Iterative evolution of digitate planktonic foraminifera. *Paleobiology*, **33**, 495–516.

DE VARGAS, C., SAEZ, A., MEDLIN, L. K. & THIERSTEIN, H. R. 2004. Super-species in the calcareous plankton. *In*: THIERSTEIN, H. R. & YOUNG, J. (eds) *Coccolithophores – from Molecular Processes to Global Impact*. Springer, Berlin, 271–298.

DE WEVER, P., DUMITRICA, P., CAULET, J. P., NIGRINI, C. & CARIDROIT, M. 2001. *Radiolarians in the Sedimentary Record*. Gordon and Breach, Amsterdam.

EMPSON-MORIN, K. M. 1984. Depth and latitude distribution of Radiolaria in Campanian (Late Cretaceous) tropical and subtropical oceans. *Micropaleontology*, **30**, 87–115.

FENNER, J. 1985. Late Cretaceous to Oligocene planktic diatoms. *In*: BOLLI, H. M., SAUNDERS, J. B. & PERCH-NIELSEN, K. (eds) *Plankton Stratigraphy*. Cambridge University Press, Cambridge, 713–762.

FENSOME, R. A., MACRAE, R. A., MOLDOWAN, J. M., TAYLOR, F. J. R. & WILLIAMS, G. L. 1996. The early Mesozoic radiation of dinoflagellates. *Paleobiology*, **22**, 329–338.

FENSOME, R. A., MACRAE, R. A. & WILLIAMS, G. L. 1997. Dinoflagellate evolution and diversity through time. Bedford Institute of Oceanography Science Review, 1994–1995, Department Fisheries and Oceans, Maritimes Region, Ottawa, CA, 45–50. http://www2.mar.dfo-mpo.gc.ca/science/review/e/pdf/dinoflagellate.pdf

FINKEL, Z. V., KATZ, M. E., WRIGHT, J. D., SCHOFIELD, O. & FALKOWSKI, P. G. 2005. Climatically driven macroevolutionary patterns in the size of marine diatoms over the Cenozoic. *Proceedings of the National Academy of Sciences USA*, **102**, 8927–8932.

FINKEL, Z. V., SEBBO, J. ET AL. 2007. A universal driver of macroevolutionary change in the size of marine phytoplankton over the Cenozoic. *Proceedings of the National Academy of Sciences USA*, **104**, 20416–20420.

FOOTE, M. & MILLER, A. 2006. *Principles of Paleontology*. Freeman, San Francisco.

FOOTE, M., CRAMPTON, J. S., BEU, A. G. & COOPER, R. A. 2008. On the bidirectional relationship between geographic range and taxonomic duration. *Paleobiology*, **34**, 421–433.

GOLL, R. M. 1976. Morphological intergradation between modern populations of Lophospyris and Phormospyris (Trissocyclidae, Radiolaria). *Micropaleontology*, **22**, 379–418.

GOULD, S. J. & ELDREDGE, N. 1993. Punctuated equilibrium comes of age. *Nature*, **6452**, 223–227.

GRADSTEIN, F. 1985. *Quantitative Stratigraphy*. D. Reidel, Dordrecht.

GRANT, V. 1971. *Plant Speciation*. Columbia University Press, New York.

HAECKEL, E. 1887. Report on the Radiolaria collected by H.M.S. Challenger during the years 1873–1876. Report on the Scientific Results of the Voyage of the H.M.S. *Challenger, Zoology*, **18**, clxxxviii + 1803.

HARPER, H. E. & KNOLL, A. H. 1975. Silica, diatoms and Cenozoic radiolarian evolution. *Geology*, **3**, 175–177.

HARWOOD, D., LAZARUS, D. B., AUBRY, M., BERGGREN, W. A., HEIDER, F., INOKUCHI, H. & MARUYAMA, T. 1992. Neogene stratigraphic synthesis, ODP Leg 120. *In*: WISE, W., SCHLICH, R., O'CONNELL, S. ET AL. (eds) *Proceedings of the Ocean Drilling Program, Scientific Results*. Ocean Drilling Program, College Station, TX, 1031–1052.

HAYS, J. D. 1970. Stratigraphy and evolutionary trends of radiolaria in North Pacific deep-sea sediments. *In*: HAYS, J. D. (ed.) *Geological Investigations of the North Pacific*. Geological Society of America Memoirs, **126**, 185–218.

HEAD, M. J. 1996. Modern dinoflagellate cysts and their biological affinities. *In*: JANSONIUS, J. & McGREGOR,

D. C. (eds) *Palynology: Principles and Applications.* American Association of Stratigraphic Palynologists, Dallas, TX, **3**, 1197–1248.

HEINZE, C., HUPE, A., MAIER-REIMER, E., DITTERT, N. & RAGUENEAU, O. 2003. Sensitivity of the marine biospheric Si cycle for biogeochemical parameter variations. *Global Biogeochemical Cycles*, **17**, 1–23 (online).

HILGEN, F. J. 2008. Recent progress in the standardization and calibration of the Cenozoic Time Scale. *Newsletters on Stratigraphy*, **43**, 15–22.

HODGE, V. J. & AUSTIN, J. 2004. A survey of outlier detection methodologies. *Artificial Intelligence Review*, **22**, 85–126.

HONJO, S., MANGANINI, S. J. & COLE, J. 1982. Sedimentation of biogenic matter in the deep ocean. *Deep Sea Research*, **29**, 609–625.

HUBER, B. T., BIJMA, J. & DARLING, K. 1997. Cryptic speciation in the living planktonic foraminifer Globigerinella siphonifera (d'Orbigny). *Paleobiology*, **23**, 33–62.

JOHNSON, D. A. 1972. Ocean-floor erosion in the equatorial Pacific. *Geological Society of American Bulletin*, **83**, 3121–3144.

JOHNSON, D. A. & NIGRINI, C. A. 1985. Synchronous and time-transgressive Neogene radiolarian datum levels in the equatorial Indian and Pacific Oceans. *Marine Micropaleontology*, **9**, 489–523.

KENNETT, J. P. 1982. *Marine Geology.* Prentice-Hall, Englewood Cliffs, NJ.

KENNETT, J. P. & INGRAM, B. L. 1995. Paleoclimatic evolution of Santa Barbara Basin during the last 20K.Y.: marine evidence for Hole 893A. *In*: KENNETT, J. P., BALDAUF, J. G. & LYLE, M. (eds) *Proceedings of the Ocean Drilling Program, Scientific Results.* Ocean Drilling Program, College Station, TX, **146**, 309–325.

KENNETT, J. P. & SRINIVASAN, M. S. 1983. *Neogene Planktonic Foraminifera, A Phylogenetic Atlas.* Hutchinson-Ross, Stroudsburg.

KIDWELL, S. M. 1998. Time-averaging in the marine fossil record: overview of strategies and uncertainties. *Geobios*, **30**, 977–995.

KNAUSS, J. A. 2005. *Introduction to Physical Oceanography.* Waveland Press, Long Grove, Illinois.

KOWALEWSKI, M. & BAMBACH, R. K. 2008. The limits of paleontological resolution. *In*: HARRIES, P. J. (ed.) *High-Resolution Approaches in Stratigraphic Paleontology.* Springer, Netherlands, 1–48.

KUCERA, M. & SCHÖNFELD, J. 2007. The origin of modern oceanic foraminiferal faunas and Neogene climate change. *In*: WILLIAMS, M., HAYWOOD, A. M., GREGORY, F. J. & SCHMIDT, D. N. (eds) *Deep-Time Perspectives on Climate Change: Marrying the Signal from Computer Models and Biological Proxies.* TMS Special Publications, **2**, 409–425.

LALLI, M. C. & GILMER, R. W. 1989. *Pelagic Snails: the Biology of Holoplanktonic Gastropods.* Stanford University Press, Stanford, CA.

LANE, A. & BENTON, M. J. 2003. Taxonomic level as a determinant of the shape of the Phanerozoic marine diversity curve. *American Naturalist*, **162**, 265–276.

LAZARUS, D. 1983. Speciation in pelagic protista and its study in the planktonic microfossil record: a review. *Paleobiology*, **9**, 327–340.

LAZARUS, D. 1989. Sampling the sediments: commentary on Jones (1988). *TREE*, **8**, 80.

LAZARUS, D. 2006. The Micropaleontological Reference Centers network. *Scientific Drilling*, **3**, 46–49.

LAZARUS, D., KOTRC, B., WULF, G. & SCHMIDT, D. N. 2009. Radiolarians decreased silicification as an evolutionary response to reduced Cenozoic ocean silica availability. *Proceedings of the National Academy of Sciences USA*, **106**, 9333–9338.

LAZARUS, D. B. 1986. Tempo and mode of morphologic evolution near the origin of the radiolarian lineage Pterocanium prismatium. *Paleobiology*, **12**, 175–189.

LAZARUS, D. B. 1994. The Neptune Project – a marine micropaleontology database. *Mathematical Geology*, **26**, 817–832.

LAZARUS, D. B. 2002. Environmental control of diversity, evolutionary rates, and taxa longevities in Antarctic Neogene Radiolaria. *Paleontologica Electronica*, **5**. http://palaeo-electronica.org

LAZARUS, D. B., HILBRECHT, H. & SPENCER-CERVATO, C. 1995*a*. Speciation and phyletic change in *Globorotalia truncatulinoides* (planktonic foraminifera). *Paleobiology*, **21**, 28–51.

LAZARUS, D. B., SPENCER-CERVATO, C., PIANKA-BIOLZI, M., BECKMANN, J. P., VON SALIS, K., HILBRECHT, H. & THIERSTEIN, H. R. 1995*b*. *Revised chronology of Neogene DSDP holes from the world ocean.* Ocean Drilling Program, Technical Note Ocean Drilling Program, Technical Notes, **24**.

LAZARUS, D., WEINKAUF, M. & DIVER, P. In press. Pacman profiling: a simple procedure to identify stratigraphic outliers in high density deep-sea microfossil data. *Paleobiology*.

LEINEN, M., CWIENK, D., HEATH, G. R., BISCAYE, P. E., KOLLA, V., THIEDE, J. & DAUPHIN, J. P. 1986. Distribution of biogenic silica and quartz in recent deep-sea sediments. *Geology*, **14**, 199–203.

LIOW, L. H. & STENSETH, N. C. 2007. The rise and fall of species: implications for macroevolutionary and macroecological studies. *Proceedings of the Royal Society, B*, **274**, 2745–2752.

LIOW, L. H., SKAUG, H. J., ERGON, T. & SCHWEDER, T. 2010. Global occurrence trajectories of microfossils: environmental volatility and the rise and fall of individual species. *Paleobiology*, **36**, 224–252.

LIPPS, J. H. 1979. The ecology and paleoecology of planktic foraminifera. *In*: LIPPS, J. H., BERGER, W. H., BUZAS, M. A., DOUGLAS, R. G. & ROSS, C. A. (eds) *Foraminiferal Ecology and Paleoecology.* Society of Economic Paleontologists and Mineralogists, Houston, Texas, 162–104.

LONGHURST, A. R. 1998. *Ecological Geography of the Sea.* Academic Press, San Diego.

LYLE, M., RAFFI, I., PÄLIKE, H., NISHI, H., GAMAGE, K. & KLAUS, A. 2009. *Integrated Ocean Drilling Program Expedition 321 Preliminary Report: Pacific Equatorial Age Transect.* http://publications.iodp.org/prelimi nary_report/321/index.html

MALMGREN, B. A. & KENNETT, J. P. 1981. Phyletic gradualism in a late Cenozoic planktonic foraminiferal lineage; DSDP Site 284, SW Pacific. *Paleobiology*, **7**, 230–240.

MAYR, E. 1963. *Animal Species and Evolution.* Harvard University, Cambridge.

McGowan, J. A. 1974. The nature of oceanic ecosystems. *In*: Miller, C. B. (ed.) *The Biology of the Oceanic Pacific – Proceedings of the 33rd Annual Biology Colloquium*. Oregon State University, Corvallis, 9–28.

McGowan, J. A. 1986. The biogeography of pelagic ecosystems. *In*: Pierrot-Bults, A. C., van der Spoel, S., Zahuranec, B. J. & Johnson, R. K. (eds) *Pelagic Biogeography*. UNESCO, Paris, 191–200.

Moore, J. T. C. 2008. Chert in the Pacific: biogenic silica and hydrothermal circulation. *Palaeogeography, Palaeoclimatology, Palaeoecology*, **261**, 87–99.

Moore, R. C., Lalicker, C. G. & Fischer, A. G. 1952. *Invertebrate Fossils*. McGraw-Hill, New York.

Moore, T. C. & Heath, G. R. 1977. Survival of deep-sea sedimentary sections. *Earth and Planetary Science Letters*, **37**, 71–80.

Moore, T. C., Andel, T. H., Sancetta, C. & Pisias, N. 1978. Cenozoic hiatuses in pelagic sediments. *Micropaleontology*, **24**, 113–138.

Moore, T. C. J., Shackleton, N. J. & Pisias, N. G. 1993. Paleoceanography and the diachrony of radiolarian events in the eastern equatorial Pacific. *Paleoceanography*, **8**, 567–586.

Morard, R., Quillévéré, F., Escarguel, G., Ujiie, Y., de Garidel-Thoron, T., Norris, R. D. & de Vargas, C. 2009. Morphological recognition of cryptic species in the planktonic foraminifer Orbulina universa. *Marine Micropaleontology*, **71**, 148–195.

Muttoni, G. & Kent, D. V. 2007. Widespread formation of cherts during the early Eocene climatic optimum. *Palaeogeography, Palaeoclimatology, Palaeoecology*, **253**, 348–362.

Nigrini, C. 1985. Radiolarian biostratigraphy in the central equatorial Pacific, Deep Sea Drilling Project Leg 85. *In*: Mayer, L. & Theyer, F. (eds) *Initial Reports of the Deep Sea Drilling Project*. United States Government Printing Office, Washington, 511–551.

Norris, R. D. 2000. Pelagic species diversity, biogeography, and evolution. *In*: Erwin, D. H. & Wing, S. L. (eds) *Deep Time: Paleobiology's Perspective*. The Paleontological Society (Allen Press), Lawrence, Kansas, 236–258.

Palumbi, S. R. 1992. Marine speciation on a small planet. *TREE*, **7**, 114–118.

Parsons, B. 1982. Causes and consequences of the relation between area and age of the ocean floor. *Journal of Geological Research*, **87**, 289–302.

Pearson, P. 1996. Cladogenetic, extinction and survivorship patterns from a lineage phylogeny: the Paleogene planktonic foraminifera. *Micropaleontology*, **42**, 179–188.

Pike, J. & Kemp, A. E. S. 1999. Diatom mats in Gulf of California sediments: implications for the paleoenvironmental interpretation of laminated sediments and silica burial. *Geology*, **27**, 311–314.

Pisias, N. & Delaney, M. L. (eds) 1999. *Complex: Conference on Multiple Platform Exploration of the Ocean*. JOI, Vancouver.

Prothero, D. & Lazarus, D. B. 1980. Planktonic microfossils and the recognition of ancestors. *Systematic Zoology*, **29**, 119–129.

Rabosky, D. L. & Sorhannus, U. 2009. Diversity dynamics of marine planktonic diatoms across the Cenozoic. *Nature*, **247**, 183–187.

Ragueneau, O., Tréguer, P. *et al.* 2000. A review of the Si cycle in the modern ocean: recent progress and missing gaps in the application of biogenic opal as a paleoproductivity proxy. *Global and Planetary Change*, **26**, 317–365.

Rosner, B. 1983. Percentage points for a generalized ESD many-outlier procedure. *Technometrics*, **25**, 165–172.

Rutherford, S., D'Hondt, S. & Prell, W. 1999. Environment controls on the geographic distribution of zooplankton diversity. *Nature*, **400**, 749–753.

Sadler, P. M. 2010. Brute-force biochronology: sequencing paleobiologic first- and last-appearance events by trial and error. *In*: Alroy, J. & Hunt, G. (eds) *Quantitative Methods in Paleobiology*. Paleontological Society Short Course, 30 October 2010. Paleontological Society Papers, **16**, 271–289.

Sanfilippo, A. & Riedel, W. R. 1985. Cretaceous radiolaria. *In*: Bolli, H., Saunders, & Perch-Nielsen, K. (eds) *Plankton Stratigraphy*. Cambridge University, Cambridge, 573–630.

Sanfilippo, A., Westberg-Smith, M. J. & Riedel, W. R. 1985. Cenozoic radiolaria. *In*: Bolli, H. M., Saunders, J. B. & Perch-Nielsen, K. (eds) *Plankton Stratigraphy*. Cambridge University Press, Cambridge, 631–712.

Sepkoski, J. J. J. 1978. A kinetic model of Phanerozoic taxonomic diversity I. analysis of marine orders. *Paleobiology*, **4**, 223–251.

Sepkoski, J. J. J., Bambach, R. K., Raup, D. M. & Valentine, J. W. 1981. Phanerozoic marine diversity and the fossil record. *Nature*, **293**, 435–437.

Shaw, A. B. 1964. *Time in Stratigraphy*. McGraw-Hill, New York.

Simon, N., Cras, A.-L., Foulon, E. & Lemée, R. 2009. Diversity and evolution of marine phytoplankton. *Comptes Rendus Biologies*, **332**, 159–170.

Smith, A. B. 2007. Marine diversity through the Phanerozoic: problems and prospects. *Journal of the Geological Society, London*, **164**, 1–15.

Sogin, M. L., Morrison, H. G. *et al.* 2006. Microbial diversity in the deep sea and the underexplored "rare biosphere". *Proceedings of the National Academy of Sciences USA*, **103**, 12 115–12 120.

Sournia, A., Chrétiennot-Dinet, M.-J. & Ricard, M. 1991. Marine phytoplankton: how many species in the world ocean? *Journal of Plankton Research*, **13**, 1093–1099.

Spencer-Cervato, C. 1999. The Cenozoic deep sea microfossil record: explorations of the DSDP/ODP sample set using the Neptune database. *Palaeontologica Electronica*, **2**. http://palaeo-electronica.org

Spencer-Cervato, C., Lazarus, D. B., Beckmann, J. P., Perch-Nielsen, K. V. S. & Biolzi, M. 1993. New calibration of Neogene radiolarian events in the North Pacific. *Marine Micropaleontology*, **21**, 261–294.

Spencer-Cervato, C., Thierstein, H. R., Lazarus, D. B. & Beckmann, J. P. 1994. How synchronous are Neogene marine plankton events? *Paleoceanography*, **9**, 739–763.

STANLEY, S. M., WHETMORE, K. L. & KENNETT, J. P. 1988. Macroevolutionary differences between the two major clades of Neogene planktonic foraminifera. *Paleobiology*, **14**, 235–249.

STEPANJANTS, S. D., CORTESE, G., KRUGLIKOVA, S. B. & BJØRKLUND, K. R. 2006. A review of bipolarity concepts: history and examples from Radiolaria and Medusozoa (Cnidaria). *Marine Biology Research*, **2**, 200–241.

STEWART, D. R. M. & PEARSON, P. N. 2000. PLANKRANGE: *A database of planktonic foraminiferal ranges.* http://palaeo.gly.bris.ac.uk/Data/plankrange.html

VALENTINE, J. W. 1995. Why no new phyla after the Cambrian? Genome and ecospace hypotheses revisited. *Palaios*, **10**, 190–194.

WAKI, K., DUTTA, S. *ET AL.* 2007. Transmembrane molecules for phylogenetic analyses of pathogenic protists: Leishmania-specific informative sites in hydrophilic loops of trans-endoplasmic reticulum N-acetylglucosamine-1-phosphate transferase. *Eukaryotic Cell*, **6**, 198–210.

YOUNG, J. R., GEISEN, M. & PROBERT, I. 2005. A review of selected aspects of coccolithophore biology with implications for paleobiodiversity estimation. *Micropaleontology*, **51**, 267–288.

Quantifying the deep-sea rock and fossil record bias using coccolithophores

GRAEME T. LLOYD*, ANDREW B. SMITH & JEREMY R. YOUNG

Natural History Museum, Cromwell Road, London SW7 5BD, UK

Corresponding author (e-mail: graemelloyd@gmail.com)

Abstract: While many studies show a correlation between observed taxonomic richness and various measures of geological sampling, all have been based on the same record of terrestrial and marine sediments collected from the land. Here we present the first analyses of how rock and fossil records vary in the deep-sea. We have developed a novel database of species occurrences of coccolithophores sampled during major drilling programs of the North Atlantic, including the Mediterranean and Caribbean. Our sampling proxy, the number of deep-sea sites sampled – perhaps the most direct measure of sampling used so far – shows an exponential rise towards the Recent. Over the same period species-richness has grown in an approximately linear fashion, but genus-level richness shows a sharp initial increase followed by a much slower decline. However, correlations between both richness measures and sampling are extremely strong and a model assuming true diversity to be constant accurately predicts much of observed richness. We conclude that the deep-sea fossil record, like its land-based counterpart, bears a rock record bias.

Fossils collected and recorded from sedimentary rocks provide the empirical evidence from which the history of life over geological time is reconstructed. However, we also know that the sedimentary rock record we can access on land does not represent a time series with uniform sampling opportunity; both the areal extent and environmental representation of rocks at outcrop vary in a non-trivial way, even over relatively short time intervals. Major sea-level cycles (for example) have driven the ratio of terrestrial to marine rocks that are preserved and accessible to palaeontologists at outcrop (Smith 2001). As a result we now know that the classic diversity curves derived from raw counts of fossils through time (e.g. Sepkoski *et al.* 1981; Benton 1995) are a product of the complex interplay between original biological diversity and sampling effort. Recent work to standardize Phanerozoic diversity curves for sampling effort (Alroy *et al.* 2001, 2008; Alroy 2010) indicates that such corrections can dramatically modify their shape, suggesting that sampling bias in the fossil record is non-negligible.

Over the last 10 years there has been significant effort first to quantify this bias in the rock record and then to compare the quality of the rock record against sampled fossil diversity curves (Peters & Foote 2001, 2002; Smith 2001; Crampton *et al.* 2003; Smith & McGowan 2007; McGowan & Smith 2008; Barrett *et al.* 2009; Butler *et al.* 2009; Marx 2009; Wall *et al.* 2009; Benson *et al.* 2010; Mannion *et al.* 2011). While these studies have

demonstrated that there is a strong correlation between the quality of the rock and fossil records, the significance of this correlation remains problematic (Smith 2007; Peters 2008; Butler *et al.* in press; Mannion *et al.* 2011; Upchurch *et al.* 2011; Benson & Butler 2011). This is because these studies have used proxies for sampling effort rather than measuring sampling effort itself, and the factors that drive the quality of the rock record also drive the fossil sampling potential. Thus sea-level change may be driving biological diversity on the land and in shallow epicontinental shelf seas (through a species-area effect), while at the same time altering availability of rock at outcrop. This is the 'common cause' hypothesis of Peters (2005). Alternatively, parts of the sampled diversity curve might be recording little more than sampling effort. To truly test the effect of sampling effort on recorded diversity we need to turn to a record where there is not such a close coupling between the area of original habitat occupied and area of preserved rock outcrop remaining. To do this we turn to the deep-sea rock and microfossil records.

Pelagic sediments accumulating in the deep-sea are composed of the calcareous or siliceous skeletons of microscopic plankton that lived and died in the overlying surface waters. These microplankton achieve extremely wide geographical distribution within ocean basins (Winter *et al.* 1994; Ziveri *et al.* 2004) and a few grams of rock sample can yield 1000s of individual microfossils for analysis. Furthermore, in some places the rock record

From: McGowan, A. J. & Smith, A. B. (eds) *Comparing the Geological and Fossil Records: Implications for Biodiversity Studies*. Geological Society, London, Special Publications, **358**, 167–177.
DOI: 10.1144/SP358.11   0305-8719/11/$15.00 © The Geological Society of London 2011.

created can be near complete for relatively long geological time spans (Wang et al. 2003; Ebra et al. 2010). Not surprisingly therefore the deep-sea microfossil record is considered by some to be amongst the best we have (Ebra 2004; McGowran 2005; Suchéras-Marx et al. 2010).

Accessing the deep-sea rock and fossil records is, of course, difficult, and it is largely through the Deep Sea (DSDP; http://www.deepseadrilling.org/) and Ocean Drilling Program (ODP; http://www-odp.tamu.edu/) that knowledge of the age distribution and fossil content of deposits in the deep-sea has been acquired. Such limited access has advantages as well as disadvantages. While the volume of rock recovered is tiny compared to the available land-based record, the numbers of cores drilled and the number of samples from which microfossils have been recovered from each core retrieved allow an accurate measure of sampling effort to be quantified. In contrast to the land-based record, therefore, we are able to measure both sampling effort and taxonomic diversity directly.

There is less reason to expect a tight correspondence between recorded taxonomic diversity and the amount of rock sampled from the deep-sea for particular time intervals. This is partly because groups such as the coccolithophores are effectively cosmopolitan in their distribution (Winter et al. 1994; Ziveri et al. 2004) and thus, unlike most shelf or terrestrial taxa, in theory a global record can be accessed from a single core. Highest biomass productivity (and hence bioclastic sediment delivery to the deep-sea) is associated with low species-richness counts, whereas highest species-richness occurs at intermediate levels of phytoplankton biomass (Irigoien et al. 2005). Although locally sections can be astonishingly complete, the global basin record of pelagic sediments in the deep-sea can be patchy and far from perfect, and in many sections hiatuses abound (Spencer-Cervato 1998). Variation in the carbonate compensation depth (CCD; Murray & Renard 1891) over time is a major factor in controlling whether calcareous oozes accumulate at specific sites and at specific times in the geological past, and the CCD is controlled by temperature and CO_2 concentration of the water (Gomitz 2009). So while the abundance of microplankton in surface waters might be expected to affect the quantity of skeletal debris being delivered to the sea floor far below, no strong link is expected connecting taxonomic diversity and extent of rock record sampling through time. The deep-sea rock and fossil records thus offer an ideal opportunity to test the strength of correlation between sampling intensity and recorded diversity over geological time.

Strategies to answer questions about how the rock and fossil records compare have evolved through time. The earliest comparisons (e.g. Raup 1972, 1976; Peters & Foote 2001; Smith 2001) involved the plotting of independently created databases of taxonomic diversity and sedimentary rock through time. More recently work has shifted to creating new, more sophisticated databases for rock measures, such as North American gap-bound packages (Peters 2006) or Western European maps (Smith & McGowan 2007). However, in order to make direct comparisons it is preferable to collect the rock and fossil data together, as many vertebrate workers have done (Frobisch 2008; Barrett et al. 2009; Butler et al. in press; Mannion et al. 2011; Benson & Butler 2011; Upchurch et al. 2011), or assign the fossils to specific rock packages a posteriori (Heim & Peters 2011). Here we adopt the former approach and create a completely novel database that houses information on both fossil occurrences and lithological information with all data coming from the published records of the DSDP and ODP. We use this data to test the correlation between deep-sea sampling effort and the taxonomic diversity of coccolithophorids recorded from Atlantic sites spanning the last 150 Ma.

Material and methods

The database

As it was not tractable to enter data from the world's oceans as a whole we limited ourselves to the North Atlantic, which we define as 90°N to 20°S and including both the Mediterranean and Caribbean. This area offers the key advantage of being relatively densely sampled by the DSDP/ODP, compared to say, the Pacific. However, our database is explicitly designed to be easily expandable in future and already incorporates the basic information on all DSDP/ODP holes (e.g. leg number, site number, depth below sea-level and latitude–longitude).

The DSDP/ODP volumes present data on several microfossil groups, but we chose to limit ourselves to the two major calcareous planktonic groups, the coccolithophores and planktic foraminifera. These are commonly occurring, frequently recorded and sufficiently abundant and speciose to be appropriate for palaeobiodiversity studies. In this paper we shall only deal with the coccolithophores.

Unlike the pre-existing NEPTUNE database (Lazarus 1994) our fundamental unit is not an individual sample, but a biozone within a specific DSDP or ODP hole. This decision was made to greatly reduce the amount of data entry required, in order to allow more sites to be entered. Here a biozone is either a nannofossil or planktic foraminifera zone. Our dates come from Gradstein et al.

(2004), and specifically the TimeScale Creator program (https://engineering.purdue.edu/Stratigraphy/tscreator/). In practice some complications arise from using zones. An initial problem is slumping which can lead to a zone occurring more than once in the same DSDP/ODP hole. This was solved by additionally defining a unit by its top and bottom as a depth in metres below sea floor (mbsf), thus allowing a zone to occur twice in the same hole. In other cases precise allocation to a single zone is not possible and instead a range is given. This is accommodated in the database by having separate fields for oldest and youngest possible zone. Finally, in some cases both nannofossil and planktic foraminiferal zones are available. Where possible, these were used to split up the units more finely based on areas of overlap that ultimately lead to more precise dating (see methods, below).

Once defined stratigraphically, other data can then be assigned to that unit. Most important amongst this data are the taxonomic occurrences that come from the distribution charts in the scientific results of the DSDP/ODP volumes. Apart from the species name, additional data recorded include the taxon's highest abundance within that unit (if recorded) and whether or not the occurrence is considered questionable. The abundance and preservation quality for the unit as a whole is also recorded where given and is based on the best sample in the unit. This is done for coccolithophores and planktic foraminifera separately, as are counts of the total number of samples within the unit as well as the number that are fossiliferous. Additional data assigned to a unit includes: the presence of other taxa, including reworked or indeterminate coccolithophores/planktic foraminifera, the reference(s) from which the data came, the taxonomist(s) responsible for the species occurrence data and lithological data, including the presence of glauconite. Reference data is linked to a separate bibliography file created in BibTeX (http://www.bibtex.org/).

Data entry proceeded with a mixture of manual entry and parsing from online resources. Several pre-existing web depositories house digitized versions of DSDP/ODP distribution charts, including NEPTUNE (http://paleodb.org/cgi-bin/bridge.pl?a=displayDownloadNeptuneForm), JANUS (http://www-odp.tamu.edu/database/) and ODSN (http://www.odsn.de/). These were parsed into the correct format for our database using custom written R (version 2.11.1; R Development Core Team 2010) code and checked for potential errors. Error checking was done by using a conservative range-through database provided to us by Paul Bown (pers. comm., 2010), the results of which are published in Bown et al. (2004). We flagged up all occurrences for shared taxa that definitively lie outside of the range of the Bown et al. (2004) data. We then consulted the original DSDP/ODP scientific results to check if the data was entered correctly (the range extension thus being considered 'real') or not. Using this procedure we found an alarmingly high error rate (c. 18% of all occurrences outside the Bown et al. 2004 range were not accurate representations of the original DSDP/ODP data). Some of these were simple misdatings of a unit and a handful were questionable occurrences that hadn't been flagged as such. However, the majority (c. 15%) were completely erroneous, bearing no resemblance to the published distribution chart. All errors were overwhelmingly concentrated in the NEPTUNE data and likely reflect the lengthy and complex history of this database. However, presently there is an effort underway to overhaul NEPTUNE (Lazarus 2011) and we have passed on our findings to them. For our database all such errors were corrected, in many cases leading to whole charts or units being re-entered.

At present the database is implemented in Microsoft Access with the data used in the analyses here coming from three separate queries: (1) a table of all species occurrences, (2) a table of units entered and, (3) a list of taxa and their statuses (see below). At present data entry is still underway, but we hope to ultimately migrate the database to MySQL and make it freely available online.

Analytical methods

Taxonomic standardization. Before analysing our data we standardize our taxonomy using a new list of valid, invalid and synonymized taxa originally based on the NEPTUNE database, but significantly overhauled by one of the authors (JRY). In the process of manual data entry we have additionally uncovered many names not included in the NEPTUNE list making our global nannofossil synonymy list the most comprehensive and up-to-date presently available. This list is stored in the main database, allowing data entry to proceed using the original names from the distribution charts. This procedure thus allows for a different future taxonomy to still be used should opinions on synonymy etc. change. For data analysis we adopt the following procedure: (1) synonyms are replaced with their senior counterparts, (2) any resulting duplicate occurrences are removed and, (3) invalid taxa, questionable occurrences, taxa whose status is presently considered unknown and cf. or aff. taxa are removed.

Creating time bins of equal duration and calculating error bars. Units are given numerical dates based on Gradstein et al. (2004) and TimeScale

Creator as follows. If only a nannofossil or planktic foraminiferal zonation is known then the top of the youngest and bottom of the oldest are used. If both zone types are present then the dates of the overlap are used, conferring greater precision. In some cases, however, the two zonations do not overlap (implying uncertainty). When this happens then the maximum possible age range is used. Finally, if the uncertainty between the maximum and minimum possible dates is large (>15 million years) then we remove that unit and its constituent taxa from the analyses.

As we are interested in counts of species-richness through time an appropriate time binning approach is required. However, nannofossil or foraminiferal zones are problematic to use as they vary considerably in length and are thus likely to give misleading results, with more taxa likely to accumulate in a longer bin than a shorter one. This problem was identified by Sepkoski & Koch (1996) who recommended using time bins of roughly equal length. Alroy *et al.* (2008) adopted such an approach by combining geological stages to get roughly 11 million-year time bins. Although this is appropriate for an overview of Phanerozoic macrofossils, or poorly time-constrained taxa such as dinosaurs (Lloyd *et al.* 2008) such coarse binning is unnecessary for the data used here. Instead we adopt the Alroy *et al.* (2008) approach, but combine biozones, instead of geological stages, to make roughly 3 million-year time bins. We made an additional modification to this approach however, which is to enforce the inclusion of major geological boundaries (the Jurassic/Cretaceous, Cretaceous/Palaeogene, Eocene/Oligocene and Palaeogene/Neogene). This is because these are often associated with major turnover events and a bin spanning such a boundary is thus likely to have artificially inflated diversity because of an extinction and recovery fauna being time-averaged together.

Even after clumping zones together it is inevitable that some units will lack the precise dating required to assign them to a single time bin. Previous workers have had diametrically opposed solutions to this quandry. For example, Alroy *et al.* (2008) simply ignore taxa that cannot be assigned to a single bin and don't count them. By contrast, vertebrate workers have tended to treat uncertainty instead as the range of a taxon, counting it in all bins it could *possibly* be in (e.g. Benton 1995; see Upchurch & Barrett 2005 for a justification of this approach). Here we regard both solutions to be somewhat extreme and prefer instead a method that is intended to better quantify this uncertainty. Firstly we assume that each unit really does belong to a single bin and assign it based on a randomization approach. This is done by picking a random number from a uniform distribution between the

oldest and youngest possible dates for the unit. We then assign the unit to a time bin based on this single date and perform all of the counts outlined below. We then repeat this procedure 1000 times and record the resulting mean and 95% confidence intervals for our counts. When plotted it can thus be clearly seen whether a rise or fall between successive bins is likely to be empirically real or within dating error. It is this procedure that is used to create the error bars on the graphs presented here and the mean values that are used in the tests below.

Picking a sampling proxy in the deep-sea. Here we use the number of DSDP/ODP sites that have yielded sediments dated to a specific time bin as a measure of sampling. Sites can be considered a good measure of sampling, as they are decided on *a priori* by the researchers on the DSDP/ODP leg. We use sites rather than holes, as although multiple holes are typically drilled at the same site appearing to represent additional sampling, these are actually just additional attempts to recover from a particular horizon. Furthermore there is zero redundancy in the database: there are no cases of entered core being drilled from the same depth at the same site. Consequently, our within bin sampling measure is the number of sites recording fossil-bearing rock of that age.

Correlation tests. In comparing our sampling measure against taxonomic richness we employ a Spearman rank test. These are performed for both the raw data (long-term correlation) and detrended data (short-term correlation). Here we use two different detrending methods, a first differences and a 5-bin moving average (following Smith & McGowan 2007). In both cases raw values were logged for plotting purposes, but this does not affect rank-based correlation. Here and elsewhere when zero-values were encountered these were treated as non-applicable, essentially removing them from the analysis. This was done to avoid the problem of a value of infinity being returned, is not expected to affect the correlations, and only occurs in the late Early Jurassic to early Middle Jurassic and once in the Upper Jurassic (9 out of 66 time-bins).

Modelling. Assuming that sampling is a major factor in producing observed taxonomic richness curves an interesting follow up question is: how much of the observed richness is unexplained by sampling? Smith & McGowan (2007) introduced a procedure to tackle this question that starts from the notion that sampling perfectly predicts observed richness. In other words, the smallest sample is matched up with the lowest observed richness, the second smallest with the second lowest and so on. A linear model is then fitted to this new data from which a function can be derived that allows us to predict the richness for a given sample. Subtracting these predicted values from those observed gives residuals that show either higher than expected (positive) or lower than expected (negative) richness. To further test if any residuals were significantly different than expected Barrett *et al.* (2009) simply looked for points outside two standard deviations of the mean.

Here we extend the method of Smith & McGowan (2007) and Barrett *et al.* (2009) as follows. Firstly, we consider more than just a linear model: we also fit logarithmic, exponential, hyperbolic, sigmoidal and polynomial models. The best model is then chosen using the sample size corrected Akaike information Criterion (AIC; Akaike 1973), the AIC_c (Johnson & Omland 2004). The AIC_c weighs both a model's fit (a close fit being best) and its complexity (a simple model being best). Second, we use the standard error and standard deviation from the model-fitting process instead of the standard deviation of the residuals (after subtracting the modelled estimate from the observed values) to assess statistically significant diversions. For further details see Lloyd (in press).

Subsampling. An alternative way to remove a sampling bias from a species-richness curve is to rarefy or subsample (e.g. Alroy *et al.* 2001, 2008). For each time bin a list is compiled by combining the species occurrences from each site that is represented. This is the occurrence list that the subsampling is based on and is different to the full inventory as we do not allow an occurrence to be counted more than once simply because multiple units from a site are present in that bin. An additional problem arises from the fact stated earlier that some units span more than one time bin, complicating the question of whether a unit (and its constituent taxa) is represented in that time bin. Previous workers (e.g. Alroy *et al.* 2001, 2008) have simply ignored such occurrences, removing them from the analysis. We dub this a 'minimum' approach. An alternative is to multiply this unit so that it occurs in every possible bin, a 'maximum' approach. Here we perform both in order to ascertain what difference it makes.

Having compiled these two sets of occurrence lists subsampling proceeded in the usual manner (see rarefaction by occurrences in Bush *et al.* 2004). Here 1000 replicates are performed in order to get a mean and 95% confidence interval that make up a rarefaction curve (samples v. species-richness). Diversity curves were then produced by recording the species-richness when the number of samples taken is equal to the bin with the fewest samples. We then removed the bin with the fewest

samples and repeated the process for the bin with the next fewest and so on until only a single bin (with the most samples) remains. The same procedure was applied to both the minimum and maximum lists.

All analyses were performed by importing the three SQL queries outlined above into the freely-available statistical programming package R 2.11.1 (R Development Core Team 2010). Custom-written code is available on request from one of the authors (GTL).

Results

Empirical pattern

Figure 1 shows our sampling proxy, number of ODP and DSDP sites, through time. Low Mesozoic values rise slightly going into the Palaeogene and then exponentially in the Neogene, with shorter-term fluctuations also apparent. Figure 2a shows species-richness through time. Intermittent Jurassic preservation is replaced by a sharp rise going into the Early Cretaceous followed by a period of large fluctuations before a latest Cretaceous drop. Richness rises sharply in the Palaeocene and then follows a plateau before rising again in the Miocene before a final Plio-Pleistocene drop. Short-term fluctuations are also evident. Genus-level richness, shown in Figure 2b, shows a different pattern. Again there is a sharp rise going into the Early Cretaceous, but this is something of a peak, with

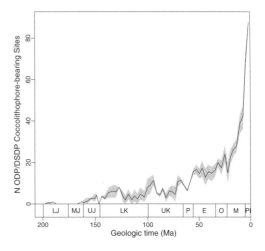

Fig. 1. Sampling proxy over time – the number of DSDP and ODP sites from which coccolithophore-bearing rock in a *c.* 3 Myr time bin have been recovered. Grey polygon shows 95% confidence interval based on 1000 iterations where uncertainly dated units are assigned at random to a time bin.

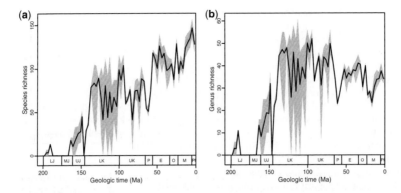

Fig. 2. Sampled coccolithophore richness through time: (**a**) species and (**b**) genera. Grey polygon shows 95% confidence interval as in Figure 1.

later richness slowly dropping off, with notable troughs in the Early Palaeocene and Early Miocene. Short-term fluctuations are again evident.

Correlation

Figure 3 shows the raw correlation between the number of sites and: (a) species-richness (Spearman $\rho = 0.94$, $P < 0.001$) and, (b) genus-level richness (Spearman $\rho = 0.37$, $P = 0.005$). Figure 4a, b shows the first differences – the absolute rise or fall over time – of number of sites (blue line) and: (a) species-richness (yellow line; Spearman $\rho = 0.87, P < 0.001$) and, (b) genus-level richness (black line; Spearman $\rho = 0.84$, $P < 0.001$). Figure 4c, d shows differences from a five-bin moving average for number of sites (blue line) and: (c) species-richness (yellow line; Spearman $\rho = 0.90$, $P < 0.001$) and, (d) genus-level richness (yellow line; Spearman $\rho = 0.87$, $P < 0.001$). Overall correlation is very high – very unlikely to be due to chance

alone – and consistent regardless of taxonomic level or detrending method. The only exception here is raw genus-level richness, which although much less strongly correlated with sampling than species-richness, is still statistically significant.

Corrected pattern

Figure 5 shows taxon richness ((a) = species, (b) = genus) estimates modelled from sampling, assuming that true diversity is constant and hence driven purely by sampling. Figure 6 shows the same data, but with observed diversity detrended by the model predictions. Figure 7 summarizes the subsampled results. Here we choose to show only two plots, both based on the maximum approach and recording the results for a relatively high sample count ($N = 109$). We chose this figure as it gives continuous results (a result for successive bins) from the Early Cretaceous to present. (Higher figures meant bins with insufficient numbers of

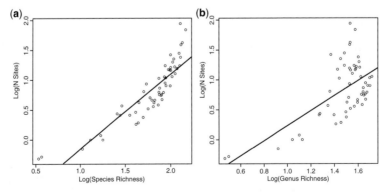

Fig. 3. Long term correlation – $\log_{10}$ number of DSDP and ODP sites against: (**a**) $\log_{10}$ species-level richness and (**b**) $\log_{10}$ genus-level richness.

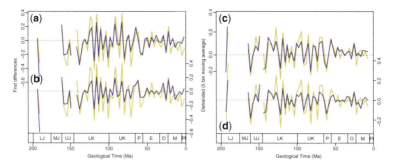

Fig. 4. Short term correlation time series – first differences of number of sites (blue) and: (**a**) species-richness (yellow) and (**b**) genus-level richness (yellow); 5-bin moving average detrended number of DSDP and ODP sites (blue) and: (**c**) species-richness (yellow) and (**d**) genus-level richness (yellow).

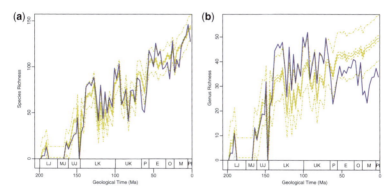

Fig. 5. Predicted richness based on a model (see Lloyd, in press) (yellow) against sampled richness (blue) for: (**a**) species and (**b**) genera. Dashed yellow line is a 95% confidence interval based on 1.96 standard errors, dash-dot yellow line is a 95% confidence interval based on 1.96 standard deviations (see Lloyd, in press).

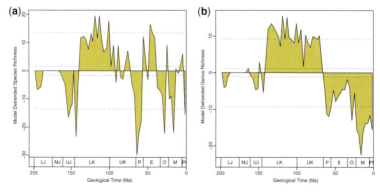

Fig. 6. Sampling model corrected richness for: (**a**) species and (**b**) genera. Confidence intervals are as in Figure 5.

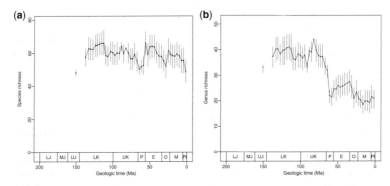

Fig. 7. Subsampling corrected richness at 109 species occurrences based on the maximum or optimistic approach (see text) for: (**a**) species and (**b**) genera. Vertical lines indicate 95% confidence interval based on 1000 iterations.

species occurrences could not be used, creating gaps that make interval-to-interval changes harder to interpret.) The pattern of roughly flat richness for species and declining richness for genera is still present in the minimum approach results, but there are a greater number of bins that return no result.

Discussion

Sampling and observed diversity are highly correlated

The clearest conclusion from our results is that our sampling proxy and observed diversity are very highly correlated. Indeed, this correlation is stronger than those obtained for marine invertebrates (Peters & Foote 2001; Smith & McGowan 2007) and even exceeds those for many terrestrial vertebrates (Barrett *et al.* 2009; Butler *et al.* 2009; Mannion *et al.* 2011). There are likely multiple explanations for this pattern. Firstly, unlike Peters & Foote (2001) and Smith & McGowan (2007) our sampling proxy and taxonomic richness are taken from the same database, and thus this is not a regional v. global comparison that is unlikely to be particularly strongly correlated. Secondly, our sampling proxy is, we contend, a much more appropriate measure of sampling (as discussed above) than those used previously and thus we are more likely to be capturing a true sampling signal rather than some other geological or anthropogenic measure. Thirdly, our data does not include Lagerstätten that can introduce outliers that confound correlation and give the false impression that correlation isn't as strong as it really is. (This is particularly true for vertebrates and future work requires methods for dealing with this such as that proposed in Cavin & Forey 2007.) The strength of this correlation, and its pervasiveness over both the short- and long-term, is the strongest evidence yet that the fossil record

should not be interpreted at face value and further that the deep-sea record is beset by the same biases as that of the land-based record.

Corrected diversity and biological signal

Despite these strong correlations our data do show evidence of sampling-independent diversity excursions. Even in the uncorrected data it is clear that in the Neogene, where sampling is exceptionally high, observed species-richness starts to become decoupled from sampling. For example, the Miocene shows a progressive rise in sampling (Fig. 1), but something of a dip in richness (Fig. 2) and in the Plio–Pleistocene where sampling rockets upwards (Fig. 1) richness shows a downturn (Fig. 2). This inversion of ups and downs is more clearly visible in the short-term correlation plots (Fig. 4). It appears then, that correlations would be diminished if the Neogene were considered separately, and this notion is supported by a Mesozoic/ Cenozoic partition. When considered separately the relatively poorly sampled Mesozoic shows a higher Spearman ρ (0.94) compared to the better-sampled Cenozoic (0.73). (The Neogene was not considered on its own as it includes too few data points for a meaningful statistical correlation.) These results are thus consistent with the notion that when sampling passes a particular threshold it no longer influences the pattern.

Additional sampling-independent signals are evident from the model-corrected measure used here. These include lower than expected (based on the standard deviation confidence interval) species-richness (Fig. 6a) in the middle Upper Jurassic, lowest Cretaceous, lowest Palaeogene, middle Oligocene, middle Miocene and in the Plio-Pleistocene. Many of these correspond to either hypothesized extinction events or periods of low speciation (see Bown *et al.* 2004; their Fig. 3) and the largest drop occurs with the largest extinction at the Cretaceous/Palaeogene boundary. Comparatively few points show significantly greater than expected diversity, however: just three in the Cretaceous and one in the Eocene. For genus-level richness the picture is notably different with a clear three-phase signal in the residuals (Fig. 6b) indicating that the model is not accurately capturing the data. The first phase, corresponding to the Jurassic, is one of constant diversity with the model fitting well. However, the second is a plateau of greater than expected diversity that lasts for almost the entire Cretaceous with several points significantly higher than expected. The third and final phase is one of lower than expected diversity, with a clear declining pattern that lasts for the entire Cenozoic. Again many points are significant excursions from the model. Taken together these results bear comparison with other studies of coccolithophore diversity (e.g. Bown et al. 2004), where diversity rises to a Late Cretaceous peak followed by a more stable Cenozoic low (their Fig. 2).

Some of these signals are also evident in the subsampling approach, although this is perhaps the most conservative correction (Alroy 2010). The species-richness curve (Fig. 7a) seems consistent with a good fit of the constant diversity model with a near flat trajectory from the Lower Cretaceous to the Plio-Pleistocene, suggesting the empirical graph of rising species-richness (Fig. 2a) is highly misleading. However, there are some medium-term features of note. These include a slight rising trend in the middle part of the Lower Cretaceous, a declining trend in the Upper Cretaceous, depressed diversity in the Palaeocene, a declining trend through the Eocene and Oligocene and finally a declining trend in the Neogene. However, in most cases the 95% confidence intervals between successive bins are overlapping. More substantial trends are apparent in the genus-level richness curve (Fig. 7b), most obviously a dichotomy between a Mesozoic plateau of high richness and a Cenozoic plateau of significantly lower richness. Interestingly though, the biggest drop between these two levels actually occurs within the Palaeocene, suggesting this shift is not directly linked to the Cretaceous–Palaeogene extinction. Finer scale changes include more volatile short-term trends in the Upper Cretaceous, a rising trend through the Palaeogene that is curtailed at the Eocene/Oligocene boundary and generally lower richness in the Neogene than at any preceding time. Comparing the two richness corrected curves suggests greater congruence at the genus- than species-level, perhaps due to the conservative nature of the subsampling.

An interesting conclusion from the corrected approaches used here is that all of them suggest that there is either no rise or a fall in both species

and genus-level richness between the Jurassic and the Pleistocene. However, there is no doubting that coccolithophores are a morphologically diverse group and consequently a strict biological interpretation of this pattern suggests that peak coccolithophore diversity must have been established very early in the clade's history. This conclusion stands in stark contrast to the results of face-value range-through curves such as those of Bown et al. (1992, 2004) that show a gradual, additive rise to a peak in species-richness in the Upper Cretaceous. Our results thus support similar findings by Alroy et al. (2008) when comparing marine invertebrate curves. This study thus suggest that the genus-level richness curve (Fig. 2b) may be a more accurate description of coccolithophore diversity trends, with a relatively rapid initial rise followed by a gradual depletion of diversity, seemingly supporting the use of higher taxa in palaeobiodiversity estimates.

The authors would like to thank A. J. McGowan and R. Benson for their helpful reviews and S. Peters for the R function used to add the geological epochs to the x-axes of Figures 1, 2 and 4–7. This research was supported by NERC grant NE/F016905/1 to ABS, JRY and P. Pearson.

References

AKAIKE, H. 1973. Information theory and an extension of the maximum likelihood principle. In: PETROV, B. N. & CSÁKI, F. (eds) 2nd International Symposium on Information Theory. Akadémiai Kiadó, Budapest, 267–281.

ALROY, J. 2010. The shifting balance of diversity among major marine animal groups. Science, 329, 1191–1194.

ALROY, J., MARSHALL, C. R. ET AL. 2001. Effects of sampling standardization on estimates of Phanerozoic marine diversification. Proceedings of the National Academy of Sciences of the United States of America, 98, 6261–6266.

ALROY, J., ABERHAN, M. ET AL. 2008. Phanerozoic trends in the global diversity of marine invertebrates. Science, 321, 97–100.

BARRETT, P. M., MCGOWAN, A. J. & PAGE, V. 2009. Dinosaur diversity and the rock record. Proceedings of the Royal Society, B, 276, 2667–2674.

BENSON, R. B. J. & BUTLER, R. J. 2011. Uncovering the diversification history of marine tetrapods: ecology influences the effect of geological sampling bias. In: MCGOWAN, A. J. & SMITH, A. B. (eds) Comparing the Geological and Fossil Records: Implications for Biodiversity Studies. Geological Society, London, Special Publications, 358, 191–207.

BENSON, R. B. J., BUTLER, R. J., LINDGREN, J. & SMITH, A. S. 2010. Mesozoic marine tetrapod diversity: mass extinctions and temporal heterogeneity in geological megabiases affecting vertebrates. Proceedings of the Royal Society, B, 277, 829–834.

BENTON, M. J. 1995. Diversity and extinction in the history of life. Science, 268, 52–58.

BOWN, P. R., BURNETT, J. A. & GALLAGHER, L. T. 1992. Calcareous nannoplankton evolution. *Memorie di Scienze Geologiche*, **43**, 1–17.

BOWN, P. R., LEES, J. A. & YOUNG, J. R. 2004. Calcareous nannoplankton evolution and diversity through time. *In*: THIERSTEIN, H. R. & YOUNG, J. R. (eds) *Coccolithophores: from Molecular Processes to Global Impact*. Springer Verlag, Berlin, Germany, 481–508.

BUSH, A. M., MARKEY, M. J. & MARSHALL, C. R. 2004. Removing bias from diversity curves: the effects of spatially organized biodiversity on sampling-standardization. *Paleobiology*, **30**, 666–686.

BUTLER, R. J., BENSON, R. J., NOWBATH, S. & UPCHURCH, P. 2009. Estimating the effects of sampling biases on pterosaur diversity patterns: implications for hypotheses of bird/pterosaur competitive replacement. *Paleobiology*, **35**, 432–446.

BUTLER, R. J., BARRETT, P. M., CARRANO, M. T., MANNION, P. D. & UPCHURCH, P. in press. Sea level, dinosaur diversity and sampling biases: investigating the 'common cause' hypothesis in the terrestrial realm. *Proceedings of the Royal Society, B*. (doi: 10.1098/rspb.2010.1754)

CAVIN, L. & FOREY, P. L. 2007. Using ghost lineages to identify diversification events in the fossil record. *Biology Letters*, **3**, 201–204.

CRAMPTON, J. S., BEU, A. G., COOPER, R. A., JONES, C. M., MARSHALL, B. & MAXWELL, P. A. 2003. Estimating the rock volume bias in paleobiodiversity studies. *Science*, **301**, 358–360.

EBRA, E. 2004. Calcareous nannofossils and Mesozoic oceanic anoxic events. *Marine Micropaleontology*, **52**, 85–106.

EBRA, E., BOTTINI, C., WEISSERT, H. J. & KELLER, C. E. 2010. Calcareous nannoplankton response to surface water acidification around Anoxic Event 1a. *Science*, **329**, 428–432.

FROBISCH, J. 2008. Global taxonomic diversity of anomodonts (Tetrapoda, Therapsida) and the terrestrial rock record across the Permo-Triassic boundary. *PLoS One*, **3**, e3733.

GOMITZ, V. (ed.) 2009. *Encyclopedia of Paleoclimatology and Ancient Environments*. Springer, Dordrecht.

GRADSTEIN, F. M., OGG, J. G. & SMITH, A. G. (eds) 2004. *A Geologic Time Scale 2004*. Cambridge University Press, Cambridge.

IRIGOIEN, X., HUISMAN, J. & HARRIS, R. P. 2005. Global biodiversity patterns of marine phytoplankton and zooplankton. *Nature*, **429**, 863–867.

JOHNSON, J. B. & OMLAND, K. S. 2004. Model selection in ecology and evolution. *Trends in Ecology and Evolution*, **19**, 101–108.

LAZARUS, D. 1994. Neptune: a marine micropaleontology database. *Mathematical Geology*, **26**, 817–832.

LAZARUS, D. B. 2011. The deep-sea microfossil record of macroevolutionary change in plankton and its study. *In*: MCGOWAN, A. J. & SMITH, A. B. (eds) *Comparing the Geological and Fossil Records: Implications for Biodiversity Studies*. Geological Society, London, Special Publications, **358**, 141–165.

LLOYD, G. T. in press. A refined modelling approach to assess the influence of sampling on palaeobiodiversity curves. *Biology Letters*.

LLOYD, G. T., DAVIS, K. E. *ET AL.* 2008. Dinosaurs and the Cretaceous Terrestrial Revolution. *Proceedings of the Royal Society, B*, **275**, 2483–2490.

MANNION, P. D., UPCHURCH, P., CARRANO, M. T. & BARRETT, P. M. 2011. Testing the effect of the rock record on diversity: a multidisciplinary approach to elucidating the generic richness of sauropodomorph dinosaurs through time. *Biological Reviews*, **86**, 157–181; doi: 10.1111/j.1469-185X.2010.00139.x

MARX, F. G. 2009. Marine mammals through time: when less is more in studying palaeodiversity. *Proceedings of the Royal Society, B*, **276**, 887–892.

MCGOWAN, B. 2005. *Biostratigraphy: Microfossils and Geological Time*. Cambridge University Press, Cambridge.

MCGOWAN, A. J. & SMITH, A. B. 2008. Are global Phanerozoic marine diversity curves truly global? A study of the relationship between regional rock records and global Phanerozoic marine diversity. *Paleobiology*, **34**, 80–103.

MURRAY, J. & RENARD, A. F. 1891. *Report on deep-sea deposits based on the specimens collected during the voyage of H.M.S. Challenger in the years 1872 to 1876*. Challenger Reports, H.M.S.O., Edinburgh.

PETERS, S. E. & HEIM, N. A. 2011. Macrostratigraphy and macroevolution in marine environments: testing the common-cause hypothesis. *In*: MCGOWAN, A. J. & SMITH, A. B (eds) *Comparing the Geological and Fossil Records: Implications for Biodiversity Studies*. Geological Society, London, Special Publications, **358**, 95–104.

PETERS, S. E. 2005. Geologic constraints on the macroevolutionary history of marine animals. *Proceedings of the National Academy of Sciences of the United States of America*, **102**, 12326–12331.

PETERS, S. E. 2006. Macrostratigraphy of North America. *Journal of Geology*, **114**, 391–412.

PETERS, S. E. 2008. Environmental determinants of extinction selectivity in the fossil record. *Nature*, **454**, 626–629.

PETERS, S. E. & FOOTE, M. 2001. Biodiversity in the Phanerozoic: a reinterpretation. *Paleobiology*, **27**, 583–601.

PETERS, S. E. & FOOTE, M. 2002. Determinants of extinction in the fossil record. *Nature*, **416**, 420–424.

R DEVELOPMENT CORE TEAM. 2010. *R: A Language and Environment for Statistical Computing*. R Foundation for Statistical Computing, Vienna, Austria. http://www.R-project.org.

RAUP, D. M. 1972. Taxonomic diversity during the Phanerozoic. *Science*, **177**, 1065–1071.

RAUP, D. M. 1976. Species diversity in the Phanerozoic: an interpretation. *Paleobiology*, **2**, 289–297.

SEPKOSKI, J. J. & KOCH, C. F. 1996. Evaluating paleontologic data relating to bioevents. *In*: WALLISER, H. (ed.) *Global Events and Event Stratigraphy*. Springer, Berlin, 21–34.

SEPKOSKI, J. J., BAMBACH, R. K., RAUP, D. M. & VALENTINE, J. W. 1981. Phanerozoic marine diversity and the fossil record. *Nature*, **293**, 435–437.

SMITH, A. B. 2001. Large-scale heterogeneity of the fossil record: implications for Phanerozoic biodiversity studies. *Philosophical Transactions of the Royal Society of London, B*, **356**, 351–367.

SMITH, A. B. 2007. Marine diversity through the Phanerozoic: problems and prospects. *Journal of the Geological Society, London,* **164**, 731–745.

SMITH, A. B. & McGOWAN, A. J. 2007. The shape of the Phanerozoic marine palaeodiversity curve: how much can be predicted from the sedimentary rock record of western Europe? *Palaeontology,* **50**, 765–774.

SPENCER-CERVATO, C. 1998. Changing depth distribution of hiatuses during the Cenozoic. *Paleoceanography,* **13**, 178–182.

SUCHÉRAS-MARX, B., MATTIALI, E., PITTET, B., ESCARGUEL, G. & SUAN, G. 2010. Astronomically-paced coccolith size variations during the early Pliensbachian (Early Jurassic). *Palaeogeography, Palaeoclimatology, Palaeoecology,* **295**, 281–292.

UPCHURCH, P. & BARRETT, P. M. 2005. Phylogenetic and taxic perspectives on sauropod diversity. *In*: CURRY ROGERS, K. A. & WILSON, J. A. (eds) *The Sauropods: Evolution and Paleobiology.* University of California Press, Berkeley, 104–124.

UPCHURCH, P., MANNION, P. D., BENSON, R. B. J., BUTLER, R. J. & CARRANO, M. T. 2011. Geological and anthropogenic controls on the sampling of the terrestrial fossil record: a case study from the Dinosauria. *In*: McGOWAN, A. J. & SMITH, A. B. (eds) *Comparing the Geological and Fossil Records: Implications for Biodiversity Studies.* Geological Society, London, Special Publications, **358**, 209–240.

WANG, P.-X., ZHAO, Q.-H. ET AL. 2003. Thirty-million year deep sea records in the South China Sea. *Chinese Science Bulletin,* **48**, 2524–2535.

WALL, P. D., IVANY, L. C. & WILKINSON, B. H. 2009. Revisiting Raup: exploring the influence of outcrop area on diversity in light of modern sample-standardization techniques. *Paleobiology,* **35**, 146–167.

WINTER, A., JORDAN, R. W. & ROTH, P. H. 1994. Biogeography of living coccolithophorids in ocean waters. *In*: WINTER, A. & SIESSER, W. G. (eds) *Coccolithophores.* Cambridge University Press, Cambridge, 161–177.

ZIVERI, P., BAUMANN, K.-H., BÖCKEL, B., BOLLMANN, J. & YOUNG, J. R. 2004. Present day coccolithophore biogeography of the Atlantic Ocean. *In*: THIERSTEIN, H. R. & YOUNG, J. R. (eds) *Coccolithophores: From Molecular Processes to Global Impact.* Springer, Berlin, 403–427.

Collateral mammal diversity loss associated with late Quaternary megafaunal extinctions and implications for the future

ANTHONY D. BARNOSKY[1,2]*, MARC A. CARRASCO[1] & RUSSELL W. GRAHAM[3]

[1]*Department of Integrative Biology and Museum of Paleontology, University of California, Berkeley, California 94720, USA*

[2]*Museum of Paleontology, University of California, Berkeley, California 94720, USA*

[3]*Department of Geosciences and Earth and Mineral Sciences Museum, The Pennsylvania State University, University Park, Pennsylvania, USA*

*Corresponding author (e-mail: barnosky@berkeley.edu)

Abstract: Using data from two palaeontological databases, MIOMAP and FAUNMAP (now linked as NEOMAP), we explore how late Quaternary species loss compared in large and small mammals by determining palaeospecies-area relationships (PSARs) at 19 temporal intervals ranging from *c.* 30 million to 500 years ago in 10 different biogeographical provinces in the USA. We found that mammalian diversity of both large and small mammals remained relatively stable from 30 million years ago up until both crashed near the Pleistocene–Holocene transition. The diversity crash had two components: the well-known megafaunal extinction that amounted to *c.* 21% of the pre-crash species, and collateral biodiversity loss due to biogeographical range reductions. Collateral loss resulted in large mammal diversity regionally falling an additional 6–31% above extinction loss, and small mammal diversity falling 16–51%, even though very few small mammals suffered extinction. These results imply that collateral losses due to biogeographical range adjustments may effectively double the regional diversity loss during an extinction event, substantially magnifying the ecological ramifications of the extinctions themselves. This is of interest in forecasting future ecological impacts of mammal extinctions, given that *c.* 8% of USA mammal species, and 22% of mammal species worldwide, are now considered 'Threatened' by the IUCN.

Previously, we determined that throughout the Holocene (from *c.* 11 000 to 500 years ago) biodiversity of mammals in the USA has been 15–42% too low, depending on biogeographical province, with respect to the pre-Holocene baseline that had existed for millions of years (Carrasco *et al.* 2009). The Holocene biodiversity decline was associated with the widely-recognized Late Quaternary Extinction (LQE) event, which affected primarily mammals (and a few birds and reptiles) >44 kg in body weight and was globally time-transgressive. The LQE began in Australia some 50 000 years ago, became most intense in temperate (our study area) and high latitudes near the Pleistocene–Holocene boundary, and then diminished into the early, middle, and late Holocene (Martin 1966; Martin & Wright 1967; Martin & Klein 1984; Barnosky *et al.* 2004; Wroe *et al.* 2004; Koch & Barnosky 2006; Wroe & Field 2006; Brook *et al.* 2007; Barnosky & Lindsey 2010). Today extinction appears to be accelerating again, as indicated by elevated extinction rates over the past few centuries, and high numbers of species threatened with extinction due to human activities (Myers 1990; Leakey & Lewin 1992; May *et al.* 1995; Pimm *et al.* 1995; Dirzo & Raven 2003; Wake & Vredenburg 2008; Barnosky *et al.* 2011).

Whether or not humans were the primary cause of the onset of LQE in various regions still engenders debate, but recent treatments tend to recognize at least some role for *Homo sapiens* as a driver, with details of timing and intensity being controlled by complex synergies between human population sizes, timing and magnitude of climate change, and ecological attributes of species (Barnosky *et al.* 2004; Wroe *et al.* 2004; Koch & Barnosky 2006; Wroe & Field 2006; Brook *et al.* 2007, 2008; Barnosky 2008; Field *et al.* 2008; Barnosky & Lindsey 2010).

Whatever the ultimate cause of the LQE, it offers a natural experiment to assess ecological effects that result from extinction (Blois *et al.* 2010). Here we examine some of those effects, by using palaeospecies-area relationships to more fully characterize the diversity loss that occurred on a continental scale and at regional scales. Exploring these details is pertinent to understanding the biotic impacts that would ensue if currently threatened species (particularly mammals), as defined by the

From: McGowan, A. J. & Smith, A. B. (eds) *Comparing the Geological and Fossil Records: Implications for Biodiversity Studies.* Geological Society, London, Special Publications, **358**, 179–189.
DOI: 10.1144/SP358.12   0305-8719/11/$15.00 © The Geological Society of London 2011.

International Union for Conservation of Nature (IUCN) (Mace *et al.* 2008; IUCN 2010), in fact did go extinct.

We focus on mammals from the lower 48 states of the USA for several reasons. First, mammals were the primary victims at the LQE near the end of the Pleistocene, and once again are at risk of extinction in significant numbers: the IUCN has assessed all 5490 known species and classified *c.* 22% worldwide and *c.* 8% of the 439 USA species as threatened (Mace *et al.* 2008; IUCN 2010). For comparison, during the LQE *c.* 5% of species worldwide, and 19% of USA species, went extinct. Third, and critically for our purposes, there is a relatively good fossil record of mammals from the USA that is accessible through palaeobiological databases for past times up to 500 years ago, and which can be adjusted for spatial and temporal sampling inconsistencies (FAUNMAP Working Group 1994; MIOMAP 2010; NEOMAP 2010; NEOTOMA 2010; PaleoDB 2010). Finally, mammal taxonomy, both fossil and modern, has been relatively stable compared to many other kinds of organisms, and different species concepts produce broadly overlapping results, thus facilitating comparison of deep-time and near-time samples.

Methods

Our general approach was to construct palaeospecies-area relationships (PSARs) for a variety of time-intervals from *c.* 30 million to 500 years ago (Table 1) and biogeographical provinces (Fig. 1). We then compared the PSARs from the Holocene time interval ('anthropogenic' interval of Carrasco *et al.* 2009) to baseline PSARs derived from pre-Holocene time intervals ('pre-anthropogenic' of Carrasco *et al.* 2009, in the sense of lacking substantial numbers of humans in our study area). We did this for all mammals in the USA, for large mammals only (as defined below), and for small mammals only, using unnested PSARs that analyzed the data using methods detailed below.

Databases

Species occurrence data were extracted from three databases: MIOMAP (Carrasco *et al.* 2007; MIOMAP 2010), FAUNMAP I (FAUNMAP Working Group 1994), and FAUNMAP II (as of August, 2010) (NEOMAP 2010). MIOMAP spans the period from 5 to 30 million years ago, FAUNMAP II from 40 000 to 5 million years ago, and FAUNMAP I from 500 to 40 000 years ago. We combined FAUNMAP I and FAUNMAP II data into a single database and served it through the Neogene Mammal Mapping Portal (NEOMAP 2010), which we

Table 1. *Temporal bins into which species occurrences were sorted*

Time interval	Age boundaries	Interval duration
Holocene	*c.* 11 500–500	*c.* 11 000
Rancholabrean	0.15 Ma–*c.* 11 500	*c.* 140 000
Irvingtonian	1.8–0.15 Ma	1.65 Ma
Blancan	4.7–1.8 Ma	2.9 Ma
Late late Hemphillian	5.9–4.7 Ma	1.2 Ma
Early late Hemphillian	6.7–5.9 Ma	0.8 Ma
Late early Hemphillian	7.5–6.7 Ma	0.8 Ma
Early early Hemphillian	9–7.5 Ma	1.5 Ma
Late Clarendonian	10–9 Ma	1.0 Ma
Middle Clarendonian	12–10 Ma	2.0 Ma
Early Clarendonian	12.5–12 Ma	0.5 Ma
Late Barstovian	14.8–12.5 Ma	2.3 Ma
Early Barstovian	15.9–14.8 Ma	1.1 Ma
Late Hemingfordian	17.5–15.9 Ma	1.6 Ma
Early Hemingfordian	18.8–17.5 Ma	1.3 Ma
Late late Arikareean	19.5–18.8 Ma	0.7 Ma
Early late Arikareean	23.8–19.5 Ma	4.3 Ma
Late early Arikareean	27.9–23.8 Ma	4.1 Ma
Early early Arikareean	30–27.9 Ma	2.1 Ma

created to facilitate uniform searches and output from MIOMAP and FAUNMAP (http://ucmp.berkeley.edu/neomap/use.html). We used the online routines in NEOMAP to generate species counts and geographical areas that were ultimately used in the analysis (a slightly different version of the FAUNMAP data also is served online as part

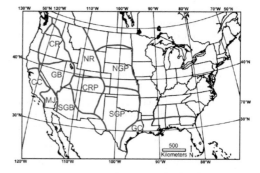

Fig. 1. Biogeographical provinces used in this study. Central California (CC); Columbia Plateau (CP); Colorado Plateau (CRP); Great Basin (GB); Gulf Coast (GC); Mojave (MJ); Northern Great Plains (NGP); Northern Rockies (NR); Southern Great Basin (SGB); Southern Great Plains (SGP).

of the NEOTOMA database effort (NEOTOMA 2010), which combines several Quaternary fossil databases).

Biogeographical provinces

Although Quaternary mammal fossils are distributed essentially continent-wide, most of the Cenozoic North American mammalian fossil record is best represented in the USA west of the Mississippi River (Tedford *et al.* 1987, 2004). Therefore, we concentrated on this region, dividing it into 10 biogeographical provinces (Fig. 1). Today, these provinces are considered biogeographically distinct from one another (Hagmeier & Stults 1964; Hagmeier 1966; Janis *et al.* 1998; Lugo *et al.* 1999), and it is likely that the same held true back through the Oligocene (Tedford *et al.* 1987; Storer 1989; FAUNMAP Working Group 1996; Barnosky & Carrasco 2002; Tedford *et al.* 2004). This is especially true in those provinces with the most complete

fossil record (e.g. Northern Great Plains), as they have undergone limited topographic change over the past 30 million years (Prothero 1998; Condon 2005).

Large v. small mammal samples

The species data from each time interval and biogeographical province were divided into two separate groups: large mammals and small mammals (Table 2). Small mammals included all members of the mammalian orders Rodentia, Insectivora, and Lagomorpha, that is, animals with body mass generally <2 kg (though a few rodents and lagomorphs exceed 2 kg). Large mammals comprised all other non-volant, terrestrial mammalian orders. Because of their limited representation in the databases, Chiropterans were eliminated from the analysis. For counts of extant species in the USA, we used IUCN data (IUCN 2010) as summarized in Table 2. Counts of extinct species and their approximate

Table 2. *Extant USA species per clade and total extinct USA species since 50 thousand years ago*

Size	Clade	Species	
Large extant	Carnivora	53	
Large extant	Cetartiodactyla	62	
Large extant	Cingulata	1	
Large extant	Didelphimorpha	1	
Small extant	Eulipotyphla	43	
Small extant	Lagomorpha	19	
Small extant	Rodentia	211	
Not included	Chiroptera	49	
Totals			
Extant	Minus Chiroptera	390	
Large extant	Minus Chiroptera	117	
Small extant	Minus Chiroptera	273	
Extant	With Chiroptera	439	
Large extant	With Chiroptera	117	
Small extant	With Chiroptera	322	
Extinct	Minus Chiroptera	106	
Large extinct	Minus Chiroptera	100	
Small extinct	Minus Chiroptera	6	
% Extinct			
Extant + extinct	With Chiroptera	545	19.45
Large extant + extinct	With Chiroptera	217	46.08
Small extant + extinct	With Chiroptera	328	1.83
Extant + extinct	Minus Chiroptera	496	21.37
Large extant + extinct	Minus Chiroptera	217	46.08
Small extant + extinct	Minus Chiroptera	279	2.15

Small 1/4 less than *c.* 2 kg body mass; Large 1/4 greater than *c.* 2 kg.
Data for extant mammals are from the IUCN Red List of Threatened Species, Version 2010.4, www.iucnredlist.org, downloaded 13 Dec. 2010 (IUCN, 2010). Not included in this table are Sirenia (2 spp.) and Primates (1 sp.). The IUCN lists 37 (includes chiropterans) or 32 (excludes chiropterans) USA species as Threatened.
Data for extinct species primarily from references in Barnosky *et al.* (2011) and Smith *et al.* (2003). The IUCN lists 2 spp. extinct in historic times; these are included in the totals, of extinct species.

body sizes were compiled from references detailed in previous publications (Smith *et al.* 2003; Barnosky *et al.* 2011).

Species counts

Species data were tabulated using the MIOMAP EstimateS (Colwell 2009) web service function via the Berkeley Mapper (http://berkeleymapper. berkeley.edu), which produced a table of the minimum number of individuals (MNI) for each taxon by locality. After exporting the EstimateS table to Microsoft Excel, the MNI data were adjusted to occurrence (presence/absence) data. Within individual localities, generic or higher level occurrences were eliminated when a more precise taxonomic assignment in that locality was present (e.g. the occurrence *Canis* sp. would be eliminated from a locality if *Canis dirus* was known from the same locality). Thus we used minimum counts of species. Occurrence data were then added across all localities for the given time period and geographical province. Only minimum counts as opposed to maximum counts (all specimens identified to genus or higher taxon assigned to a new species) were employed here as previous work showed little difference among these two counting methods (Barnosky *et al.* 2005).

Geographical area calculation

For each time interval, the geographical area encompassed by the sample was determined by using the routines in NEOMAP to zoom in to a scale that included all pertinent localities, trace the minimum convex polygon that would include all the localities of interest, and calculate the area enclosed by the polygon. Geographical areas were calculated using the Berkeley Mapper mapping interface.

Temporal binning

Methods based on taxon co-occurrences have been developed to sort fossil occurrences into one million year intervals (Alroy 1992, 1996, 1998, 1999, 2003). However, these methods were not appropriate for our study because they reduce the number of localities per time slice such that not enough data exists for many time slices when dividing the record into discrete biogeographical regions, and also can introduce false precision for localities that are not well dated and that have few taxa. Therefore, we assigned fossil occurrences to one of 19 subdivisions of the North American Land Mammal Ages (Table 2) as specified in Tedford *et al.* (2004) for pre-Blancan time intervals and FAUNMAP I (FAUNMAP Working Group 1994) for post-Blancan temporal bins.

Durations of temporal bins are not equal, but we determined this has little influence on diversity counts per time period because (i) there is no correlation between bin length and number of localities (Barnosky *et al.* 2005) or species richness (Carrasco *et al.* 2009); (ii) the localities do not span the entire time represented by each bin but instead subsample discrete times within bins, thus correcting for number of localities as described below also to some extent corrects for temporal variations; and (iii) bins of the sort used here, based on maximum taxon associations, are best suited to comparisons of diversity through time, as they produce a series of biologically meaningful groupings that do not change much within each bin (Tedford *et al.* 2004).

We limited our Holocene sample to contain localities older than 500 years, in order to use only fossil data and therefore make the Holocene sample comparable to the pre-Holocene sample. Therefore, both Holocene and pre-Holocene data were assembled in the same way: primarily through fossils reported over the past century by scientists employing similar collecting methodologies. Nevertheless, Holocene samples were often larger, which would be expected to result in a greater number of total species in each biogeographical province. Because prior statistical analyses have demonstrated a significant correlation within each temporal bin between the total number of species and the rarefied species richness as well as the total number of occurrences (Carrasco *et al.* 2009), the palaeospecies-area relationships from the Holocene bin should have higher diversity if this bias were present. Therefore, any results that reveal a *reduction* in species diversity during the Holocene should be particularly robust.

Sampling biases and sample standardization

There are well-recognized sampling problems that must be adjusted for when using fossil data to assess diversity, including differences in sampling intensity and geographical areas sampled for each time slice (Barnosky *et al.* 2005). To standardize samples, we computed species richness values per time slice and per geographical area (for both small and large mammals) by rarefaction using a richness value of 75 taxon occurrences. Rarefaction of the raw minimum species counts was accomplished with S. Holland's analytical rarefaction software (http://www.uga.edu/~strata/software/). A review of the development of rarefaction methodology can be found in Tipper (1979) while the programs we used were ultimately based on the rarefaction work of (Raup 1975), and originally derived by Hurlbert (1971) and Heck *et al.* (1975). The data was rarefied by occurrences instead of the number of individual specimens to remove the effect of high-graded localities and missing data (Barnosky *et al.* 2005). We set the rarefaction occurrence value at 75 because that value provided an adequate number of data points while at the same time eliminating samples that were based on spotty data.

Constructing palaeospecies-area Relationships (PSARs)

The rarefied species richness data were plotted against sampled area to determine palaeospecies-area relationships (PSARS) (Barnosky *et al.* 2005). PSARs were determined at two different geographical scales: continental and provincial. The continental analysis plots continental species richness per time interval against area per time interval for all intervals for which data existed (Figs 2–4). The provincial analysis plots species richness per biogeographical province against sampled area within the province for each time interval for which data were available (Figs 5 & 6). Thus each data point in the graphs represents a single time slice for either all 10 provinces combined (Figs 2–4) or for individual provinces (Figs 5 & 6).

PSARs in this paper correspond to Type IV unnested species-area relationships (SARs) (Scheiner 2003). They simply plot species richness against geographical area for each area sampled. These differ from nested Type I SARs, in that they do not represent species accumulation curves, an important distinction in interpreting the data. In a nested Type I SAR, the expectation is that adding more area will always result in adding more species, because as new species are encountered as one expands the sampling into new habitats, the species are added to the list that has already accumulated by previous sampling efforts. Thus, as beta diversity (the difference in species composition between sites) increases, so does the slope of the nested SAR. In contrast, in unnested Type IV curves, which we use here, the relationship shows how many species are in a given sampling area, but the species in one area may be entirely the same as those in another, entirely different, or somewhere in between. Therefore the slope of an unnested Type IV curve does not reflect beta diversity in a straightforward way, because there is no embedded information about species identity. Likewise, the slope of the unnested SAR may deviate from positive, if, for example, different biogeographical provinces are being compared, and diversity is markedly different within each because of provincial environmental constraints. One might imagine, for example, equally sized areas, one in a productive, topographically diverse mountainous area

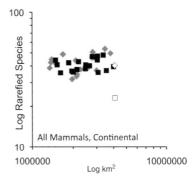

Fig. 2. Type IV palaeospecies-area relationship for all mammals at the continental scale. Squares = large mammals, black from pre-Holocene time intervals, white from Holocene. Diamonds = small mammals, grey from pre-Holocene time intervals, white from Holocene. Each point represents one time interval and encompasses all fossil localities known from that time interval.

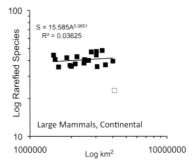

Fig. 3. Type IV palaeospecies-area relationship for large mammals at the continental scale. Black symbols are pre-Holocene time intervals, white symbol is Holocene. Each point represents one time interval and encompasses all fossil localities known from that time interval.

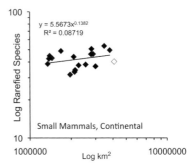

Fig. 4. Type IV palaeospecies-area relationship for small mammals at the continental scale. Black symbols are pre-Holocene time intervals, white symbol is Holocene. Each point represents one time interval and encompasses all fossil localities known from that time interval.

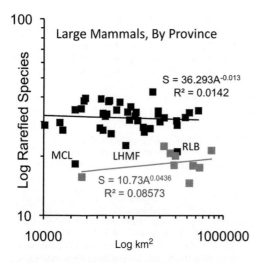

Fig. 5. Type IV palaeospecies-area relationship for large mammals per biogeographical province. Black symbols are pre-Holocene time intervals, grey symbols are Holocene. Each point represents one species richness value from a single time interval within a single biogeographical province. MCL, Middle Clarendonian from Central California; LHMF, Late Hemingfordian from the Northern Rocky Mountains; RLB, Rancholabraean from the Colorado Plateau.

with overall high species richness, and the other in an adjacent flat desert, with overall low species richness. In that case, a larger area in the desert may still sample fewer species than a smaller area in the

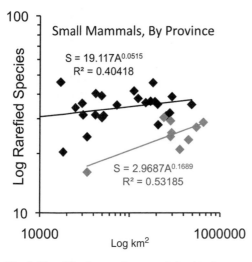

Fig. 6. Type IV palaeospecies-area relationship for small mammals per biogeographical province. Black symbols are pre-Holocene time intervals, grey symbols are Holocene. Each point represents one species richness value from a single time interval within a single biogeographical province.

mountains. For that reason the key information in unnested analyses that compare different provinces, as we do in some cases, is not the slope of the SAR, but whether the diversity values per area for a given time slice are higher or lower than for comparative time slices.

For Type IV SARS within a biogeographical province, one would expect an increase of species with an increase of area, up until a large enough part of the province was sampled to include most species, at which point adding area no longer adds species and the curve flattens. Therefore, a series of samples all from areas that are larger than that critical value, even within a province, would be expected to produce a flat SAR, even if there was variation in sampling area.

We plotted separate PSARs for the Holocene data; these are the PSARs that characterize the time humans were abundant in our study area. We then compared the Holocene PSARs to corresponding ones for pre-Holocene (pre-anthropogenic) times. This separation of the data is justified by previous work (Carrasco *et al.* 2009), which through analysis of nested Type I and unnested Type IV PSARS demonstrated that PSARs for various pre-Holocene time intervals did not differ significantly from one another, but Holocene PSARs plotted significantly lower than pre-Holocene PSARs.

Results

The continental-scale analysis highlights that prior to the Holocene, PSARS for both small and large mammals were similar (Figs 2–4). However, while Holocene large mammals demonstrate a prominent diversity crash (Fig. 3) at this scale, the small mammals maintain diversity similar to that predicted from pre-Holocene PSARs (Fig. 4). This is consistent with previous knowledge about the LQE, long recognized to have caused extinctions primarily of the megafauna (Martin 1966; Martin & Wright 1967; Martin & Klein 1984; Barnosky *et al.* 2004, 2008; Koch & Barnosky 2006; Barnosky & Lindsey 2010).

The LQE worldwide took place largely between *c.* 50 000 and 8000 years ago and in our study area seems concentrated mostly between 13 500– 11 000 years ago. During the LQE the USA lost *c.* 106 of *c.* 545 total species (*c.* 19%) estimated to have inhabited the continent prior to the extinction episode (Table 2). This estimation of total species is simply 106 extinct by the Pleistocene–Holocene transition (*c.* 11 000 years ago), plus 439 extant today (including Chiropterans), which are presumed also to have been on the continent during the late Pleistocene. However, the fossil sample excludes Chiropterans; extracting them from the extant sample as well results in an extinction percentage of c. 21% (Table 2).

Looking only at the large mammal subset of our data (that is, larger than rabbits) there are 100 species that went extinct during the LQE out of 217 total species (extinct plus extant). Thus, c. 46% of large mammal species suffered extinction during the LQE. The PSARs for the large mammal subset of the total data indicates that diversity dropped by 45% then (Table 3; this is the percentage difference between the species richness value expected from the pre-Holocene PSAR v. the observed Holocene value). These two different ways of calculating extinction intensity agree well, and indicate that essentially all of the diversity drop recognized from the large-mammal PSAR analysis can be accounted for by actual extinction of large mammals during the LQE.

However, the large-mammal PSARs constructed at the provincial level (Fig. 5) suggest loss by extinction did not evenly affect biogeographic provinces across the continent. Depending on the province, large mammal diversity is 27% to 52% (average 39%) lower than one would expect based on the pre-Holocene PSARs (Table 3). This is 6–31% more reduction, depending on province, than the loss predicted by extinction alone.

The pre-Holocene large-mammal PSAR at this scale exhibits flattening, probably for reasons elaborated in the Methods section. It is also notable that points for the MCL (see Fig. 5 for abbreviations) in Central California, the RLB in the Colorado Plateau, and to a lesser extent the LHMF in the Northern Rockies are more characteristic of the Holocene PSAR than the pre-Holocene PSAR. The LHMF previously was shown to have both a relatively small sampling area, and low overall diversity, in both the Northern Rockies and Great Plains (Barnosky et al. 2005). Also the boundary between the LHMF and the preceding Early Hemingfordian marks one of the most impressive faunal turnover events in the northern Rockies in the last 30 million years, including high extinction, immigration, and emigration (Barnosky 2001). It is unclear whether the LHMF decline in large mammal diversity is related to this faunal turnover event, or to the small sampling area. Sampling area cannot account for the low large-mammal diversity in the

Table 3. *Expected v. observed rarefied species richness values for large and small mammals per province and for the continent*

Province	Expected	Actual	% Decline
Small mammals			
CC	32.7	16.1	50.8
Colorado	36.6	25.5	30.3
CP	36.5	24.4	33.2
GB	36.5	29.3	19.8
GC	37.0	21.0	43.2
NGP	37.4	23.5	37.1
NR	38.1	28.8	24.4
SGB	36.2	30.5	15.7
SGP	37.8	27.2	28.0
Continental	45.7	40.3	11.8
Province average			31.4
Province regression	$S = 19.117A^{0.0515}$		
Continental regression	$S = 5.5673A^{0.1382}$		
Large mammals			
CC	31.8	15.7	50.6
Colorado	30.8	18.0	41.6
CP	30.8	20.1	34.8
GB	30.9	20.7	32.9
GC	30.7	14.6	52.4
NGP	30.6	17.9	41.6
NR	30.4	21.4	29.7
SGB	30.9	22.5	27.3
SGP	30.6	17.5	42.7
Continental	42.0	23.1	45.0
Province average			39.3
Province regression	$S = 36.293A^{-0.013}$		
Continental regression	$S = 15.585A^{0.0651}$		

MCL of Central California, because this time is characterized by one of the largest sampling areas, yet one of the lowest diversity values. Whether this anomalously low point indicates a real, biologically significant provincial diversity decline, or some sort of taphonomic bias, remains to be determined. While the RLB low point could indicate that large mammal diversity declined in the Colorado Plateau slightly earlier than the Holocene decline elsewhere, it is more likely due to an anomalous sampling situation, in which many localities consist of a single occurrence, generally of a large mammal such as *Bison* or *Mammuthus* (Carrasco *et al.* 2009).

Only six small mammal species went extinct during the LQE (1.1% of the small plus large species and 2.2% of the small species subset). The loss by extinction is so small that in the PSARs a diversity drop is barely apparent (and not statistically below expectations) at the continental scale (Figs 2 & 4), yet the diversity drop at the provincial level is dramatic, between 16% and 51% (average 31%), depending on province (Fig. 6, Table 3). This suggests that geographical range changes accompanying extinction episodes may dramatically reduce biodiversity provincially even in mammal groups that are little culled by extinction itself, as anticipated from\ previous studies (Graham 1976, 1985).

Discussion

Spurious results introduced by sampling problems are always an issue in using palaeontological data as we have. In our study, perhaps the largest red flag is that most of the pre-Holocene time intervals are so much longer than the Holocene one. If species were accumulating through evolutionary replacement in the time averaged by the samples, one might expect higher diversity in the longer time intervals. For reasons outlined in the methods, this does not seem to be the case for the composite large-mammal plus small-mammal sample (Carrasco *et al.* 2009). To further explore this possibility for the separated large- and small-mammal samples, we examined the correlation between temporal duration and diversity for the non-Holocene continental sample and for the provincial sample in each of the two body-size classes, using a power function (given the order-of-magnitude differences in lengths of time bins), for time intervals ranging from 4.3 million to 140 000 years. We found no significant correlation between diversity and bin length for either large or small mammals at the continental or provincial scale, with *P*-values of 0.12 and 0.28 for large mammals (continental and provincial scales, respectively) and 0.99 (continental) and 0.49 (provincial) for small mammals. Further verifying that temporal interval is not the best

explanation for our results is the much more comprehensive sample and more highly resolved taxonomy for Holocene mammals; as noted in the Methods section, one would expect this to elevate Holocene diversity, so depressed Holocene diversity is a particularly robust result.

Given these considerations, it seems likely that geographical range shifts, especially contractions, explain most of the apparent diversity crash of small mammals at the provincial scale. In effect, the diversity loss becomes apparent at the provincial scale because even relatively small range shifts (relative to the continental scale) can retract a species distribution outside the province. This finding is consistent with other studies that have demonstrated dramatic geographical-range shifts of mammals at the Pleistocene–Holocene transition (Graham 1976; Graham & Grimm 1990; Graham 1997; Lyons 2003, 2005; Lyons *et al.* 2010), and Holocene diversity declines in small mammals at the local scale (Blois *et al.* 2010) and continental scale (FAUNMAP Working Group 1996). At the local scale, such declines have been related to abundance increases in 'weedy' species as ecologically restricted species move out of a given area (Blois *et al.* 2010), whereas at the continental scale, the explanations have revolved around decreasing environmental heterogeneity (Graham & Lundelius 1984; Guthrie 1984). It is also tempting to speculate a feedback between the removal of large mammals and the distributional patterns of small mammals, because large mammals act as ecosystem engineers to help maintain mosaic habitats in landscapes where they are abundant (Owen-Smith 1987).

Conclusions

Previous work (Carrasco *et al.* 2009) showed that observed Holocene mammalian diversity (e.g. the diversity baseline traditionally assumed to be the 'normal' one) in the USA is actually between 15% and 42% too low, depending on biogeographical province, with respect to diversity expected from PSARs. That study did not differentiate how much of the diversity decline was due to actual extinction, and how much resulted from reducing the average size of geographical ranges such that intra-provincial diversity fell even though species have survived. Our results shed light on that distinction through separating the diversity reductions by body size and by looking at them on both a continental and provincial biogeographical scale.

Our analyses suggest that at extinction events such as those that took place during the LQE, diversity declines in two ways. There is loss by extinction itself, but there also is collateral diversity loss within biogeographical provinces as the surviving fauna rearrange their geographical ranges. For small mammals, that collateral diversity loss reduced modern mammalian diversity in every biogeographical province we analyzed to levels far below the pre-Holocene baseline, coincident with the late Pleistocene megafaunal extinctions, even though very few small mammals went extinct. The result is that contemporary mammal communities have between 16 and 51% fewer small-bodied species than was typical before the large-mammal extinction event (Table 3), as well as 6–31% fewer large-mammal species than can be accounted for by extinction alone.

Besides shedding light on how mammalian diversity reduced in the USA during the LQE, our data suggest that similar collateral biodiversity loss should probably be anticipated as a by-product of future mammal extinctions – a very real possibility inasmuch as 8% of USA species and 22% of species worldwide are currently regarded as threatened. Should these species actually be lost, the diversity decline at the scale of biogeographical provinces might well be much more than the percentage of species-reduction caused by the extinctions themselves, if the LQE is any guide. The fact that future extinctions will take place on a landscape much more fragmented by human modifications than was the case at the LQE would probably exacerbate collateral losses on local and regional scales. We qualify our conclusions by noting they are based on a single large clade, mammals, from a single large geographical region, the lower 48 states of the USA. It will be of interest to learn whether these inferences from our data agree with future studies on other taxa from other regions.

We thank A. Smith and A. J. McGowan for making it possible for us to participate in this symposium. Research funding was provided in part by the National Science Foundation (DEB-0543641 and EAR Geoinformatics Program 0622349 and 0622289), the University of California Museum of Paleontology (UCMP), The Pennsylvania State University, and the Illinois State Museum. This is UCMP Contribution 2027.

References

ALROY, J. 1992. Conjunction among taxonomic distributions and the Miocene mammalian biochronology of the Great Plains. *Paleobiology*, **18**, 326–343.

ALROY, J. 1996. Constant extinction, constrained diversification, and uncoordinated stasis in North American mammals: new perspectives on faunal stability in the fossil record. *Palaeogeography, Palaeoclimatology, Palaeoecology*, **127**, 285–311.

ALROY, J. 1998. Equilibrial diversity dynamics in North American mammals. *In:* MCKINNEY, M. L. & DRAKE, J. A. (eds) *Biodiversity Dynamics, Turnover of Populations, Taxa, and Communities.* Columbia University Press, New York, 232–287.

ALROY, J. 1999. The fossil record of North American mammals: evidence for a Paleocene evolutionary radiation. *Systematic Biology*, **48**, 107–118.

ALROY, J. 2003. Taxonomic inflation and body mass distributions in North American fossil mammals. *Journal of Mammalogy*, **84**, 431–443.

BARNOSKY, A. D. 2001. Distinguishing the effects of the Red Queen and Court Jester on Miocene mammal evolution in the northern Rocky Mountains. *Journal of Vertebrate Paleontology*, **21**, 172–185.

BARNOSKY, A. D. 2008. Megafauna biomass tradeoff as a driver of Quaternary and future extinctions. *Proceedings of the National Academy of Sciences USA*, **105**, 11543–11548.

BARNOSKY, A. D. & CARRASCO, M. A. 2002. Effects of Oligo-Miocene global climate changes on mammalian species richness in the northwestern quarter of the USA. *Evolutionary Ecology Research*, **4**, 811–841.

BARNOSKY, A. D. & LINDSEY, E. L. 2010. Timing of Quaternary megafaunal extinction in South America in relation to human arrival and climate change. *Quaternary International*, **217**, 10–29.

BARNOSKY, A. D., KOCH, P. L., FERANEC, R. S., WING, S. L. & SHABEL, A. B. 2004. Assessing the causes of late Pleistocene extinctions on the continents. *Science*, **306**, 70–75.

BARNOSKY, A. D., CARRASCO, M. A. & DAVIS, E. B. 2005. The impact of the species-area relationship on estimates of paleodiversity. *PLoS Biology*, **3**, 1356–1361.

BARNOSKY, A. D., MATZKE, N. ET AL. 2011. Has the Earth's sixth mass extinction already arrived? *Nature*, **471**, 51–57.

BLOIS, J. L., MCGUIRE, J. L. & HADLY, E. A. 2010. Small mammal diversity loss in response to late-Pleistocene climatic change. *Nature*, **465**, 771–774.

BROOK, B. W., BOWMAN, D. M. J. S. ET AL. 2007. Would the Australian megafauna have become extinct if humans had never colonised the continent? *Quaternary Science Reviews*, **26**, 560–564.

BROOK, B. W., SODHI, N. S. & BRADSHAW, C. J. A. 2008. Synergies among extinction drivers under global change. *Trends in Ecology & Evolution*, **23**, 453–460.

CARRASCO, M. A., BARNOSKY, A. D., KRAATZ, B. P. & DAVIS, E. B. 2007. The Miocene Mammal Mapping Project (MIOMAP): an online database of Arikareean through Hemphillian fossil mammals. *Bulletin of the Carnegie Museum of Natural History*, **39**, 183–188.

CARRASCO, M. A., BARNOSKY, A. D. & GRAHAM, R. W. 2009. Quantifying the extent of North American mammal extinction relative to the pre-anthropogenic baseline. *PLoS One*, **4**, Article No.: e8331, doi: 10.1371/journal.pone.0008331.

COLWELL, R. K. 2009. *EstimateS: Statistical estimation of species richness and shared species from samples.* Version 8.2. User's Guide and application, http://viceroy.eeb.uconn.edu/estimates

CONDON, S. M. 2005. Geological studies of the Platte River, south-central Nebraska and adjacent areas – geologic maps, subsurface study, and geologic history. *U. S. Geological Survey Professional Paper*, **1706**, 1–63.

DIRZO, R. & RAVEN, P. H. 2003. Global state of biodiversity and loss. *Annual Review of Environment and Resources*, **28**, 137–167.

FAUNMAP WORKING GROUP. 1994. FAUNMAP: A Database Documenting Late Quaternary Distributions of Mammal Species in the United States. *Illinois State Museum Scientific Papers*, **25**, 690.

FAUNMAP WORKING GROUP. 1996. Spatial response of mammals to late Quaternary environmental fluctuations. *Science*, **272**, 1601–1606.

FIELD, J., FILLIOS, M. & WROE, S. 2008. Chronological overlap between humans and megafauna in Sahul (Pleistocene Australia–New Guinea): a review of the evidence. *Earth-Science Reviews*, **89**, 97–115.

GRAHAM, R. W. 1976. Late Wisconsin mammalian faunas and environmental gradients of the eastern United States. *Paleobiology*, **2**, 343–350.

GRAHAM, R. W. 1985. Diversity and community structure of the late Pleistocene mammal fauna of North America. *Acta Zoologica Fennica*, **170**, 181–192.

GRAHAM, R. W. 1997. The spatial response of mammals to Quaternary climate changes. *In*: HUNTLEY, B., CRAMER, W., MORGAN, A. V., PRENTICE, H. C. & SOLOMON, A. M. (eds) *Past and Future Rapid Environmental Changes: the Spatial and Evolutionary Responses of Terrestrial Biota*. NATO ASI Series 1: Global Environmental Change, **47**. Springer, Berlin, 153–162.

GRAHAM, R. W. & GRIMM, E. C. 1990. Effects of global climate change on the patterns of terrestrial biological communities. *Trends in Ecology and Evolution*, **5**, 289–292.

GRAHAM, R. W. & LUNDELIUS, E. L. JR. 1984. Coevolutionary disequilibrium and Pleistocene extinction. *In*: MARTIN, P. S. & KLEIN, R. G. (eds) *Quaternary Extinctions: a Prehistoric Revolution*. University of Arizona Press, Tucson, 223–249.

GUTHRIE, R. D. 1984. Mosaics, allelochemics, and nutrients: an ecological theory of Late Pleistocene megafaunal extinctions. *In*: MARTIN, P. S. & KLEIN, R. G. (eds) *Quaternary Extinctions: a Prehistoric Revolution*. University of Arizona Press, Tucson, 259–298.

HAGMEIER, E. M. 1966. A numerical analysis of the distributional patterns of North American mammals. II. Re-evaluation of the provinces. *Systematic Zoology*, **15**, 279–299.

HAGMEIER, E. M. & STULTS, C. D. 1964. A numerical analysis of the distributional patterns of North American mammals. *Systematic Zoology*, **13**, 125–155.

HECK, K. L. J., VAN BELLE, G. & SIMBERLOFF, D. 1975. Explicit calculation of the rarefaction diversity measurement and the determination of sufficient sample size. *Ecology*, **56**, 1459–1461.

HURLBERT, S. 1971. The nonconcept of species diversity: a critique and alternative parameters. *Ecology*, **52**, 577–586.

IUCN. 2010. *International Union for Conservation of Nature Red List*. International Union for Conservation of Nature, http://www.iucn.org/about/work/programmes/species/red_list/.

JANIS, C. M., SCOTT, K. M. & JACOBS, L. L. 1998. *Evolution of Tertiary Mammals of North America. Volume 1: Terrestrial Carnivores, Ungulates, and Ungulatelike Mammals*. Cambridge University Press, Cambridge.

KOCH, P. L. & BARNOSKY, A. D. 2006. Late Quaternary extinctions: state of the debate. *Annual Review of Ecology Evolution and Systematics*, **37**, 215–250.

LEAKEY, R. & LEWIN, R. 1992. *The Sixth Extinction: Patterns of Life and the Future of Humankind*. Doubleday, New York.

LUGO, A. E., BROWN, S. L., DODSON, R., SMITH, T. S. & SHUGART, H. H. 1999. The Holdridge life zones of the conterminous United States in relation to ecosystem mapping. *Journal of Biogeography*, **26**, 1025–1038.

LYONS, S. K. 2003. A quantitative assessment of the range shifts of Pleistocene mammals. *Journal of Mammalogy*, **84**, 385–402.

LYONS, S. K. 2005. A quantitative model for assessing the community dynamics of Pleistocene mammals. *The American Naturalist*, **165**, E1–E18.

LYONS, S. K., WAGNER, P. J. & DZIKIEWICZ, K. 2010. Ecological correlates of range shifts of Late Pleistocene mammals. *Philosophical Transactions of the Royal Society B*, **365**, 3681–3693.

MACE, G. M., COLLAR, N. J. *ET AL.* 2008. Quantification of extinction risk: IUCN's system for classifying threatened species. *Conservation Biology*, **22**, 1424–1442.

MARTIN, P. S. 1966. African and Pleistocene overkill. *Nature*, **212**, 339–342.

MARTIN, P. S. & KLEIN, R. D. 1984. *Quaternary Extinctions: a Prehistoric Revolution*. University of Arizona Press, Tucson.

MARTIN, P. S. & WRIGHT, H. E., JR. 1967. *Pleistocene Extinctions: the Search for a Cause. Volume 6 of the Proceedings of the VII Congress of the International Association for Quaternary Research*. Yale University Press, New Haven.

MAY, R. M., LAWTON, J. H. & STORK, N. E. 1995. Assessing extinction rates. *In*: LAWTON, J. H. & MAY, R. M. (eds) *Extinction Rates*. Oxford University Press, Oxford, 1–24.

MIOMAP. 2010. *Miocene Mammal Mapping Project*, http://www.ucmp.berkeley.edu/.

MYERS, N. 1990. Mass extinctions: what can the past tell us about the present and future? *Palaeogeography Palaeoclimatology Palaeoecology*, **82**, 175–185.

NEOMAP. 2010. *Neogene Mammal Mapping Portal*, http://www.ucmp.berkeley.edu/neomap/

NEOTOMA. 2010. *Paleoecology Database, Plio-Pleistocene to Holocene*, http://www.neotomadb.org/

OWEN-SMITH, N. 1987. Pleistocene extinctions: the pivotal role of megaherbivores. *Paleobiology*, **13**, 351–362.

PaleoDB. 2010. *The Paleobiology Database*, http://paleodb.org/cgi-bin/bridge.pl

PIMM, S. L., RUSSELL, G. J., GITTLEMAN, J. L. & BROOKS, T. M. 1995. The future of biodiversity. *Science*, **269**, 347–350.

PROTHERO, D. R. 1998. The chronological, climatic, and paleogeographic background to North American mammalian evolution. *In*: JANIS, C. M., SCOTT, K. M. & JACOBS, L. L. (eds) *Evolution of Tertiary Mammals of North America. Volume 1: Terrestrial Carnivores, Ungulates, and Ungulatelike Mammals*. Cambridge University Press, Cambridge, **1**, 9–36.

RAUP, D. M. 1975. Taxonomic diversity estimation using rarefaction. *Paleobiology*, **1**, 333–342.

SCHEINER, S. M. 2003. Six types of species-area curves. *Global Ecology & Biogeography*, **12**, 441–447.

SMITH, F. A., LYONS, S. K. *ET AL.* 2003. Body mass of late Quaternary mammals. *Ecology/Ecological Archives*, **84/E084-094-D1** 3403/3401–3417, doi:10.1890/02-9003

STORER, J. E. 1989. Rodent faunal provinces, Paleocene–Miocene of North America. *In*: BLACK, C. C. & DAWSON, M. R. (eds) *Papers on Fossil Rodents in Honor of Albert Elmer Wood*. Natural History Museum of Los Angeles County, Los Angeles, **33**, 17–29.

TEDFORD, R. H., ALBRIGHT, L. B., III *ET AL.* 2004. Mammalian biochronology of the Arikareean through Hemphillian interval (Late Oligocene through Early Pliocene Epochs). *In*: WOODBURNE, M. O. (ed.) *Late Cretaceous and Cenozoic Mammals of North America*. Columbia University Press, New York, 169–231.

TEDFORD, R. H., SKINNER, M. F. *ET AL.* 1987. Faunal succession and biochronology of the Arikareean through Hemphillian interval (late Oligocene through earliest Pliocene epochs) in North America. *In*: WOODBURNE,

M. O. (ed.) *Cenozoic Mammals of North America*. University of California Press, Berkeley, CA, 153–210.

TIPPER, J. C. 1979. Rarefaction and rarefiction – the use and abuse of a method in paleoecology. *Paleobiology*, **5**, 423–434.

VIÉ, J.-C., HILTON-TAYLOR, C. & STUART, S. N. (eds.), 2009. *Wildlife in a Changing World – An Analysis of the 2008 IUCN Red List of Threatened Species*. IUCN, Gland, Switzerland.

WAKE, D. B. & VREDENBURG, V. T. 2008. Are we in the midst of the sixth mass extinction? A view from the world of amphibians. *Proceedings of the National Academy of Sciences of the United States of America*, **105**, 11466–11473.

WROE, S. & FIELD, J. 2006. A review of the evidence for a human role in the extinction of Australian megafauna and an alternative interpretation. *Quaternary Science Reviews*, **25**, 2692–2703.

WROE, S., FIELD, J., FULLAGAR, R. & JERMIIN, L. S. 2004. Megafaunal extinction in the late Quaternary and the global overkill hypothesis. *Alcheringa*, **28**, 291–331.

Uncovering the diversification history of marine tetrapods: ecology influences the effect of geological sampling biases

ROGER B. J. BENSON[1]* & RICHARD J. BUTLER[2]

[1]*Department of Earth Sciences, University of Cambridge, Downing Street, Cambridge CB2 3EQ, UK*

[2]*Bayerische Staatssammlung für Paläontologie und Geologie, Richard-Wagner-Straße 10, 80333 Munich, Germany*

*Corresponding author (e-mail: rbb27@cam.ac.uk)

Abstract: Mesozoic terrestrial vertebrates gave rise to sea-going forms independently among the ichthyosaurs, sauropterygians, thalattosaurs, crocodyliforms, turtles, squamates, and other lineages. Many passed through a shallow marine phase before becoming adapted for open ocean life. This allows quantitative testing of factors affecting our view of the diversity of ancient organisms inhabiting different oceanic environments. We implemented tests of correlation using generalized difference transformed data, and multiple regression models. These indicate that shallow marine diversity was driven by changes in the extent of flooded continental area and more weakly influenced by uneven fossil sampling. This is congruent with studies of shallow marine invertebrate diversity and suggests that 'common cause' effects are influential in the shallow marine realm. In contrast, our view of open ocean tetrapod diversity is strongly distorted by temporal heterogeneity in fossil record sampling, and has little relationship with continental flooding. Adaptation to open ocean life allowed plesiosaurs, ichthyosaurs and sea turtles to 'escape' from periodic extinctions driven by major marine regressions, which affected shallow marine taxa in the Late Triassic and over the Jurassic–Cretaceous boundary. Open ocean taxa declined in advance of the end-Cretaceous extinction. Shallow marine taxa continued diversifying in the terminal stages due to increasing sea-level.

Supplementary material: The data series and full analytical results are available at http://www.geolsoc.org.uk/SUP18486

Documenting the diversification history of extinct organisms is a fundamental goal of palaeobiology (e.g. Valentine 1969; Sepkoski 1981, 1982), but recent work has raised serious concerns that uneven temporal sampling of the fossil record may limit our ability to distinguish genuine and artifactual patterns (e.g. Raup 1972, 1976; Peters & Foote 2001, 2002; Smith 2001, 2007; Smith *et al.* 2001; Crampton *et al.* 2003; Peters 2005; Smith & McGowan 2005, 2007; Alroy *et al.* 2008). Most work examining the influence of sampling on perceived diversity patterns has focused on the record of shallow marine invertebrates. Vertebrates have been relatively neglected, with the significant exceptions of Permo-Triassic Russian tetrapods (Benton *et al.* 2004), Cenozoic mammals (e.g. Alroy 2000; Uhen & Pyenson 2007; Marx 2008; Marx & Uhen 2010), anomodont therapsids (Fröbisch 2008), pterosaurs (Butler *et al.* 2009, 2011*b*) and dinosaurs (Wang & Dodson 2006; Lloyd *et al.* 2008; Barrett *et al.* 2009; Mannion *et al.* 2010; Butler *et al.* 2011*a*). As a result, studies of vertebrate palaeodiversity do not routinely consider sampling biases (Slack *et al.* 2006; Benton & Emerson 2007; Sahney *et al.* 2010) despite the fact that the vertebrate record (which is dominantly terrestrial) is generally thought to be less complete than that of shallow marine invertebrates (but see Benton 2001) and thus might be subject to more severe biases.

Vertebrates have the potential to provide a unique perspective on heterogeneity (temporal, spatial and ecomorphological) in the nature of sampling biases because they occupy a broad range of habitats, and thus depositional environments, and are ecomorphologically diverse. For example, contemporaneous groups of Mesozoic vertebrates occupied open ocean (e.g. plesiosaurians, chelonioid turtles), shallow marine (e.g. thalattosaurs, placodonts, see below), coastal and fully terrestrial (e.g. pterosaurs, dinosaurs) habitats. In addition, the marine tetrapods of the Mesozoic formed at least twelve independent radiations into the marine realm from terrestrial ancestors, often passing through a shallow marine phase early in their adaptation to open ocean life (Fig. 1; Storrs 1993*a*; Rieppel 2000; Bell & Polcyn 2005; Motani 2005, 2009).

From: McGowan, A. J. & Smith, A. B. (eds) *Comparing the Geological and Fossil Records: Implications for Biodiversity Studies.* Geological Society, London, Special Publications, **358**, 191–208.
DOI: 10.1144/SP358.13   0305-8719/11/$15.00 © The Geological Society of London 2011.

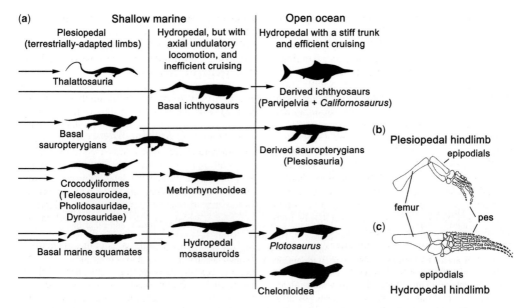

Fig. 1. The evolution of Mesozoic marine tetrapods. (**a**) schematic representation of transitions (represented by arrows) from the terrestrial to shallow marine and open ocean habitats by independent tetrapod groups. (**b, c**) diagrams showing plesiopedal (terrestrially-adapted), (b) (modified from Caldwell 1997, fig. 2K: *Serpianosaurus*, a basal sauropterygian), and hydropedal, (c) (modified from Caldwell 1997, fig. 3D: *Hydrorion*, a plesiosaurian), hindlimb morphologies.

The occurrence of multiple radiations means that inferences drawn from marine tetrapods may be independent of phylogenetic relationships. Although vertebrate fossils are less abundant than invertebrate fossils, exceptional professional and public interest in fossil vertebrates means that their taxonomy and spatiotemporal distributions are often well-understood, and ecological inferences based on their preserved anatomy are often well-constrained.

Here, we focus on the relationship between sampling and diversity for Mesozoic marine reptiles, building on a previous study (Benson *et al.* 2010). Most other previous studies of marine tetrapod diversity only considered sampling biases qualitatively (Bardet 1992, 1994; Pierce *et al.* 2009a (thalattosuchians); Young *et al.* 2010 (metriorhynchoids)). A study of mosasauroids using rarefaction to standardize sample size by (Ross 2009) is one exception, and Storrs (1993a) attempted to quantify the quality of the Triassic sauropterygian record. Our aim here is to test for and examine the nature of ecomorphological heterogeneity in sampling biases. This forms the basis for a refined understanding of trends in marine tetrapod diversity. Nearly all previous work on sampling biases and vertebrate diversity has ignored (or not explicitly considered) ecomorphological variation in the study taxa. Consequently, most hypotheses of Phanerozoic animal diversification are predominantly based on shallow marine invertebrates that make up the bulk of the

fossil record (e.g. Sepkoski 1981, 1982; Alroy *et al.* 2008). This may be problematic if open ocean or terrestrial taxa follow a different diversification trajectory.

Ecology of marine reptiles

Based on Carrier's (1987) observations of running lizards, Cowen (1996) recognized that rapid axial undulatory locomotion in aquatic vertebrates would have impaired the ability to breathe using paired, bilateral lungs. This would have limited the stamina of aquatic vertebrates propelled by axial undulation. This hypothesis is consistent with Massare's (1988) observation that axial undulatory swimmers such as mosasauroids and crocodyliforms have proportionally long, narrow bodies. In principle, this allowed them rapid bursts of acceleration but low sustained swimming speeds that limited them to 'ambush' predation in marginal and shallow marine environments. The axial undulatory mode was inherited from the terrestrial ancestors of most marine tetrapod groups and is therefore plesiomorphic (Cowen 1996; marine mammals, aquatic birds, and turtles are exceptions). Axial undulatory locomotion is inferred for all marine tetrapods with 'plesiopedal' (i.e. terrestrially-proportioned; Fig. 1b; Bell & Polcyn 2005) limbs, including thalattosaurs (e.g. Liu & Rieppel 2001; Jiang *et al.*

2004), basal mosasauroids and related squamate groups (Bell & Polcyn 2005), and most basal sauropterygians (Carroll 1985; Caldwell 1997; Rieppel 2000), as well as many hydropedal taxa (with limbs forming flippers; Fig. 1c) such as basal ichthyosaurs (McGowan 1991; Motani et al. 1996; Sander 2000; Motani 2005), derived mosasauroids (Massare 1988; Cowen 1996; other than *Plotosaurus*) and thalattosuchian crocodyliforms (Massare 1988).

In contrast, parvipelvian ichthyosaurs (e.g. Motani 2002a, b, 2005), plesiosaurians (Storrs 1993b; Rieppel 2000), turtles and the mosasauroid *Plotosaurus* (Lindgren et al. 2007) had appendicular- or caudally-driven locomotion and rigid trunks that in principle allowed efficient, cruising locomotion over long distances, sometimes at relatively high speeds (Massare 1988). These locomotory functional inferences, combined with facies data, have led several authors to suggest that basal representatives of most clades were limited to shallow water environments on the flooded continental shelf, whereas parvipelvian ichthyosaurs, plesiosaurians, turtles and *Plotosaurus* could cross the open ocean (Rieppel 2000; Motani 2002a, b, 2005; Lindgren et al. 2007).

Prediction

We predict that if the ecomorphological inferences discussed above are correct, then the actual taxic diversity of 'shallow marine' tetrapods should be strongly influenced by the extent of continental flooding, which provided habitable area for these taxa. Thus, the origination and extinction of shallow marine tetrapods may be controlled by eustatic sea-level changes. However, because of their inferred capacity to survive in the open ocean, the actual diversity of 'non-undulating' taxa should be less dependent upon shallow marine area and potentially show a stronger correlation with fossil sampling metrics.

Methods

Data series

All data series were assigned to stage-level time bins. The stratigraphic age and total duration of stages was taken from Walker & Geismann (2009). Mesozoic sea-level estimates were drawn from Miller et al. (2005, supplementary data) who provided two data series: one for the curve of Haq et al. (1987) covering the time period of 0–244 Ma and a novel one spanning 0–172 Ma. Because the points within these data series are not distributed evenly in time, we interpolated equally spaced data points onto linear segments spanning adjacent

points in the original data series at 0.1 million year intervals. To do this we used a freeware function (XlXtrFun) for Microsoft Excel that interpolates data using a third-order piecewise polynomial. We then calculated the mean sea-level for each of our time bins. Data on reconstructed total non-marine surface area were derived from Smith et al. (1994, table 3), who generated their data from a series of reconstructed global palaeogeographical maps. Non-marine surface area exactly corresponds to the continental area that is not flooded by shallow seas, and thus varies inversely with the amount of shallow marine habitat available. Non-marine area and sea-level estimates are used here as proxies for the area of shallow marine habitat available.

Taxic diversity counts were extracted from a modified version of the dataset of Benson et al. (2010), which is available as an online appendix to Benson et al. (2010) and on request from the authors. This includes species occurrences of chelonioid turtles, ichthyosaurs, mosasauroids and other marine squamates, sauropterygians, thalattosaurs and thalattosuchian, dyrosaurid and pholidosaurid crocodyliforms. These were compiled from recent taxonomic compendia or systematic assessments (e.g. Steel 1973; Hirayama 1997; Rieppel 2000; O'Keefe 2001; McGowan & Motani 2003; Bell & Polcyn 2005; Druckenmiller & Russell 2008; Hill et al. 2008; Jouve et al. 2008; Pierce et al 2009a, b; Young & Andrade 2009; Ketchum & Benson 2010; Young et al. 2010) and a review of the wider literature conducted by Benson et al. (2010). This data was modified by the removal of Jurassic 'plesiochelyid' turtles, the ecology of which is poorly understood (J. Anquetin pers. comm. 2010; although they may be marginal marine; Billione-Bruyat et al. 2005; Fuente & Fernandez 2011). Due to the small number of Jurassic turtle occurrences (one Oxfordian; two Tithonian), this is not expected to have a major impact on the data. This resulted in a total of 570 occurrences by stage. The total taxic diversity of marine tetrapods was divided into two non-overlapping subsets comprising 'open ocean' and 'shallow marine' taxa, identified by locomotor inferences (above). This resulted in three data series: total taxic diversity (TDE$_{total}$: Benson et al. 2010, fig. 2a), shallow marine taxic diversity (TDE$_{shallow\ marine}$: Fig. 2) and open ocean taxic diversity (TDE$_{open\ ocean}$: Fig. 3).

Counts of fossiliferous marine formations were used as a proxy for geological sampling of marine depositional environments and were downloaded from The Paleobiology Database (Benson et al. 2010; accessed 12th May 2009). Use of formation counts as a sampling proxy does not assume that all formations are equal, only that variation in weathering rates, outcrop area, thickness, lithostratigraphic research, and palaeontological sampling

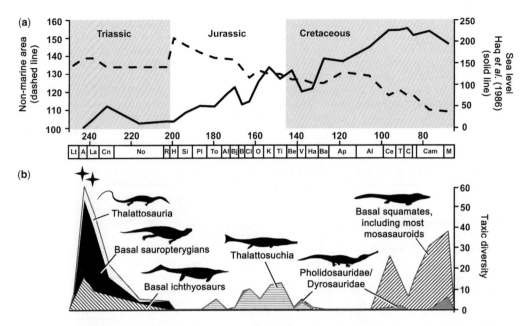

Fig. 2. Continental flooding proxies (**a**) and shallow marine taxic (**b**) diversity plotted against geological time (Ma; stage names abbreviated). Stars indicate shallow marine Lagerstätten stages explained in the text.

effort are distributed randomly and do not introduce systematic biases. The fossil record has been sampled over more than two centuries and historic collections are often sparsely documented. Thus, sampling effort can only rarely be measured directly. Although sampling proxies only provide an estimate of sampling effort, they are necessary in quantitative studies of ancient biotic diversity such as this one.

Taxon records and fossiliferous marine formations that span multiple stages were considered to occur within their entire range, even when the range represented uncertainty in dating or provenance (following the arguments of Upchurch & Barrett (2005, pp. 111–112)).

Lagerstätten

Benson *et al.* (2010) identified Lagerstätten within five Triassic–Jurassic intervals. Greater than half of marine tetrapod taxa from these stages were collected from a single formation, or a restricted geographical region characterized by intensive local sampling. Marine tetrapod Lagerstätten include the Anisian (55%) and Ladinian (66%) formations of central Europe (values in parentheses indicate the percentage of taxa from localized deposits), the primarily Sinemurian Lower Lias Group of the UK (100%), the Toarcian Posidonienschiefer Lagerstätte of Germany (52%), the Callovian Peterborough Member of the Oxford Clay Formation

(73%) and the primarily Kimmeridgian Kimmeridge Clay Formation of the UK (55%). Lagerstätten effects present a major challenge to palaeodiversity studies. Despite representing the most extreme form of uneven fossil sampling, Lagerstätten weaken the apparent correlation between most sampling proxies and observed palaeodiversity. They also spuriously inflate taxic diversity estimates for the intervals in which they occur. A basic approach to removing Lagerstätten effects from palaeodiversity data series is to simply exclude Lagerstätten data. Here we propose an alternative approach whereby Lagerstätten are coded as present or absent using a binary variable in multiple regression models (described below) as a coarse attempt to account for their presence without discarding data.

Pairwise tests of correlation

As a preliminary survey, we conducted multiple pairwise tests of correlation between our data series using Pearson's product moment (r). Non-parametric tests (Spearman's ρ, and Kendall's τ) were also performed to corroborate these analyses (Supplementary Material). Tests of correlation were implemented using the computer program PAST (Hammer *et al.* 2001), applied using both the raw data series, and modified versions of the data series after the application of generalized differencing (e.g. McKinney 1990; Alroy 2000; see below). Because of the strong influence of Jurassic

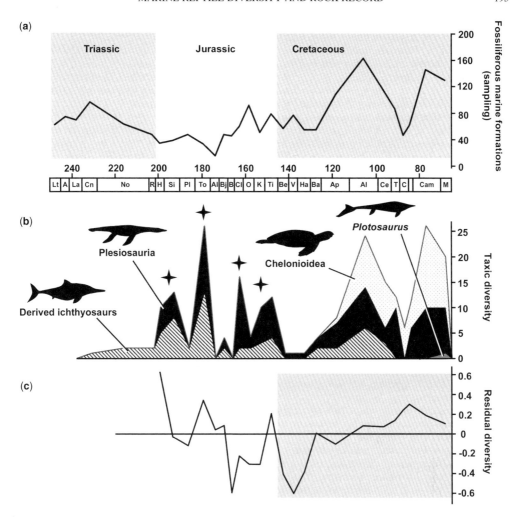

Fig. 3. Fossiliferous marine formations (**a**; sampling proxy), open ocean taxic diversity (**b**), and residual diversity after subtraction of a GLS multiple regression model including stage duration, deep water Lagerstätten, and sampling (**c**; Table 3) plotted against geological time (Ma; stage names abbreviated). Stars indicate deep water Lagerstätten stages explained in the text.

Lagerstätten on total taxic diversity and open ocean taxic diversity, tests of correlation with these data series were also performed excluding Lagerstätten stages. Because each data series was compared to a maximum of four other data series, the threshold for statistical significance was made more stringent ($\alpha = 0.05/4 = 0.0125$; Bonferroni correction) to avoid the increased risk of detecting spurious correlations as a result of making multiple comparisons. Triassic data were not analysed because our objective was to test the strength of correlations between taxic diversity and proxies representing sampling (fossiliferous marine formations) and the area of shallow marine shelf available (sea-level, non-marine area). The Triassic comprises only five

time bins of a total 28 in the Mesozoic and is marked by several events that could confound underlying correlations, including (1) the initial diversification of marine tetrapods, incorporating a necessary rise from zero diversity independent of external factors; (2) the invasion of the open ocean by thunniform or intermediate-grade ichthyosaurs from the Carnian onwards (Motani 2005) and plesiosaurians at least as early as the Rhaetian (e.g. Storrs 1994); (3) the early Late Triassic mass extinction among marine tetrapods (e.g. Bardet 1994; Rieppel 2000; Benson *et al.* 2010) incorporating a dramatic decline in diversity. Note that the Triassic data were included in our generalized least squares multiple regression analyses described below.

One assumption of tests of correlation is that values in each data series are drawn independently from one another. However, for time series data, such as those analysed here, the presence of long-term trend (i.e. a directed change in the mean value through time) or short-term autocorrelation (correlation between successive data points) may violate this assumption and cause overestimation of correlation coefficients. Generalized differencing is a two-stage approach in which data series are first detrended, then corrected for autocorrelation by differencing of successive values, modulated by the observed strength of autocorrelation between successive time bins (McKinney 1990; no differencing is applied if statistically significant autocorrelation is not detected). This technique was applied to fossil vertebrate diversity data by Alroy (2000) and Butler et al. (2011a), but most recent authors examining the link between fossil sampling and palaeodiversity in vertebrates have left the data uncorrected (Fröbisch 2008; Barrett et al. 2009; Butler et al. 2009; Mannion et al. 2010) or applied first differencing (Uhen & Pyenson 2007; Marx 2008). In first differencing, each data point is transformed by subtraction of the preceding datum. This results in a data series describing the change in an observation through time (e.g. Chatfield 2003). Benson et al. (2010, appendix S2) voiced concerns that first differencing may be an overcorrection of the data and result in loss of signal. However, application of generalized differencing addresses this concern (McKinney 1990; Alroy 2000).

The Jarque–Bera test indicated that all detrended and generalized difference transformed data series were distributed normally. The raw data series for sea-level estimates, non-marine area and fossiliferous marine formations were also distributed normally. However, the raw taxic diversity data series were significantly non-normal. Thus, the raw taxic diversity series were $\log_{10}$-transformed prior to tests of correlation, resulting in fully normal distributions.

Generalized differencing: implementation

Generalized differencing was implemented manually following the protocol described by McKinney (1990). Concomitant statistical procedures were carried out in PAST (Hammer et al. 2001). Initially, values of each data series were regressed against the midpoint ages of the time bins. In those cases where the Durbin–Watson test statistic indicated a significant fit of the least squares regression line, long-term trend was inferred. This was removed by subtraction of the regression slope from values in the data series, resulting in detrended versions of all series other than time bin duration and open ocean taxic diversity (which did not

show statistically significant evidence for long-term trend).

The presence of Lagerstätten, yielding high values of taxic diversity, and concentrated in the Triassic and Jurassic may affect estimates of trend in taxic diversity estimates (Benson et al. 2010). Thus, the regression slope (trend line) was also calculated excluding Jurassic stages in which Lagerstätten effects were observed by Benson et al. (2010; Sinemurian, Toarcian, Callovian, Kimmeridgian). In most cases, the regression slope calculated excluding Lagerstätten was closely similar to the slope calculated from the complete data series, as expected for sea-level, non-marine surface area and fossiliferous marine formations (these variables are not oversampled in Lagerstätten). Correspondingly, data series calculated using the two methods were strongly correlated (Pearson's $r > 0.99$) in most cases, including those for shallow marine taxic diversity (N.B. Jurassic Lagerstätten represent deep water facies; Hudson et al. 1991; Röhl et al. 2001; Martill et al. 2006). The correlation was lower for total taxic diversity ($r = 0.97$), and for open ocean taxic diversity a significant regression slope, indicating long-term trend, was only detected when Lagerstätten were excluded. This indicates a strong impact of Lagerstätten on the inference of long-term trend in total and open ocean taxic diversities. Thus, the trend slope estimated excluding Lagerstätten was used exclusively for these latter two data series (although pairwise correlation test were applied both to the full data series, and to the data series excluding Lagerstätten; see Results). Next, the autocorrelation coefficient at a time lag of one interval was estimated by regressing the values of each data series (t_i) against a series comprising values from the immediately preceding time bins (t_{i-1}). A significant fit of the least squares regression line was absent for most data series, including both open ocean and total taxic diversity (both including or excluding Lagerstätten stages). A significant fit was indicated for non-marine area and shallow marine taxic diversity, for both of which generalized differences were thus calculated. The slope of the regression line represents the autocorrelation coefficient (a). This was used to remove autocorrelation via the following equation yielding the generalized differenced values (t_{GD}):

$$t_{GD} = t_i - at_{i-1}$$

Generalized least squares

Generalized least squares (GLS) is a multiple regression technique that does not assume independence of data series or points within a data series. For instance, the problem of autocorrelation described above was accounted for by an underlying autoregressive model (e.g. Box *et al.* 1994; Chatfield 2003) in our GLS analyses. GLS was previously applied to palaeontological time series data by Hunt *et al.* (2005), Marx & Uhen (2010), Benson & Mannion (2011) and Butler *et al.* (2011*b*). One advantage of GLS over pairwise tests of correlation is that it allows multiple explanatory variables to be examined simultaneously, and the effect of adding additional variables to be assessed quantitatively. We used GLS to examine the relationship between sampling (fossiliferous marine formations), shallow marine area (using sea-level estimates or the amount of non-marine area as proxies), taxic diversity of shallow marine and open ocean tetrapods, and the total taxic diversity. For each taxic diversity series, the best combinations of explanatory variables were identified using an information criterion (AICc; Sugiura 1978). This is a measure that rewards goodness fit of the regression model (combination of explanatory variables; Table 1) but penalizes models incorporating higher numbers of variables. Thus, the best model is deemed to be one that explains the highest proportion of variation in taxic diversity using the fewest explanatory variables. GLS and associated statistical tests were implemented in R version 2.10.1 (R Development Core Team 2009) with the packages lmtest (Zeilis & Hothorn 2002), nlme version 3.1–96 (Pinheiro *et al.* 2009), qpcR version 1.2–7 (Spiess & Ritz 2010) and tseries version 0.10–22 (Trapletti & Hornik 2009). Because it was possible to take account of multiple explanatory variables (e.g. Lagerstätten, continental flooding) that otherwise confound pairwise correlation tests (described above; also see Benson & Mannion 2011), Triassic data were included in our GLS analyses. GLS analyses excluding Triassic data can be found in the supplementary material.

Table 1. *List of regression models including various factors that may explain observed taxic diversity*

Regression models (combinations of explanatory variables)

Stage duration [null model]
Stage duration + sampling*
Stage duration + shallow marine area[†]
Stage duration + sampling* + shallow marine area[†]
Stage duration + Lagerstätten + sampling*
Stage duration + Lagerstätten + shallow marine area[†]
Stage duration + Lagerstätten + sampling* + shallow marine area[†]

---

*Sampling is measured by a proxy, the number of fossiliferous marine formations.
[†]Shallow marine area is measured by one of three proxies described in the text.

Due to the importance of Lagerstätten stages (as detected during application of generalized differencing and preliminary tests of correlation, see above, and Results), Lagerstätten stages observed by Benson *et al.* (2010) were coded for presence or absence by a binary variable. The Triassic (Anisian–Ladinian) deposits that have yielded the majority of discoveries are European formations (Benson *et al.* 2010), in which marine reptiles are abundant in shallow and marginal facies with carbonate and terrigenous input (Hagdorn & Rieppel 1999). These stages were scored for a 'shallow marine Lagerstätten' variable used in comparison with shallow marine taxic diversity (Fig. 2b). Contrastingly, the Jurassic Lagerstätten represent deeper water facies with fully marine fauna and virtually no terrigenous input (Hudson *et al.* 1991; Röhl *et al.* 2001; Martill *et al.* 2006). These stages were scored for a 'deep marine Lagerstätten' variable used in comparison with open ocean taxic diversity (Fig. 3b).

Because both taxic diversity and fossiliferous marine formations may accumulate through longer stages (Sepkoski & Koch 1996), uneven stage duration may cause a spurious increase in correlation between these variables. To counter this effect, stage duration was included as a non-optional explanatory variable in all models, including the null model. Sea-level estimates are only available from the Anisian (Middle Triassic) onwards (Haq *et al.* 1987) and the Bajocian (Middle Jurassic) onwards (Miller *et al.* 2005). Therefore, we implemented GLS on the full duration of each taxic diversity series (i.e. Triassic–Cretaceous for shallow marine and total taxic diversity; Jurassic–Cretaceous or open ocean taxic diversity (excluding a small number of Triassic open ocean ichthyosaurs)) only when comparing with non-marine area (Smith *et al.* 1994), shorter subsets of each data series were compared with the sea-level estimates (Haq *et al.* 1987; Miller *et al.* 2005). However, because the sea-level estimates showed substantially worse fit to taxic diversity estimates than did non-marine area in all cases, these results are presented in the supplementary material. Because open ocean marine tetrapods only appeared late in the Triassic and were not well-established until the Jurassic, we analysed open ocean taxic diversity only for the Jurassic–Cretaceous and shorter time intervals.

Autoregressive models (e.g. Box *et al.* 1994; Chatfield 2003) of order zero, one or two were fit to combinations of explanatory variables used to predict taxic diversity (Table 1). This was implemented using the 'GLS' function of nlme. The modified version of Akaike's information criterion introduced by Sugiura (1978) for small sample sizes (AIC$_c$; 'AICc' function of qpcR) was used to calculate Akaike weights (Burnham & Anderson

2001) to identify the best combination of explanatory variables. The generalized coefficient of determination (R^2) introduced by Cox & Snell (1989; Magee 1990; Nagelkerke 1991) was calculated manually from the output of the GLS analysis. This coefficient indicates the proportion of variance in taxic diversity explained by the combination of variables in the regression model. The Jarque–Bera ('jarque.bera.test' function of tseries) and Breusch–Pagan ('bptest' function of lmtest) tests were used to assess the normality and homoscedasticity of residuals. The residuals were normally distributed in all cases after $\log_{10}$ transformation of the data series prior to analysis (only stage duration and the presence or absence of Lagerstätten were not transformed). In a few cases the Breusch–Pagan test indicated heteroskedasticity and this could not be removed by prior transformation of the dependent data series. Heteroskedasticity may cause overestimation of regression fit (e.g. Burnham & Anderson 2001). However, in most cases it was only present in the residuals from regression models

with low Akaike weights, in which case it should not affect the interpretation of our results.

Results

Pairwise tests of correlation

Almost all pairwise tests of correlation between the raw, untransformed data series recovered statistically significant results with correlation coefficients ranging from 0.5–0.9 (Table 2). Thus, non-marine surface area is correlated negatively with both sea-level estimates, counts of fossiliferous marine formations are correlated negatively with non-marine surface area and positively with both sea-level estimates, although the correlation with the sea-level estimate of Miller et al. (2005) is rendered non-significant after correction for multiple comparisons (Table 2).

Shallow marine taxic diversity correlates negatively with non-marine area, and positively with

Table 2. *Summary of pairwise tests of correlation over the Jurassic–Cretaceous interval*

		Fossiliferous marine formations (sampling) $N = 23$ (19)	Non-marine area $N = 23$ (19)	Sea-level (Haq et al. 1987) $N = 23$ (19)	Sea-level (Miller et al. 2005) $N = 18$ (16)
'Raw' data comparisons					
Taxic diversity ($\log_{10}$ transformed)	Shallow marine	0.366^{ns}	-0.706^{**}	0.503^{*}	0.543^{*}
	Open ocean	0.513^{**}	-0.261^{ns}	0.380^{ns}	0.709^{**}
		(0.735^{**})	(-0.458^{*})	(0.655^{**})	(0.804^{**})
	Total	0.574^{**}	-0.480^{*}	0.462^{*}	0.739^{**}
		(0.733^{**})	(-0.626^{**})	(0.654^{**})	$(0.819 ^{**})$
Shallow marine area proxy	Sea-level (Haq et al. 1987)	0.631^{**}	-0.852^{**}		
	(Miller et al. 2005)	0.568^{*}	-0.825^{**}		
	Non-marine area (Smith et al. 1994)	-0.609^{**}			
Generalized differenced (or detrended) data comparisons					
Taxic diversity	Shallow marine	0.220^{ns}	-0.545^{**}	-0.160^{ns}	0.570^{*}
	Open ocean	0.449^{*}	-0.0467^{ns}	-0.0359^{ns}	0.409^{ns}
		(0.721^{**})	(0.164^{ns})	(-0.0319^{ns})	(0.388^{ns})
	Total	0.369^{ns}	-0.338^{ns}	-0.125^{ns}	0.546^{*}
		(0.499^{*})	(-0.256^{ns})	(-0.132^{ns})	(0.514^{*})
Shallow marine area proxy	Sea-level (Haq et al. 1987)	-0.0964^{ns}	0.113^{ns}		
	(Miller et al. 2005)	0.167^{ns}	-0.335^{ns}		
	Non-marine area (Smith et al. 1994)	0.160^{ns}			

Correlation coefficients are derived from Pearson's product moment (r), suffixed by^{ns}, non-significant; *significant at $\alpha = 0.05$; **, significant at $\alpha = 0.0125$ (i.e. incorporating a correction for multiple comparisons). Results in brackets were calculated excluding Lagerstätten stages. Shaded cells contain results that were significant after correction for multiple comparisons.

both sea-level estimates, but not with fossiliferous marine formations, and the correlations with sea-level are rendered non-significant after correction for multiple comparisons. Total taxic diversity and open ocean taxic diversity both correlate positively with counts of fossiliferous marine formations and the sea-level estimate of Miller *et al.* (2005). After exclusion of Lagerstätten stages, both of these series also correlate positively with the sea-level estimate of Haq *et al.* (1987) and total taxic diversity correlates negatively with non-marine area. This is consistent with the strong Lagerstätten effect on these data series over the Jurassic–Cretaceous interval for which correlations were tested.

Many of these pairwise tests of correlation are rendered insignificant after the application of generalized differencing. A few significant correlations remain (Table 2): diversity of shallow marine taxa is negatively correlated with non-marine area ($\rho = -0.545$); diversity of open ocean taxa is positively correlated with counts of fossiliferous marine formations ($\rho = 0.721$), but only when Lagerstätten stages are excluded. The intermediate strength of these correlations suggests that neither non-marine area nor our sampling proxy provide a complete explanation of either taxic diversity series. Neither sea-level series is correlated with either non-marine area or counts of fossiliferous marine formations, and non-marine area is not correlated with counts of fossiliferous marine formations.

Generalized least squares

Among generalized least squares regression models, neither sea-level estimate (Haq *et al.* 1987; Miller *et al.* 2005) fits any taxic diversity data as well as the estimates of non-marine area calculated from palaeogeographical maps by Smith *et al.* (1994) (Supplementary Material). Thus, sea-level is not included in any well-supported explanatory model of taxic diversity, and only analyses that use non-marine area as a proxy for the extent of habitable shallow marine area are presented here (Table 3).

Shallow marine taxic diversity across the whole Mesozoic (Induan–Maastrichtian) is best explained by a regression model including the presence or absence of shallow marine Lagerstätten, and the amount of non-marine surface area (Table 3). Including the number of fossiliferous marine formations in the model results in a slight improvement of fit (Table 3), yielding an approximately equivalent, but slightly lower Akaike weight. Excluding the presence or absence of shallow marine Lagerstätten from either model results in a lower, but non-negligible Akaike weight (Table 3). However, residuals of the model including fossiliferous marine formations and non-marine area, but excluding Lagerstätten, are heteroskedastic,

suggesting that the model fit is overestimated. Furthermore, within this model, none of the explanatory variables has a statistically significant slope (Table 4). Among the best models, only the intercept, Lagerstätten, and non-marine area have a statistically significant slope (Table 4). These results suggest that non-marine surface area and shallow marine Lagerstätten are the key determinants of observed shallow marine tetrapods palaeodiversity, and the influence of sampling is correspondingly weakened.

Open ocean taxic diversity through the Jurassic and Cretaceous is best explained by a model including the presence or absence of deep water Lagerstätten, a count of fossiliferous marine formations, and the amount of non-marine surface area. Models including Lagerstätten and one of either fossiliferous marine formations or non-marine area also have high Akaike weights, although residuals from the latter (weaker) model show heteroskedasticity, suggesting that its fit is overestimated (Table 3). Within the stronger two models, Lagerstätten and fossiliferous marine formations have statistically significant, positive slopes. Within the weaker model (Lagerstätten + non-marine area), all explanatory variables, including stage duration, have statistically significant slopes (Table 4). Despite this conflicting signal from the weakest (and heteroskedastic) model, these results suggest that deep water Lagerstätten and fossiliferous marine formations are the key determinants of observed open ocean tetrapod diversity and any fluctuations in underlying biological diversity driven by continental flooding are very weak.

Total taxic diversity is best explained by regression models minimally including the presence or absence of Lagerstätten (both shallow and deep water as a single variable) and a count of fossiliferous marine formations. Including the amount of non-marine area yields equivalent, though slightly higher AICc scores (Table 3). Within these models, only Lagerstätten and fossiliferous marine formations have statistically significant (positive) slopes. These results suggest that fossiliferous marine formations and Lagerstätten are the primary determinants of the observed palaeodiversity of all Mesozoic marine tetrapods and are consistent with results from pairwise correlation analyses (Table 2). This is similar to the pattern observed among open ocean marine tetrapods and suggests that the strength of the signal is strongest among open ocean taxa, despite the fact that these represent a lower proportion of the total data (40.7%). This is congruent with the slightly higher correlation coefficients (Table 2) and R^2 values (Table 3) recovered from analyses of open ocean taxic diversity when compared to those from shallow marine taxic diversity.

Table 3. *Summary of GLS multiple regression models for taxic diversity during the Mesozoic (shallow marine taxic diversity, total taxic diversity) and Jurassic–Cretaceous (open ocean taxic diversity). All models include stage duration. AR order indicates the order of the autoregressive model. 'Model rank' indicates the rank order of models based on Akaike weights. Models with Akaike weight less than 1/8th that of the best model are not ranked. Model ranks in brackets indicate heteroscedasticity of model residuals, suggesting that the model's fit is overestimated*

Dependent variable	Regression model	AR order	R^2	Log likelihood	AICc	AIC weight	Model rank
TDE$_{\text{shallow marine}}$	Null	1	–	−21.918	52.797	0.00324	
	Sampling	1	0.135	−19.816	51.298	0.00687	
	Non-marine area	1	0.286	−17.038	45.742	0.110	3
	Sampling + non-marine area	1	0.349	−15.697	46.003	0.0970	(4)
	Lagerstätten	1	0.0860	−20.614	52.894	0.00309	
	Lagerstätten + sampling	1	0.226	−18.209	51.026	0.00787	
	Lagerstätten + non-marine area	1	0.415	−14.151	42.910	0.455	1
	Lagerstätten + non-marine area + sampling	1	0.463	−12.911	43.641	0.316	2
TDE$_{\text{open ocean}}$	Null	0	–	−15.639	38.541	0.0148	
	Sampling	0	0.129	−14.046	38.313	0.0166	
	Non-marine area	0	0.204	−13.0175	36.257	0.0465	
	Sampling + non-marine area	0	0.279	−11.883	37.296	0.0276	
	Lagerstätten	0	0.111	−14.288	38.799	0.0130	
[residuals: Fig. 3C]	Lagerstätten + sampling	1	0.458	−8.590	32.710	0.274	2
	Lagerstätten + non-marine area	0	0.377	−10.197	33.923	0.149	(3)
	Lagerstätten + non-marine area + sampling	1	0.554	−6.349	31.948	0.401	1
TDE$_{\text{total}}$	Null	1	–	−20.717	50.394	<0.0001	
	Sampling	0	0.298	−15.579	40.825	0.000498	
	Non-marine area	0	0.144	−18.465	46.596	<0.0001	
	Sampling + non-marine area	1	0.397	−13.377	39.929	0.000780	
	Lagerstätten	2	0.321	−15.108	43.883	0.000108	
	Lagerstätten + sampling	1	0.629	−6.347	27.302	0.430	2
	Lagerstätten + non-marine area	0	0.405	−13.187	38.984	0.00125	
	Lagerstätten + non-marine area + sampling	1	0.674	−4.466	26.750	0.567	1

Discussion

Determinants of marine reptile palaeodiversity

Lagerstätten effects have an influential role in determining the observed palaeodiversity of marine tetrapods. This is indicated by (1) the confounding effect of Lagerstätten on the ability to estimate long-term trend in diversity (calculated during the process of generalized differencing) and (2) on pairwise correlation of open ocean and total taxic diversity with other data series (data for detrending and correlation tests spanned only the Jurassic–Cretaceous and thus excluded shallow water Lagerstätten so pairwise correlations involving shallow marine taxa are not affected), and (3) the universal inclusion of a variable describing the presence or absence of

Lagerstätten in the best predictive models for taxic diversity (Table 3).

Pairwise tests of correlation on log$_{10}$- and generalized difference-transformed data indicate a strong, statistically significant negative correlation between the taxic diversity of shallow marine tetrapods and non-marine area (Table 2). This relationship is confirmed by the strong and statistically significant fit of non-marine area within generalized least squares multiple regression models (Tables 3 & 4). A statistically significant correlation with our sampling proxy, the number of fossiliferous marine formations, is absent (Table 2). However, multiple regression models including sampling as an additional variable are approximately as good as those excluding it (Table 3). This suggests that sampling still has a weak influence on the observed taxic diversity of shallow marine tetrapods. The quantity of

Table 4. *Summary of explanatory variables within the best GLS multiple regression models for taxic diversity (indicated in Table 3)*

	Slope	SE	P			Slope	SE	P
TDE$_{shallow\ marine}$ ($N = 29$ stages)					TDE$_{open\ ocean}$ ($N = 23$)			
1 Intercept	25.547	7.881	0.0034*	1	Intercept	−6.295	7.432	0.408
Stage duration	−0.0107	0.0147	0.472		Stage duration	−0.0121	0.0260	0.646
Lagerstätten	0.882	0.328	0.0126*		Lagerstätten	0.725	0.144	0.0001*
Non-marine area	−11.689	3.730	0.0044*		Non-marine area	2.0250	3.362	0.0555
					Sampling	1.586	0.459	0.0028*
2 Intercept	21.428	8.637	0.0205*					
Stage duration	−0.0167	0.0146	0.264	2	Intercept	−1.775	0.686	0.0180*
Lagerstätten	0.827	0.319	0.0160*		Stage duration	−0.00568	0.0254	0.8254
Non-marine area	−10.314	3.966	0.0157*		Lagerstätten	0.704	0.146	0.0001*
Sampling	0.687	0.452	0.1412		Sampling	1.426	0.407	0.0024*
3 Intercept	22.420	9.389	0.0245*	3	Intercept	11.142	4.757	0.0302*
Stage duration	−0.0135	0.0155	0.3927		Stage duration	0.0586	0.0255	0.0329*
Non-marine area	−10.172	4.443	0.0304*		Lagerstätten	0.578	0.205	0.0109*
					Non-marine area	−5.0786	2.251	0.0360*
4 Intercept	17.861	9.669	0.0766					
Stage duration	−0.0197	0.0157	0.222					
Non-marine area	−8.642	4.463	0.0642					
Sampling	0.745	0.492	0.1424					
TDE$_{total}$ ($N = 29$)					TDE$_{total}$			
1 Intercept	0.976	4.997	0.847	2	Intercept	−1.901	0.513	0.0011*
Stage duration	−0.0142	0.0110	0.207		Stage duration	−0.0152	0.0106	0.165
Lagerstätten	0.632	0.111	<0.0001*		Lagerstätten	0.626	0.108	<0.0001*
Non-marine area	−1.295	2.236	0.568		Sampling	1.655	0.285	<0.0001*
Sampling	1.576	0.319	<0.0001*					

*Indicates statistical significance of slope or intercept at $\alpha = 0.05$. SE = standard error.

non-marine area is inversely correlated with flooding of the continental shelf, and therefore provides a proxy for the areal extent of shallow marine habitats. Decreases in non-marine area result from increased continental flooding and are strongly correlated with increases in shallow marine taxic diversity, providing strong support for a species diversity-area relationship in this ecomorphological grouping.

In contrast, non-marine area and sea-level have little influence on the diversity of open ocean taxa. Although some tests detect a statistically significant correlation with sea-level (Table 2), correlation with non-marine surface area is absent. Furthermore, after generalized difference transformation, only fossiliferous marine formations show a statistically significant (positive) correlation with the taxic diversity of open ocean marine tetrapods (Table 2). This relationship is confirmed by the strong and statistically significant fit of fossiliferous marine formations within generalized least squares multiple regression models (Tables 3 & 4). Counts of fossiliferous marine formations form a proxy for (1) the amount of rock available for fossil sampling, (2) the geographical extent of sampled

formations (different sedimentary basins have different formations), (3) the heterogeneity of facies available for fossil sampling, and (4) the amount of geological study that has been undertaken (e.g. Raup 1976; Peters & Foote 2001; Peters & Heim 2010). Thus, a strong relationship with fossiliferous marine formations, and weak relationship with non-marine area confirms that observed open ocean taxic diversity is profoundly influenced by heterogeneous temporal sampling of the fossil record. Any underlying fluctuations in genuine biological diversity are evidently too weak to obscure the relationship between open ocean taxic diversity and sampling.

Our multiple regression models explain a higher proportion of variance in total taxic diversity than they do in either of the other taxic diversity data series (Table 3). The best explanatory models yield statistically significant (positive) slopes for Lagerstätten and fossiliferous marine formations, and the P-values of these slopes are the most strongly significant of any in the present study. This confirms the conclusions of Benson *et al.* (2010), who suggested that fossil sampling was influential in determining observed palaeodiversity of marine reptiles, and that Lagerstätten in the Jurassic and Middle Triassic confounded quantitative analyses of palaeodiversity.

Correlation between sea-level and sampling data series

Many previous studies have predicted or recovered a relationship between sea-level or continental flooding, and fossil sampling metrics (e.g. Sepkoski 1976; Peters & Foote 2001; Smith *et al.* 2001; Peters 2005, 2006*a*; Benton & Emerson 2007). However, we only found a correlation between our sampling metric (fossiliferous marine formations) and estimates of non-marine area or eustatic sea-level, prior to generalized differencing of the data series (Table 2). This indicates that the data series share general features such as a long-term trend of increase over the Jurassic–Cretaceous, but that the pattern of peaks and troughs differs. This lack of detailed correspondence was also observed for the terrestrial Mesozoic record by Butler *et al.* (2011*a*), and may be genuine or reflect inadequacy of either the estimates of sea-level/continental flooding (Haq *et al.* 1987; Smith *et al.* 1994; Miller *et al.* 2005) or our sampling proxy. Given that many studies of the highly abundant record of marine invertebrates suggest a correlation between sampling and sea-level (see above), its absence from our data is an issue that requires further investigation. Purported 'eustatic' sea-level estimates such as those presented by Miller *et al.* (2005) have been criticized as they may be strongly influenced by localized mantle flow induced topographic changes (Moucha *et al.* 2008). Detailed macrostratigraphic data have only been compiled for North America (Peters 2006*b*, 2008; Peters & Heim 2010), but their assembly over wider geographical areas presents one avenue by which more accurate estimates of fossil sampling and continental flooding might be quantified.

Implications for the 'common cause' hypothesis

The concept of fossil sampling as a direct explanation of observed palaeodiversity has been challenged on the grounds that both may instead be driven by a common, external driving mechanism (Sepkoski 1976; Peters 2005, 2006*a*; Benton & Emerson 2007; Smith 2007). The most commonly cited mechanism is sea-level, which drives continental flooding and has been proposed for both the marine (e.g. Sepkoski 1976; Peters 2005, 2006*a*) and terrestrial (Benton & Emerson 2007; Benton 2009) settings. In the marine

setting, increased submerged continental shelf area may result in increased deposition of fossiliferous rock, as well as an increased habitable area for shallow marine organisms, thus inflating estimates of correlation between sampling proxies and observed palaeodiversity. This principle is termed the 'common cause' hypothesis.

The strong relationship between taxic diversity of shallow marine tetrapods and continental flooding recovered in our study is congruent with the results of studies of the Phanerozoic fossil record (Sepkoski 1976; Peters 2005, 2006*a*), the majority of which comprises shallow marine invertebrates. These studies have yielded the strongest evidence in support of continental flooding as the driver of 'common cause'. However, because a strong relationship with continental flooding is absent (or very weak), the relationship between taxic diversity of open ocean tetrapods and fossil sampling cannot be explained by the 'common cause' hypothesis. Instead, our data support a direct, causal relationship between observed taxic diversity of open ocean tetrapods and temporal heterogeneity in fossil sampling. This contrasts with results recovered by Marx (2008) and Marx & Uhen (2010), which suggest that sampling is not an important determinant of the observed palaeodiversity of open ocean mammals (cetaceans; which would be classified as open ocean taxa using our approach).

Diversification and extinctions of Mesozoic marine tetrapods

Our previous analysis of total taxic diversity of marine tetrapods suggested a link between diversity and sampling, obscured by pronounced Lagerstätten effects (Benson *et al.* 2010). This allowed the construction of a 'sampling corrected' curve of residual diversity remaining after subtraction of the value of diversity expected given measured sampling within each geological stage. The present study shows a more complex picture in which the diversity of shallow marine taxa is more strongly tied to flooding of the continental shelves (a similar pattern was observed by Hagdorn & Rieppel (1999) in a detailed study of marine tetrapods from the Triassic sediments of the Germanic Basin). This hypothesis was reviewed and discussed in detail by Hallam & Cohen (1989) and Smith (2007). Both studies proposed that factors other than the literal extent of habitable area (e.g. ocean bottom anoxia, perturbation of primary productivity) might be important. In our study, high values in taxic diversity of shallow marine tetrapods span the Anisian–Carnian (Triassic: basal sauropterygians, non-parvipelvian ichthyosaurs, thalattosaurs), Bathonian–Tithonian (Jurassic: thalattosuchian crocodyliforms) and

Cenomanian–Maastrichtian (Cretaceous: dyrosaurid crocodyliforms and marine squamates, including mosasauroids). These high values approximately correspond to sea-level high-stands and are separated by major regression events in the Late Triassic–Early Jurassic and earliest Cretaceous (Fig. 2; Hallam 1978; Haq et al. 1987; Miller et al. 2005). During these regressions, taxic diversity of shallow marine tetrapods declined to low or zero observed values, and higher clades of shallow marine taxa became extinct. The Late Triassic regression, potentially representing the minimum extent of Mesozoic continental flooding (Smith et al. 1994), had a catastrophic effect on shallow marine taxa: thalattosaurs, non-plesiosaurian sauropterygians (Storrs 1993a; Hagdorn & Rieppel 1999; Rieppel 2000) and non-parvipelvian ichthyosaurs (Motani 2005) became extinct (see also Bardet 1992, 1994). The earliest Cretaceous regression corresponds to a drop in thalattosuchian diversity, and morphological disparity (Pierce et al. 2009a; Young et al. 2010) that may have occurred later in the southern hemisphere than it did in the northern hemisphere (Hallam 1986; Bardet 1994; Pierce et al. 2009a). Finally, a dramatic drop in sea-level, and corresponding reduction in the extent of continental flooding at the Cretaceous–Palaeogene boundary may have contributed to the extinction of abundant shallow marine squamates, including mosasauroids. Note, however, that the plesiopedal dyrosaurid crocodyliforms (Hill et al. 2008; Jouve et al. 2008) were not strongly affected by this event, and may have taken refuge in terrestrial freshwater ecosystems (Buffetaut 1990). Note also that numerous other factors have been invoked to explain this catastrophic extinction event that affected a wide range of terrestrial and marine organisms (e.g. Bardet 1994; Archibald et al. 2010; Schulte et al. 2010). Our previous results suggested declining diversity during the terminal stages of the Cretaceous based on sampling-corrected diversity estimates (Benson et al. 2010). However, the present study suggests that these estimates are not appropriate for shallow marine tetrapods, which may therefore have been undergoing a major diversification right up until the end of the Cretaceous, as suggested by Bardet (1992, 1994) and Ross (2009; for mosasauroids). This is especially likely given a slight decrease in fossil sampling, coincident with an increase in observed palaeodiversity in the final stage of the Cretaceous (Maastricthian). One remaining question regarding shallow marine-tetrapod diversity is the presence or absence of end-Cenomanian crash in diversity. This was suggested by Bardet (1992, 1994) based on observation of taxic diversity and origination and extinction rates among marine tetrapods. However, Benson et al. (2010) suggested that it coincided with

substantially low sampling of Turonian/Coniacian marine fossils (Fig. 3). Although temporal heterogeneity in fossil sampling has been shown not to have a strong influence on observed palaeodiversity of shallow marine tetrapods for most stages of the Mesozoic, the magnitude of the post-Cenomanian sampling low is such that it is difficult to dismiss as a possible explanation.

The relationship between open ocean tetrapod diversity and continental flooding is weak or absent. Thus, the evolution of highly pelagic forms among ichthyosaurs and plesiosaurians, which occurred by the Late Triassic (Storrs 1994; Motani 2005), released these animals from dependence on shallow marine habitats. This may explain their 'escape' from extinction events driven by marine regressions such as that in the Late Triassic. Thus, plesiosaurians and ichthyosaurs were diverse in the earliest Jurassic (Fig. 3; e.g. McGowan & Motani 2003; Ketchum & Benson 2010), and are represented by high levels of residual diversity after regression against our sampling proxy and a variable coding the presence or absence of Lagerstätten (Fig. 3c). The effect of the Jurassic–Cretaceous extinction on open ocean tetrapods is difficult to determine as this boundary also marks a transition out of the Lagerstätten-dominated Jurassic sampling regime. However, Bakker (1993) suggested that Jurassic plesiosaur lineages were truncated at this boundary. Unfortunately, the problem of deep water Lagerstätten is pervasive in the Jurassic. Extreme cases were identified, in which a single or small number of formations have yielded a high proportion of fossil discoveries (Benson et al. 2010). However, other, less extreme cases were not identified or accounted for in our analyses and it is difficult to interpret the meaning of our residual diversity plot confidently (Fig. 3c). We do not advocate a literal interpretation of high frequency oscillations in residual diversity observed in the Jurassic: the Jurassic record of marine tetrapods has been sampled by palaeontologists in an extremely heterogeneous fashion and this presents a fundamental obstacle to interpreting patterns in diversity. In contrast, the Cretaceous record is more evenly sampled (Fig. 3). This may allow a more confident interpretation of patterns in diversity. Residual diversity of open ocean tetrapods shows a similar pattern to that of total marine tetrapod diversity recovered by Benson et al. (2010): a progressive increase in diversity from the Early Cretaceous to a Santonian (middle Late Cretaceous) peak. This is interrupted by a local peak in the Barremian. An extended period of low ichthyosaur diversity preceded their final disappearance after the Cenomanian (e.g. Sander 2000; McGowan & Motani 2003), which does not coincide with reduced diversity of other clades and thus did not occur during a catastrophic episode of mass extinction. One possible explanation is the rise of fast-swimming actinopterygian fish as prey and marcopredaceous sharks as competitors (Lingham-Soliar 2003). The terminal stages of the Cretaceous, the Campanian and Maastrichtian, show a slight decrease in diversity, suggesting a gradual decline prior to the Cretaceous–Palaeogene mass extinction event (Fig. 3c). Highly pelagic chelonioid turtles (and continental, freshwater turtles; Hutchison & Archibald 1986) were largely unaffected by the terminal Cretaceous extinction event, but plesiosaurians and open ocean mosasauroids like *Plotosaurus* became extinct. This is congruent with observed high levels of selectivity in terminal Cretaceous extinctions (e.g. Buffetaut 1990). One possibility is that because turtles are oviparous with zero parental care, they survived the acme of highly-stressed terminal Cretaceous environments as eggs (which is also possible for crocodyliformes). Other marine tetrapods were likely viviparous (Caldwell & Lee 2001; Cheng *et al.* 2004). This is analogous to the situation among open ocean planktonic organisms, among which diatoms suffered proportionally little extinction (e.g. Thierstein 1982). This is hypothesized to result from a meroplanktic life cycle, incorporating a dormant resting cyst phase (Kitchell *et al.* 1986).

Timing of the 'Late Triassic' marine tetrapod extinction

Extinction of shallow marine tetrapod lineages in the early Late Triassic is demonstrated here and by previous studies (Bardet 1992, 1994; Rieppel 2000). The last appearances of many higher clades occur early in the Late Triassic (Benson *et al.* 2010), although the basal sauropterygian placodonts (Pinna & Mazin 1993; Storrs 1994) and some non-parvipelvian ichthyosaurs (Motani 2005) are represented in the latest Triassic (Rhaetian). It is possible that all, or most, of these extinctions coincide with a wider, but controversial, global extinction event at the end of the Triassic, affecting terrestrial tetrapods, plants and marine invertebrates (e.g. Benton 1995; Tanner *et al.* 2004). However, this is not currently consistent with the last appearance data of most marine tetrapod clades, or our results. If extinction among shallow marine tetrapods was driven by reduction in flooded continental area then it should precede the end of the Triassic as the sea-level minimum may have occurred as early as the Norian (Haq *et al.* 1987; Miller *et al.* 2005). Unfortunately, this time interval is poorly resolved in the non-marine area estimates of Smith *et al.* (1994) (Fig. 2a).

Conclusions

Vertebrates offer unique insights into the relationships between observed palaeodiversity, sampling of the fossil record, and continental flooding, because they are well-studied and occupy a range of habitats indicated by clearly established ecomorphotypes. During the Mesozoic, shallow marine tetrapod diversity was strongly tied to the extent of flooded continental area. Decreases in diversity in the Late Triassic, earliest Cretaceous and latest Cretaceous coincide with major regressions and may have been driven by a reduction in habitable shallow marine area, or concurrent effects such as a break in primary productivity or ocean bottom anoxia. In contrast, open ocean marine tetrapod diversity shows a weak or absent relationship with shallow marine habitat area. Instead, open ocean palaeodiversity has a stronger relationship with temporal heterogeneity in fossil sampling.

Tetrapods gave rise to multiple independent radiations adapted for shallow marine life. Representatives of these radiations were vulnerable to major regressions, which drove extinction events. Invasion of the open ocean by parvipelvian ichthyosaurs, plesiosaurians and chelonioid turtles freed these lineages from their dependence on shallow marine environments and made them less vulnerable to extinction driven by regression. The existence of different diversification trajectories among shallow marine and open ocean tetrapods suggests that diversity curves predominantly based on shallow marine invertebrates should not be generalized across all animals.

We thank A. Smith and A. J. McGowan for inviting us to contribute to this volume. RJB is funded by an Alexander von Humboldt Research Fellowship. This is Paleobiology Database official publication 128 and benefits from the data contributed by numerous individuals. We are grateful to J. Lindgren and A. Smith, who helped to compile the taxonomic diversity dataset. We also thank J. Alroy, F. Marx and P. Mannion for advice. G. Lloyd and an anonymous reviewer provided constructive comments that improved the manuscript.

References

ALROY, J. 2000. Successive approximations of diversity curves: ten more years in the library. *Geology*, **28**, 1023–1026.

ALROY, J., ABERHAN, M. ET AL. 2008. Phanerozoic trends in the global diversity of marine invertebrates. *Science*, **321**, 97–100.

ARCHIBALD, J. D., CLEMENS, W. A. ET AL. 2010. Cretaceous extinctions: multiple causes. *Science*, **328**, 973.

BAKKER, R. T. 1993. Plesiosaur extinction cycles – events that mark the beginning, middle and end of the

Cretaceous. *Geologial Association of Canada, Special Papers*, **39**, 641–664.

BARDET, N. 1992. Evolution et extinction des reptiles marins au cours du Mésozoique. *Paleovertebrata*, **24**, 177–283.

BARDET, N. 1994. Extinction events among Mesozoic marine reptiles. *Historical Biology*, 7, 313–324.

BARRETT, P. M., McGOWAN, A. J. & PAGE, V. 2009. Dinosaur diversity and the rock record. *Proceedings of the Royal Society, B*, **276**, 2667–2674.

BELL, G. L., JR. & POLCYN, M. J. 2005. *Dallasaurus turneri*, a new primitive mosasauroid from the Middle Turonian of Texas and comments on the phylogeny of Mosasauridae (Squamata). *Geologie en Mijnbouw*, **84**, 177–194.

BENSON, R. B. J. & MANNION, P. D. 2011. Multivariate models are essential for understanding vertebrate diversification. *Biology Letters*, doi: 10.1098/rsbl.2011.0460.

BENSON, R. J. B., BUTLER, R. J., LINDGREN, J. & SMITH, A. S. 2010. Mesozoic marine tetrapod diversity: mass extinctions and temporal heterogeneity in geological megabiases affecting vertebrates. *Proceedings of the Royal Society, B*, **277**, 829–834.

BENTON, M. J. 1995. Diversification and extinction in the history of life. *Science*, **268**, 52–58.

BENTON, M. J. 2001. Biodiversity on land and in the sea. *Geological Journal*, **36**, 211–230.

BENTON, M. J. 2009. The fossil record: biological or geological signal? *In*: SEPKOSKI, D. & RUSE, M. (eds) *The Paleobiological Revolution*. University of Chicago Press, Chicago, 43–59.

BENTON, M. J. & EMERSON, B. J. 2007. How did life become so diverse? The dynamics of diversification according to the fossil record and molecular phylogenetics. *Palaeontology*, **50**, 23–40.

BENTON, M. J., TVERDOKHLEBOV, V. P. & SURKOV, M. V. 2004. Ecosystem remodelling among vertebrates at the Permian–Triassic boundary in Russia. *Nature*, **432**, 97–100.

BILLIONE-BRUYAT, J.-P., LÉCUYER, C., MARTINEAU, F. & MAZIN, J.-M. 2005. Oxygen isotope compositions of Late Jurassic vertebrate remains from lithographic limestones of western Europe: implications for the ecology of fish, turtles, and crocodilians. *Palaeogeography Palaeoclimatology Palaeoecology*, **216**, 359–375.

BOX, G., JENKINS, G. M. & REINSEL, G. 1994. *Time Series Analysis: Forecasting and Control*, 3rd edn. Prentice-Hall, New Jersey.

BUFFETAUT, E. 1990. Vertebrate extinctions and survival across the Cretaceous–Tertiary boundary. *Tectonophysics*, **171**, 337–345.

BURNHAM, K. P. & ANDERSON, D. 2001. *Model Selection and Multi-Model Inference: A Practical Information-Theoretic Approach*. 2nd edn. Springer, New York.

BUTLER, R. J., BARRETT, P. M., NOWBATH, S. & UPCHURCH, P. 2009. Estimating the effects of sampling biases on pterosaur diversity patterns: implications for hypotheses of bird/pterosaur competitive replacement. *Paleobiology*, **35**, 432–446.

BUTLER, R. J., BENSON, R. B. J., CARRANO, M. T., MANNION, P. D. & UPCHURCH, P. 2011a. Sea level, dinosaur diversity, and sampling biases: investigating the 'common cause' hypothesis in the terrestrial

realm. *Proceedings of the Royal Society, B*, **278**, 1165–1170, doi: 10.1098/rspb.2010.1754.

BUTLER, R. J., BRUSATTE, S. L., ANDRES, B. & BENSON, R. B. J. 2011b. How do geological sampling biases affect studies of morphological evolution in deep time? A case study of pterosaur (Reptilia: Archosauria) disparity. *Evolution*, doi: 10.1111/j.1558-5646.2011.01415.x.

CALDWELL, M. W. 1997. Limb osteology and ossification patterns in *Cryptoclidus* (Reptilia: Plesiosauroidea) with a review of sauropterygian limbs. *Journal of Vertebrate Paleontology*, **17**, 295–307.

CALDWELL, M. W. & LEE, M. S. Y. 2001. Live birth in Cretaceous marine lizards (mosasauroids). *Proceedings of the Royal Society, B*, **268**, 2397–2401.

CARRIER, D. R. 1987. The evolution of locomotor stamina in tetrapods: circumventing a mechanical constraint. *Paleobiology*, **13**, 326–341.

CARROLL, R. L. 1985. Evolutionary constraints in aquatic diapsid reptiles. *Special Papers in Palaeontology*, **33**, 145–155.

CHATFIELD, C. 2003. *The Analysis of Time-Series: An Introduction*. Chapman & Hall, London.

CHENG, Y.-N., WU, X.-C. & JI, Q. 2004. Triassic marine reptiles gave birth to live young. *Nature*, **432**, 383–385.

COWEN, R. 1996. Locomotion and respiration in marine air-breathing vertebrates. *In*: JABLONSKI, D., ERWIN, D. H. & LIPPS, J. H. (eds) *Evolutionary Biology*. University of Chicago Press, Chicago, 337–353.

COX, D. R. & SNELL, E. J. 1989. *The Analysis of Binary Data*. 2nd edn. Chapman & Hall, London.

CRAMPTON, J. S., BEU, A. G., COOPER, R. A., JONES, C. M., MARSHALL, B. & MAXWELL, P. A. 2003. Estimating the rock volume bias in paleobiodiversity studies. *Science*, **301**, 358–360.

DRUCKENMILLER, P. S. & RUSSELL, A. P. 2008. A phylogeny of Plesiosauria (Sauropterygia) and its bearing on the systematic status of *Leptocleidus* Andrews, 1922. *Zootaxa*, **1863**, 1–120.

FRÖBISCH, J. 2008. Global taxonomic diversity of anomodonts (Tetrapoda, Therapsida) and the terrestrial rock record across the Permian–Triassic boundary. *PLoS ONE*, **3**, 1–14.

DE LA FUENTE, M. S. & FERNANDEZ, M. S. 2011. An unusual pattern of limb morphology in the Tithonian marine turtle *Neusticemys neuquina* from the Vaca Muerta Formation, Neuquén Basin, Argentina. *Lethaia*, **44**, 15–25.

HAGDORN, H. & RIEPPEL, O. 1999. Stratigraphy of marine reptiles in the Triassic of Central Europe. *Zentralbatt für Geologie und Paläontologie, Teil I*, **1998**, 651–678.

HALLAM, A. 1978. Eustatic sea level cycles in the Jurassic. *Palaeogeography Palaeoclimatology Palaeoecology*, **23**, 1–32.

HALLAM, A. 1986. The Pliensbachian and Tithonian extinction events. *Nature*, **319**, 765–768.

HALLAM, A. & COHEN, J. M. 1989. Mass extinction of marine invertebrates [and discussion]. *Philosophical Transactions of the Royal Society of London B: Biological Sciences*, **325**, 437–455.

HAMMER, Ø., HARPER, D. A. T. & RAYN, P. D. 2001. PAST: Palaeontological statistics software package for education and data analysis. *Palaeontologica Electronica*, **4**, 1–9.

HAQ, B. U., HARDENBOL, J. & VAIL, P. R. 1987. Chronology of fluctuating sea levels since the Triassic. *Science*, **235**, 1156–1167.

HILL, R. V., McCARTNEY, J. A., ROBERTS, E., BOUARÉ, M., SISSOKO, F. & O'LEARY, M. 2008. Dyrosaurid (Crocodyliformes: Mesoeucrocodylia) fossils from the Upper Cretaceous and Paleogene of Mali: implications for phylogeny and survivorship across the K/T boundary. *American Museum Novitates*, **3631**, 1–19.

HIRAYAMA, R. 1997. Distribution and diversity of Cretaceous chelonioids. *In*: CALLAWAY, J. M. & NICHOLLS, E. L. (eds) *Ancient Marine Reptiles*. Academic Press, San Diego, 225–241.

HUDSON, J. D., MARTILL, D. M. & PAGE, K. N. 1991. *Fossils of the Oxford Clay*. Palaeontological Association Field Guide to Fossils, **4**.

HUNT, G., CRONIN, T. M. & ROY, K. 2005. Species-energy relationship in the deep sea: a test using the Quaternary fossil record. *Ecology Letters*, **8**, 739–747.

HUTCHISON, J. H. & ARCHIBALD, J. D. 1986. Diversity of turtles across the Cretaceous/Tertiary boundary in northeastern Montana. *Palaeogeography Palaeoclimatology Palaeoecology*, **55**, 1–22.

JIANG, D.-Y., MAISCH, M. W., SUN, Y.-L., MATZKE, A. T. & HAO, W.-C. 2004. A new species of *Xinpusaurus* (Thalattosauria) from the Upper Triassic of China. *Journal of Vertebrate Paleontology*, **24**, 80–88.

JOUVE, S., BOUYA, B. & AMAGHAZAZ, M. 2008. A long-snouted dyrosaurid (Crocodyliformes, Mesoeucrocodylia) from the Paleocene of Morocco: phylogenetic and palaeobiogeographic implications. *Palaeontology*, **51**, 281–294.

KETCHUM, H. F. & BENSON, R. B. J. 2010. Global interrelationships of Plesiosauria (Reptilia, Sauropterygia) and the pivotal role of taxon sampling in determining the outcome of phylogenetic analyses. *Biological Reviews*, **85**, 361–392.

KITCHELL, J. A., CLARK, D. L. & GOMBOS, A. M. JR. 1986. Biological selectivity of extinction: a link between background and mass extinction. *Palaios*, **1**, 504–511.

LINDGREN, J., JAGT, J. W. M. & CALDWELL, M. W. 2007. A fishy mosasaur: the axial skeleton of *Plotosaurus* (Reptilia, Squamata) reassessed. *Lethaia*, **40**, 153–160.

LINGHAM-SOLIAR, T. 2003. Extinction of ichthyosaurs: a catastopic or evolutionary paradigm. *Neues Jahrbuch für Geologie und Paläontologie, Abhandlungen*, **228**, 421–452.

LIU, J. & RIEPPEL, O. 2001. Restudy of *Anshunsaurus huangguoshuensis* (Reptilia: Thalattosauria) from the Middle Triassic of Guizhou, China. *American Museum Novitates*, **3488**, 1–34.

LLOYD, G. T., DAVIS, K. E. ET AL. 2008. Dinosaurs and the Cretaceous Terrestrial Revolution. *Proceedings of the Royal Society, B*, **275**, 2483–2490.

MAGEE, L. 1990. R^2 measures based on Wald and likelihood ratio joint significance tests. *American Statistician*, **44**, 250–253.

MANNION, P. D., UPCHURCH, P., CARRANO, M. T. & BARRETT, P. M. 2010. Testing the effect of the rock record on diversity: a multidisciplinary approach to elucidating the generic richness of sauropodomorph dinosaurs through time. *Biological Reviews*, **86**, 157–181, doi: 10.1111/j.1469-185X.2010.00139.x.

MARTILL, D. M., EARLAND, S. & NAISH, D. 2006. Dinosaurs in marine strata: evidence from the British Jurassic, including a review of the allochthonous vertebrate assemblage from the marine Kimmeridge Clay Formation (Upper Jurassic) of Great Britain. *III Jornadas Internacionales sobre Paleontología de Dinosaurios y su Entorno, Salas de los Infantes, Burgos*, 1–31.

MARX, F. G. 2008. Marine mammals through time: when less is more in studying palaeodiversity. *Proceedings of the Royal Society, B*, **276**, 887–892.

MARX, F. G. & UHEN, M. D. 2010. Climate, critters, and cetaceans: Cenozoic drivers of the evolution of modern whales. *Science*, **327**, 993–996.

MASSARE, J. A. 1988. Swimming capabilities of Mesozoic marine reptiles: implications for method of predation. *Palaeobiology*, **14**, 187–205.

McGOWAN, C. 1991. *Dinosaurs, Spitfires, and Sea Dragons*. Harvard University Press, Cambridge, Massachusetts.

McGOWAN, C. & MOTANI, R. 2003. Ichthyopterygia. *Handbuch der Paläoherpetologie*, **8**, 1–173.

McKINNEY, M. L. 1990. Classifying and analysing evolutionary trends. *In*: McNAMARA, K. J. (ed.) *Evolutionary Trends*. University of Arizona Press, Tuscon, 28–58.

MILLER, K. G., KOMINZ, M. A. ET AL. 2005. The Phanerozoic record of global sea-level change. *Science*, **310**, 1293–1298.

MOTANI, R. 2002a. Scaling effects in caudal fin kinematics and the speeds of ichthyosaurs. *Nature*, **415**, 309–312.

MOTANI, R. 2002b. Swimming speed estimation of extinct marine reptiles. I. Energetic approach revisited. *Paleobiology*, **28**, 251–262.

MOTANI, R. 2005. Evolution of fish-shaped reptiles (Reptilia: Ichthyopterygia) in their physical environments and constraints. *Annual Reviews in Earth and Planetary Sciences*, **33**, 395–420.

MOTANI, R. 2009. The evolution of marine reptiles. *Evolution: Education and Outreach*, **2**, 224–235.

MOTANI, R., YOU, H. & McGOWAN, C. 1996. Eel-like swimming in the earliest ichthyosaurs. *Nature*, **382**, 347–348.

MOUCHA, R., FORTE, A. M., MITOVICA, J. X., ROWLEY, D. B., QUÉRE, S., SIMMONS, N. A. & GRAND, S. P. 2008. Dynamic topography and long-term sea-level variations: there is no such thing as a stable continental platform. *Earth and Planetary Science Letters*, **271**, 101–108.

NAGELKERKE, N. J. D. 1991. A note on a general definition of the coefficient of determination. *Biometrika*, **78**, 691–692.

O'KEEFE, F. R. 2001. A cladistic analysis and taxonomic revision of the Plesiosauria (Reptilia: Sauropterygia). *Acta Zoologica Fennica*, **213**, 1–63.

PETERS, S. E. 2005. Geologic constraints on the macroevolutionary history of marine animals. *Proceedings of the National Academy of Science of the USA*, **102**, 12 326–12 331.

PETERS, S. E. 2006a. Genus extinction, origination, and the durations of sedimentary hiatuses. *Paleobiology*, **32**, 387–407.

PETERS, E. 2006b. Macrostratigraphy of North America. *Journal of Geology*, **114**, 391–412.

PETERS, S. E. 2008. Macrostratigraphy and it promise for paleobiology. *Paleontological Society Papers*, **14**, 205–232.

PETERS, S. E. & FOOTE, M. 2001. Biodiversity in the Phanerozoic: a reinterpretation. *Paleobiology,* **27**, 583–601.

PETERS, S. E. & FOOTE, M. 2002. Determinants of extinction in the fossil record. *Nature*, **416**, 420–424.

PETERS, S. E. & HEIM, N. A. 2010. The geological completeness of paleontological sampling in North America. *Paleobiology*, **36**, 61–79.

PIERCE, S. E., ANGIELCZYK, K. D. & RAYFIELD, E. J. 2009a. Morphospace occupation in thalattosuchian crocodylomorphs: skull shape variation, species delineation and temporal patterns. *Palaeontology*, **52**, 1057–1097.

PIERCE, S. E., ANGIELCZYK, K. D. & RAYFIELD, E. J. 2009b. Shape and mechanics in thalattosuchian (Crocodylomorpha) skulls: implications for feeding behaviour and niche partitioning. *Journal of Anatomy*, **215**, 555–576.

PINHEIRO, J., BATES, D., DEBROY, S., SARKAR, D. & R DEVELOPMENT CORE TEAM. 2009. *nlme: Linear and Nonlinear Mixed Effects Models.* R package version 3.1–93. http://cran.r-project.org/web/packages/nlme/index.html

PINNA, G. & MAZIN, J.-M. 1993. Stratigraphy and paleobiogeography of the Placodontia. *Paleontologia Lombarda, New Series*, **2**, 125–130.

R DEVELOPMENT CORE TEAM. 2009. *A language and environment for statistical computing.* http://www.R-project.org.

RAUP, D. M. 1972. Taxonomic diversity during the Phanerozoic. *Science*, **177**, 1065–1071.

RAUP, D. M. 1976. Species diversity in the Phanerozoic: an interpretation. *Paleobiology*, **2**, 289–297.

RIEPPEL, O. 2000. Sauropterygia. *Handbuch Paläoherpetologie*, **12A**, 1–134.

RÖHL, H.-J., SCHMID-RÖHL, A., OSCHMANN, W., FRIMMEL, A. & SCHWARK, L. 2001. The Posidonia Shale (Lower Toarcian) of SW-Germany: an oxygen-depleted ecosystem controlled by sea level and palaeoclimate. *Palaeogeography Palaeoclimatology Palaeoecology*, **165**, 27–52.

ROSS, M. R. 2009. Charting the Late Cretaceous seas: mosasaur richness and morphological diversification. *Journal of Vertebrate Paleontology*, **29**, 409–416.

SAHNEY, S., BENTON, M. J. & FERRY, P. A. 2010. Links between global taxonomic diversity, ecological diversity and the expansion of vertebrates on land. *Biology Letters*, **6**, 544–547.

SANDER, P. M. 2000. Icthyosauria: their diversity, distribution, and phylogeny. *Paläontologische Zeitschrift*, **74**, 1–35.

SCHULTE, P., ALEGRET, L. ET AL. 2010. The Chicxulub asteroid impact and mass extinction at the Cretaceous Paleogene boundary. *Science*, **327**, 1214–1218.

SEPKOSKI, J. J. JR. 1976. Species diversity in the Phanerozoic; species-area effects. *Paleobiology*, **2**, 298–303.

SEPKOSKI, J. J. JR. 1981. A factor analytic description of the Phanerozoic marine fossil record. *Palaeobiology*, **7**, 6–53.

SEPKOSKI, J. J. JR. 1982. A compendium of fossil marine families. *Milwaukee Public Museum, Contributions in Biology and Geology*, **51**, 1–125.

SEPKOSKI, J. J. JR. & KOCH, C. F. 1996. Evaluating paleontologic data relating to bio-events. *In*: WALLISER, O. H. (ed.) *Global Events and Event Stratigraphy.* Springer, Berlin, 21–34.

SLACK, K. E., JONES, C. M., ANDO, T., HARRISON, G. L., FORDYCE, R. E., ARNASON, U. & PENNY, D. 2006. Early penguin fossils, plus mitochondrial genomes, calibrate avian evolution. *Molecular Biology and Evolution* **23**, 1144–1155.

SMITH, A. B. 2001. Large-scale heterogeneity of the fossil record: implications for Phanerozoic biodiversity studies. *Philosophical Transactions of the Royal Society of London B: Biological Sciences*, **356**, 351–367.

SMITH, A. B. 2007. Marine diversity through the Phanerozoic: problems and prospects. *Journal of the Geological Society, London*, **164**, 1–15.

SMITH, A. B. & MCGOWAN, A. J. 2005. Cyclicity in the fossil record mirrors rock outcrop area. *Biology Letters*, **1**, 443–445.

SMITH, A. B. & MCGOWAN, A. J. 2007. The shape of the Phanerozoic marine palaeodiversity curve: how much can be predicted from the sedimentary rock record of western Europe? *Palaeontology*, **50**, 765–774.

SMITH, A. B., GALE, A. S. & MONKS, N. E. A. 2001. Sea-level change and rock-record bias in the Cretaceous: a problem for extinction and biodiversity studies. *Paleobiology*, **27**, 241–253.

SMITH, A. G., SMITH, D. G. & FUNNELL, B. M. 1994. *Atlas of Mesozoic and Cenozoic Coastlines.* Cambridge University Press, Cambridge.

SPIESS, A. N. & RITZ, C. 2010. *qpcR: Modelling and analysis of real-time PCR data.* R package version 1.2–7. http://CRAN.R-project.org/package=qpcR.

STEEL, R. 1973. Crocodylia. *Handbuch der Paläoherpetologie*, **16**, 1–116.

STORRS, G. W. 1993a. The quality of the Triassic sauropterygian fossil record. *Revue Paléobiologique, Volume Speciale*, **7**, 217–228.

STORRS, G. W. 1993b. Function and phylogeny in sauropterygian (Diapsida) evolution. *American Journal of Science*, **293A**, 63–90.

STORRS, G. W. 1994. Fossil vertebrate faunas from the British Rhaetian (latest Triassic). *Zoological Journal of the Linnean Society*, **112**, 217–259.

SUGIURA, N. 1978. Further analysis of the data by Akaike's Information Criterion and the Finite Corrections. *Communications in Statistics, Theory and Methods*, **7**, 13–26.

TANNER, L. H., LUCAS, S. G. & CHAPMAN, M. G. 2004. Assessing the record and causes of Late Triassic extinctions. *Earth Science Reviews*, **65**, 103–139.

THIERSTEIN, H. R. 1982. Terminal Cretaceous plankton extinctions: a critical assessment. *Geological Society of America Special Paper*, **190**, 385–399.

TRAPLETTI, A. & HORNIK, K. 2009. *tseries: Time Series Analysis and Computational Finance.* R package

version   0.10–22.   http://cran.r-project.org/web/
packages/tseries/index.html

UHEN, M. D. & PYENSON, N. D. 2007. Diversity estimates,
biases, and historiographic effects: resolving cetacean
diversity in the Tertiary. *Palaeontologia Electronica*,
10.

UPCHURCH, P. & BARRETT, P. M. 2005. Sauropodomorph
diversity through time. *In*: CURRY ROGERS, K. &
WILSON, J. (eds) *The Sauropods: Evolution and Paleobiology*. University of California Press, Berkeley,
104–124.

VALENTINE, J. W. 1969. Niche diversity and niche size
patterns in marine fossils. *Journal of Paleontology*,
43, 905–915.

WALKER, J. D. & GEISMANN, J. W. 2009. *Geologic Time
Scale*. Geological Society of America, doi: 10.1029/
2007JB005407.

WANG, S. C. & DODSON, P. 2006. Estimating the diversity
of dinosaurs. *Proceedings of the National Academy of
Science*, **37**, 13601–13605.

YOUNG, M. T. & DE ANDRADE, M. B. 2009. What is
Geosaurus? Redescription of *Geosaurus giganteus*
(Thalattosuchia: Metriorhynchidae) from the Upper
Jurassic of Bayern, Germany. *Zoological Journal of
the Linnean Society*, **157**, 551–585.

YOUNG, M. T., BRUSATTE, S. L., RUTA, M. & ANDRADE,
M. B. 2010. The evolution of Metriorhynchoidea
(Mesoeucrocodylia, Thalattosuchia): an integrated
approach using geometric morphometrics, analysis of
disparity and biomechanics. *Zoological Journal of
the Linnean Society*, **158**, 801–859.

ZEILIS, A. & HOTHORN, T. 2002. Diagnostic checking
in regression relationships. *Rochester News*, **2**,
7–10.

Geological and anthropogenic controls on the sampling of the terrestrial fossil record: a case study from the Dinosauria

P. UPCHURCH[1]*, P. D. MANNION[1], R. B. J. BENSON[2],
R. J. BUTLER[3] & M. T. CARRANO[4]

[1]*Department of Earth Sciences, University College London, Gower Street,
London, WC1E 6BT, UK*

[2]*Department of Earth Sciences, University of Cambridge, Downing Street,
Cambridge CB2 3EQ, UK*

[3]*Bayerische Staatssammlung für Paläontologie und Geologie, Richard-Wagner-Straße 10,
80333 Munich, Germany*

[4]*Department of Paleobiology, National Museum of Natural History, Smithsonian Institution,
P.O. Box 37012, Washington, DC 20013-7012, USA*

*Corresponding author (e-mail: p.upchurch@ucl.ac.uk)

Abstract: Dinosaurs provide excellent opportunities to examine the impact of sampling biases on the palaeodiversity of terrestrial organisms. The stratigraphical and geographical ranges of 847 dinosaurian species are analysed for palaeodiversity patterns and compared to several sampling metrics. The observed diversity of dinosaurs, Theropoda, Sauropodomorpha and Ornithischia, are positively correlated with sampling at global and regional scales. Sampling metrics for the same region correlate with each other, suggesting that different metrics often capture the same signal. Regional sampling metrics perform well as explanations for regional diversity patterns, but correlations with global diversity are weaker. Residual diversity estimates indicate that sauropodomorphs diversified during the Late Triassic, but major increases in the diversity of theropods and ornithischians did not occur until the Early Jurassic. Diversity increased during the Jurassic, but many groups underwent extinction during the Late Jurassic or at the Jurassic/Cretaceous boundary. Although a recovery occurred during the Cretaceous, only sauropodomorphs display a long-term upward trend. The Campanian–Maastrichtian diversity 'peak' is largely a sampling artefact. There is little evidence for a gradualistic decrease in diversity prior to the end-Cretaceous mass extinction (except for ornithischians), and when such decreases do occur they are small relative to those experienced earlier in dinosaur evolution.

Supplementary material: The full data set and details of analyses are available at www.geolsoc.org.uk/SUP18487
The same materials (in the form of an Excel workbook) are also available from the first author on request.

Fluctuations in taxic diversity through time form a key aspect of evolutionary history. The reconstruction and analysis of such fluctuations depends, almost exclusively, on data from the fossil record, either directly in the form of counts of fossil species through time, or indirectly when insights into past diversity are obtained from time-calibrated morphological or molecular phylogenies. Palaeontologists universally acknowledge the fact that the fossil record is incomplete, even if debate continues over the extent of this incompleteness and the evenness of sampling over different temporal and spatial scales (e.g. Smith & McGowan 2007; Alroy *et al.* 2008; Butler *et al.* 2009, 2011; Dyke *et al.* 2009; Benson *et al.* 2010; Heads 2010; Goswami &

Upchurch 2010; Sahney *et al.* 2010; Mannion *et al.* 2011; Benton *et al.* 2011). Concerns about the impact of uneven sampling of the fossil record on estimates of diversity arose soon after the first large-scale quantitative studies of diversity were carried out in the 1960s and 1970s (e.g. Raup 1972, 1976). These concerns have persisted to the current time and, during the past decade, have resulted in numerous studies of the impact of uneven sampling on diversity reconstruction, and the development of several methods for measuring and 'removing' the effects of sampling (Peters & Foote 2001; Smith 2001; Smith *et al.* 2001; Crampton *et al.* 2003; Peters 2005; Upchurch & Barrett 2005; Smith & McGowan 2007; Uhen & Pyenson 2007; Alroy

From: McGowan, A. J. & Smith, A. B. (eds) *Comparing the Geological and Fossil Records: Implications for Biodiversity Studies*. Geological Society, London, Special Publications, **358**, 209–240.
DOI: 10.1144/SP358.14   0305-8719/11/$15.00 © The Geological Society of London 2011.

et al. 2008; Fröbisch 2008; McGowan & Smith 2008; Barrett *et al.* 2009; Butler *et al.* 2009, 2011; Wall *et al.* 2009; Benson *et al.* 2010; Mannion *et al.* 2011). Today, there are three broad approaches to the latter (Mannion *et al.* 2011): (1) rarefaction or subsampling approaches (e.g. Fastovsky *et al.* 2004; Alroy *et al.* 2008); (2) phylogenetic corrections ('ghost ranges') (Norell & Novacek 1992*a*, *b*; Lane *et al.* 2005; Upchurch & Barrett 2005; Lloyd *et al.* 2008); and (3) sampling metrics or proxies coupled to 'residual' diversity estimation (Smith & McGowan 2007; Fröbisch 2008; McGowan & Smith 2008; Barrett *et al.* 2009; Butler *et al.* 2009, 2011; Benson *et al.* 2010; Mannion *et al.* 2011). Recently, this situation has become more complex because several authors have proposed that the sampling of the fossil record, and the genuine diversity of taxa, might be correlated because they are both controlled by a third 'common cause' factor such as sea-level (Peters & Foote 2001; Smith *et al.* 2001; Peters 2005, 2006; Benton & Emerson 2007; Lloyd *et al.* 2008; Wall *et al.* 2009). This creates a dilemma for anyone wishing to reconstruct the diversity of a particular group of organisms: the decision to correct for apparent uneven sampling might only be justified when the absence of a common cause can be confirmed (see discussion in Butler *et al.* 2011).

The majority of studies of palaeodiversity patterns and fossil record sampling have been based on marine invertebrates (e.g. Crampton *et al.* 2003; Smith & McGowan 2007; Alroy *et al.* 2008; McGowan & Smith 2008; Wall *et al.* 2009). There have also been several studies of marine vertebrate diversity (Uhen & Pyenson 2007; Marx 2009; Benson *et al.* 2010; Marx & Uhen 2010; Benson & Butler 2011). Although 'sampling-corrected' studies of terrestrial organismal diversity (typically vertebrates) remain scarce, a number of analyses have appeared recently, focusing on Permo-Triassic vertebrates (Benton *et al.* 2004), anomodont synapsids (Fröbisch 2008), dinosaurs (Sereno 1997, 1999; Weishampel & Jianu 2000; Fastovsky *et al.* 2004; Carrano 2005, 2008*a*; Upchurch & Barrett 2005; Wang & Dodson 2006; Lloyd *et al.* 2008; Barrett *et al.* 2009; Butler *et al.* 2011; Mannion *et al.* 2011), pterosaurs (Butler *et al.* 2009) and Cenozoic mammals (Alroy 2000). These analyses have varied in their conclusions with regard to the impact of sampling in the terrestrial realm. Typically, sampling seems to have had significant effects on observed diversity, but it is usually possible to tease apart such effects from genuine diversity change (Wang & Dodson 2006; Lloyd *et al.* 2008; Barrett *et al.* 2009; Mannion *et al.* 2011). However, there are occasions, such as the study of pterosaur diversity by Butler *et al.* (2009), where the apparent peaks and troughs in diversity are almost entirely controlled by sampling, especially the presence or absence of Lagerstätten (though see Dyke *et al.* 2009 for a contrary view). At present there remains considerable work still to be carried out on a number of issues, including reconstruction of diversity patterns for a wider range of organisms (e.g. terrestrial plants), tests of common cause hypotheses, and evaluations of the relative performances of competing methods for estimating and correcting for sampling biases.

All of these areas are being explored at present, but one aspect of this field has received little attention. Virtually all previous studies have been either 'global' (in the sense that they deal with nearly all available data for a particular clade), or regional (e.g. the study of North American Cenozoic mammals by Alroy (2000) and New Zealand molluscs by Crampton *et al.* (2003)). To date, few studies have looked at the relationships between global and regional diversity and global and regional sampling. This subject represents an important set of issues. One danger with global analyses of diversity is that they can obscure local or regional variations that might be highly significant for our understanding of the patterns and processes relating to radiations and extinctions. Diversity changes could occur simultaneously across the globe (e.g. a sudden mass extinction) or alternatively could display a different pattern in each region (e.g. a group might originate in one area and then subsequently disperse to other regions or along latitudinal gradients). There are also dangers associated with regional studies of diversity. If regional variations in extinction tempo and selectivity have occurred, then extrapolation from a single regional pattern to the global scale will oversimplify the diversity pattern and will make it difficult to obtain an accurate understanding of evolutionary history and the causes of extinction events. Such concerns have been raised with regard to the end-Cretaceous mass extinction because, until recently, much of our understanding of the tempo and selectivity of this event was dominated by data from North America (Alroy *et al.* 2001; Jackson & Johnson 2001; Wang & Dodson 2006). Problems also occur when comparing fossil record sampling at global and regional scales (e.g. see the discussion of competing explanations for correlations between diversity and sampling metrics in Smith & McGowan (2007) and the 'Discussion' here). Thus, without detailed examination of the relationships between global and regional diversity and sampling, our understanding of both genuine diversity change and artefacts created by sampling biases must remain incomplete.

In this paper, we use a recently compiled dinosaur data set to address the following questions that touch upon global v. regional diversity and sampling patterns in the terrestrial realm:

(i) At a global scale, is the observed diversity of dinosaurs (and the clades Theropoda, Sauropodomorpha and Ornithischia) correlated with sampling metrics (numbers of dinosaur-bearing formations and collections)?

(ii) Do different regions display similar or different observed diversity patterns through time?

(iii) Are similarities or differences in region-to-region diversity patterns created by genuine evolutionary events or do they reflect artefacts generated by sampling?

(iv) How well do regional sampling metrics predict observed diversity at regional and global scales?

Although dinosaurs form the focus of our analyses, the abundance and ecological dominance of this group during the Mesozoic means that this study has the potential to produce insights into the more general relationships between global and regional diversity and sampling in the terrestrial realm.

Methods and materials

Data set

The Dinosauria provides a particularly suitable group for a case study of palaeodiversity and sampling in the terrestrial realm. This group produced the dominant elements of global terrestrial faunas from the Late Triassic to the end of the Cretaceous, a time span of approximately 160 million years (Weishampel et al. 2004; Lloyd et al. 2008). Their fossil record has received considerable attention from taxonomists and phylogeneticists, accurate geographical and stratigraphical data are available, and most of the major clades of dinosaurs were globally distributed (Upchurch et al. 2002; Weishampel et al. 2004; The Paleobiology Database (PaleoDB)). Moreover, previous studies suggest that dinosaur evolutionary history exhibits some important diversity events, including a gradual rise in diversity throughout much of the Mesozoic, and extinctions at the Triassic/Jurassic, Jurassic/Cretaceous and Cretaceous/Palaeogene boundaries (Fastovsky et al. 2004; Wang & Dodson 2006; Lloyd et al. 2008; Barrett et al. 2009; Mannion & Upchurch 2010a; Mannion et al. 2011).

Data on the stratigraphical occurrences and geological settings of dinosaurian species were taken from Butler et al. (2011), which draws heavily upon the sauropodomorph data set of Mannion et al. (2011) and information on theropods and ornithischians in the PaleoDB (data compiled primarily by MT Carrano [Carrano 2008b]). We have also included the sauropod species Haplocanthosaurus

delfsi (now recognized by us as probably not congeneric with the type species H. priscus), as well as Mesozoic birds (which were omitted by Butler et al. 2011) based on data downloaded from the PaleoDB. All species were reviewed in order to eliminate synonyms, nomina dubia and nomina nuda. The final data set comprises information on 847 valid dinosaurian species, and can be regarded as up-to-date as of March 2010. This represents the largest species-level data set currently available on terrestrial Mesozoic animals. In order to generate data for regional analyses, this data set has been partitioned into five geographical areas based on the continents of Europe, North America, South America, Africa and Asia. Indo-Madagascan, Australasian and Antarctic taxa were excluded from the regional (but not the global analyses) because these areas have yielded very small numbers (4.3%) of dinosaurian species. The number of dinosaur species present in each of the regions is shown in Table 1.

Two global sampling metrics (numbers of dinosaur-bearing formations (DBFs) and collections (DBCs)) have been derived from the PaleoDB (downloaded March 2010). These estimated numbers of DBFs and DBCs include all formations and collections where dinosaur remains have been recovered, irrespective of the ability of systematists to assign the remains to a lower-level taxonomic group or particular species. DBC bin counts range from 53 to 1486 and DBF bin counts range from 22 to 172 (see Table 1 and supplementary material). Regional sampling metrics were generated by calculating the numbers of DBCs and DBFs for a given region (i.e. each of the five continental areas specified above – see Table 1). In addition, we have utilized the numbers of terrestrial collections of fossils and geological units for North America presented by Peters & Heim (2010) and the outcrop area for

Table 1. *Summary of the number of dinosaurian species, dinosaur-bearing collections (DBCs) and dinosaur-bearing formations (DBFs) in each of the five geographical regions used in the regional analyses (and the Indo-Madagascar, Australasia and Antarctica regions ('Other areas') that contributed to the global analyses)*

Region	No. species	No. DBCs	No. DBFs
Africa	58	431	61
Asia	308	1125	224
Europe	116	1079	200
North America	224	2449	204
South America	104	397	86
Other areas	37	242	45
Total	847	5723	820

terrestrial sedimentary rock for western Europe estimated by Smith & McGowan (2007). These various global and regional sampling metrics potentially capture information on two key aspects of fossil record sampling. Numbers of terrestrial units, numbers of DBFs and outcrop area potentially measure geological controls on sampling, whereas numbers of terrestrial collections and numbers of DBCs attempt to estimate anthropogenic controls relating to variation in collecting effort by palaeontologists. There are numerous possible alternatives to DBFs in the sense that we could count all vertebrate-bearing formations, or all terrestrial formations irrespective of whether or not they preserve fossils, or we could limit the taxonomic scope (e.g. by comparing sauropodomorph diversity with the number of sauropodomorph-bearing formations). Upchurch & Barrett (2005), and especially Barrett *et al.* (2009), summarized several arguments for using the number of geological formations as a sampling proxy in general, and DBFs in particular:

(i)   Peters (2005) demonstrated that the number of geological formations generally correlates strongly with other measures of fossil record sampling such as numbers of sedimentary rock sections and estimates of total rock volume.

(ii) Peters & Foote (2001) suggested that the number of formations captures variability in the range of habitats present during each time bin.

(iii) There may be little benefit in counting geological formations that do not preserve any fossils (or fossils of the groups under investigation): ecological surveys of extant organisms generally do not devote major search effort in environments that are not inhabited by the target organisms (Barrett *et al.* 2009). However, it is important to establish thoroughly that a formation genuinely lacks fossils as opposed to lacking reports of fossils merely because collecting effort has been insufficient to date.

(iv) Smith & McGowan (2007) have shown that taking into account unfossiliferous formations does not significantly affect the relationship between rock outcrop area and diversity.

(v)   Dinosaurs had a global distribution and their fossils are recovered from all terrestrial facies. Therefore, absence of dinosaurs in a particular formation is more likely to arise from taphonomic factors (dinosaurs were not preserved), or anthropogenic factors (the formation has not been sampled for dinosaurs) than their genuine absence. If the former explanation (taphonomy) is correct, then it is not appropriate to include non-dinosaur-bearing formations within our sampling proxies, because such

formations do not offer a genuine opportunity to sample dinosaurs. However, if it is the case that certain formations have not been inspected for dinosaurs then these might be capable of preserving dinosaur material and ideally should be added into the list of DBFs. Since we cannot be sure which of these explanations is correct (and in reality both are likely to be true), then some arbitrary cut-off point must be established in order to determine which formations are included/excluded in the sampling proxy. Here, we define a DBF as any formation that has produced dinosaur material of any type, even if the specimens are so fragmentary that they can only be identified as 'Dinosauria indet'. In future analyses, it would be interesting to evaluate the relationships between observed dinosaur diversity and a wider sampling proxy such as all terrestrial vertebrate-bearing formations.

Therefore, we follow Upchurch & Barrett (2005), Wang & Dodson (2006), Barrett *et al.* (2009), Butler *et al.* (2011) and Mannion *et al.* (2011) in using DBFs as a proxy for geological controls on observed dinosaur diversity.

Methods

Choice of time bins. Butler *et al.* (2011) based the time bins for their study on Standard European stages and the absolute dates provided by Gradstein *et al.* (2005). These authors examined the effects of variable time bin duration by assessing the statistical correlation between bin length and taxic diversity, and bin length and geological sampling. They also removed the influence of bin duration by calculating partial correlations for pairwise comparisons for parameters such as taxic diversity counts and sampling metrics. The results of this work demonstrated that diversity of dinosaurian species does not correlate with time bin duration (see also Mannion *et al.* 2011).

Fastovsky *et al.* (2004), Barrett *et al.* (2009) and Mannion *et al.* (2011) analysed dinosaur diversity using substage time bins. However, here we do not attempt this level of temporal resolution. The stratigraphical ages of most dinosaurian genera and species can only be determined accurately to stage level at best (and there are several instances where a taxon's age cannot be estimated more precisely than epoch level). In the current data set, for example, only 26% of 212 sauropodomorph species can be dated with reasonable confidence to the substage level (see also Wang & Dodson 2006). Previous analyses that have employed substages as time bins have had to make the assumption that many dinosaurian taxa known from a given stage (e.g.

the Kimmeridgian) occurred throughout the entirety of that stage (i.e. such taxa are assigned to the early, middle and late Kimmeridgian time bins). Given that the majority of taxa cannot be dated with more precision than the stage level, this means that genuine fluctuations in diversity from one substage to the next within a single stage are likely to be overwhelmed by the 'noise' generated by the large number of taxa whose stratigraphical ranges have been 'smeared' across the entire stage. Moreover, this poor temporal resolution is likely to have the effect of increasing the similarity between scores of adjacent time bins, increasing temporal autocorrelation and potentially leading to artificial inflation of correlation coefficients. Thus, although substage level temporal resolution is important, especially when examining key events such as the lead up to the end-Cretaceous mass extinction (see below), considerable caution should be exercised when interpreting the results. Since the current study is largely based on the data set of Butler *et al.* (2011), we therefore simply use Standard European stages as time bins.

Transformation of the data and statistical comparisons. In order to deal with potentially spurious or inflated correlations caused by trend and temporal autocorrelation in time series data, we applied the method of generalized differencing, which incorporates detrending and differencing but modulates the differences by the strength of the correlation between successive data points (McKinney 1990; Alroy 2000). A full description of this approach is provided by Benson & Butler (2011). All of the data series are detrended; however, only series that show a serial correlation at a time lag of one stage are differenced (with the strength of 'differencing' modulated by observed autocorrelation). Comparisons between the detrended data series were made using the non-parametric Spearman's rank and Kendall's tau, as implemented in PAST (Hammer *et al.* 2001). Statistical significance was determined using an alpha value of 0.05. However, our analyses form sets or 'families' in which the same data (or overlapping portions of the data set) are analysed several times (e.g. the 16 analyses that include comparisons of global DBFs with other data series). For each such family of analyses, the *P*-value used to determine statistical significance has been adjusted for multiple comparisons using the Bonferroni correction (Rice 1989; Waite 2000; Mannion & Upchurch 2010*b*; Butler *et al.* 2011). For example, the global DBF data series is utilized 16 times in our analyses (see supplementary material), so the *P*-value for statistical significance for analyses involving global DBFs is $0.05/16 = 0.0031$. Because each pairwise comparison involves two data series (i.e. members of two families of analyses), the most

stringent *P*-value is used: for example, global dinosaur diversity is analysed 20 times (giving a *P*-value cut-off of 0.0025), so this more stringent *P*-value is applied to the comparison of global dinosaur diversity with global DBFs. However, there are logical inconsistencies inherent to the Bonferroni correction. For example, one researcher carrying out two analyses should apply a *P*-value of 0.025, whereas two researchers each running a single analysis should apply a *P*-value of 0.05. Moreover, we have good reason to suspect a correlation between sampling and diversity and are not simply applying numerous tests of correlation on the off chance that statistically significant relationships are detected. This means that the risk of detecting spurious correlations by making multiple comparisons is not of primary concern. Here, therefore, we discuss the results of our analyses in terms of those that pass or fail when a *P*-value of 0.05 is used, and then comment on the sensitivity of their statistical significance to the application of the Bonferroni correction.

'Correction' of taxic diversity and sampling estimates. Differenced taxic diversity counts were 'corrected' for both of the global geological sampling metrics (DBCs and DBFs) using the residuals method of Smith & McGowan (2007). This approach first reorganizes the data in two data series so that each has its values ranked from low to high. The relationship between the two data series is then expressed as a regression equation, which allows the estimation of the diversity score that would be predicted if observed diversity is entirely controlled by sampling. Residual diversity values are then calculated by subtracting the predicted diversity from the observed diversity (i.e. residuals represent the amount of diversity that cannot be explained by sampling).

Long-term trends. In order to explore putative long-term trends in dinosaur diversity, we have applied a non-parametric runs test using PAST (Hammer *et al.* 2001), which tests for non-random trends in time series data (Hammer & Harper 2006; see also Mannion & Upchurch 2010*a*).

Analyses and results

The results of all analyses (labelled A to Q) are presented in the supplementary material (see also Table 2 for a key to the A–Q analysis labels).

Global diversity patterns

Global dinosaur diversity and sampling. Analyses A1 and A2 compare transformed observed global dinosaur diversity with global DBCs and DBFs

Table 2. *Key to analysis labels*

Label	Description of analysis
A	Tests of correlation between the observed diversity of dinosaurs, sauropodomorphs, theropods and ornithischians with global DBCs and DBFs
B	Tests of correlation between the observed 'Rest of the world' dinosaur diversity and regional diversity
C	Tests of correlation between the observed dinosaur diversity in two regions
D	Tests of correlation between global sampling and regional sampling
E	Tests of correlation between sampling metrics in two regions
F	Tests of correlation between different sampling metrics for the same region (e.g. DBCs and the numbers of terrestrial units for North America)
G	Tests of correlation between global observed dinosaur diversity and regional sampling metrics
H	Tests of correlation between regional observed dinosaur diversity and a regional sampling metric
I	Tests of correlation between regional observed sauropodomorph diversity and a regional sampling metric
J	Tests of correlation between regional observed theropod diversity and a regional sampling metric
K	Tests of correlation between regional observed ornithischian diversity and a regional sampling metric
L	Tests of correlation between regional observed dinosaur diversity and global sampling metrics
M	Tests of correlation between the DBC-based RDEs of the three dinosaurian subclades
N	Tests of correlation between 'rest of the world' residual diversity estimates and a regional RDE
O	Tests of correlation between the DBC-based RDEs for two regions
P	Tests of correlation between the DBF-based RDEs of the three dinosaurian subclades
Q	Runs tests of long-term trend in the RDEs of Dinosauria, Sauropodomorpha Theropoda and Ornithischia

respectively. There are strong positive correlations between the two sampling metrics and observed global dinosaur diversity (see Fig. 1 for plots of the untransformed data series). Support for these correlations persists even when the Bonferroni correction is applied.

Analyses A3–A8 compare the transformed global observed diversity of sauropodomorphs, theropods and ornithischians with global DBCs and DBFs (see Fig. 2a–c for plots of the untransformed

data series). All analyses demonstrate the presence of strong positive correlations between the observed global diversity of each clade and both sampling metrics, except for the non-significant relationship between sauropodomorph diversity and global DBFs. Application of the Bonferroni correction slightly weakens the support for these positive correlations: all results remain statistically significant except that the comparison of sauropodomorph diversity with global DBCs fails narrowly.

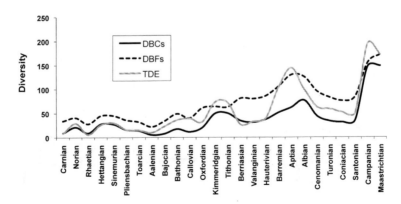

Fig. 1. Graphs showing the fluctuations in observed dinosaur species diversity (TDE), the number of dinosaur-bearing collections (DBCs) and the number of dinosaur-bearing formations (DBFs). N.B. the number of DBCs has been divided by 10 so it can be plotted on the same *y*-axis as the number of dinosaur species and the numbers of DBFs.

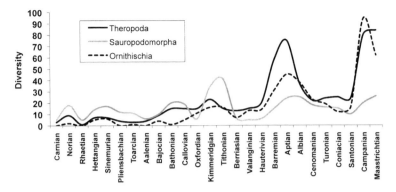

Fig. 2. Graphs of global observed diversity for Ornithischia, Sauropodomorpha and Theropoda.

Global v. regional diversity patterns and regional variations

Global v. regional observed diversity. In Analyses B1–B5, 'global' diversity is compared with regional diversity by calculating a 'rest of the world' diversity (e.g. the diversity counts for Africa + Asia + Europe + North America) and comparing this with the fifth region (in this case South America). Analyses B1–B4 do not produce statistically significant correlations, although B2 and B3 are relatively narrow fails (P-values are just above 0.05). Analysis B5 ('rest of the world' dinosaur diversity compared with African dinosaur diversity) produces statistically significant positive correlations (Fig. 3), but support for this result disappears once the Bonferroni correction is applied.

Region-to-region comparisons of observed diversity. Analyses C1–C10 are a set of pairwise comparisons between the transformed observed diversities of each of the five regions (e.g. European dinosaur diversity v. Asian dinosaur diversity). With

the exception of North American v. South American dinosaur diversity (analysis C2, see Fig. 4 for plots of the untransformed data), all comparisons fail to produce statistically significant results. Moreover, the positive correlation between the dinosaur diversities of North and South America is not supported after the Bonferroni correction is applied.

Comparisons of sampling metrics at global and regional scales

Global sampling v. regional sampling. Analyses D1–D4 compare global sampling metrics (i.e. DBCs and DBFs) with regional sampling metrics (i.e. the North American terrestrial collections and units of Peters & Heim (2010), the western European rock outcrop area of Smith & McGowan (2007), and regional DBC and DBF counts). D5–D14 use 'rest of the world' sampling metrics (e.g. the DBCs of Africa + Asia + Europe + North America) to represent global sampling in comparisons with a fifth region (i.e. South America in this example). Six of these comparisons produce

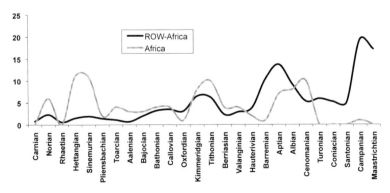

Fig. 3. Graphs of 'rest of the world' observed dinosaur diversity (minus Africa) and African dinosaur diversity (see analysis B5). N.B. 'Rest of the world' observed diversity has been divided by 10 so that it can be plotted on the same y-axis as African observed diversity.

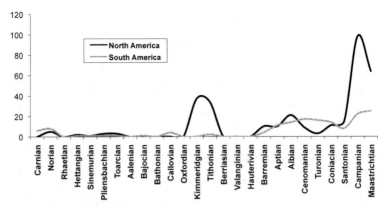

Fig. 4. Graphs of observed dinosaur diversity in North and South America.

statistically significant positive correlations between global (or 'rest of the world') and regional sampling metrics. These statistical 'passes' include: D1, DBCs v. North American terrestrial collections; D5, rest of the world DBCs v. North American DBCs; D9, rest of the world DBCs v. African DBCs; D11, rest of the world DBFs v. European DBFs; D12, rest of the world DBFs v. South American DBFs (narrow failure according to the Spearman's rank test, but a pass according to Kendal's tau); and D14, rest of the world DBFs v. African DBFs. Support for these correlations disappears for D1, D5, D11 and D12 when the Bonferroni correction is applied, but D9 (Fig. 5) and D14 remain statistically significant. Thus, African DBCs and DBFs are apparently positively correlated with 'rest of the world' DBCs and DBFs.

Region-to-region comparisons of sampling metrics. Analyses E1–E20 represent pairwise comparisons between sampling metrics from different regions (e.g. North American DBFs v. European

DBFs). Two analyses (E4 and E7) produce statistically significant positive correlations, and E10 and E17 fail narrowly. All four of these analyses are based on comparisons with African DBCs or DBFs. The statistical significance of analyses E4 and E7 disappears when the Bonferroni correction is applied.

Comparisons of different sampling metrics within the same region. Analyses F1–F4 compare different types of sampling metric within the same region (e.g. European DBCs v. western European terrestrial rock outcrop area). All four of these analyses demonstrate the presence of significant positive correlations between the different types of regional metric. For example, analysis F4 demonstrates that European DBFs are positively correlated with the western European rock outcrop area estimates of Smith & McGowan (2007) (Fig. 6a). Support for the significance of F1, F2 and F4 disappears when the Bonferroni correction is applied (F1 and F2 only fail very narrowly), but persists for F3 (North

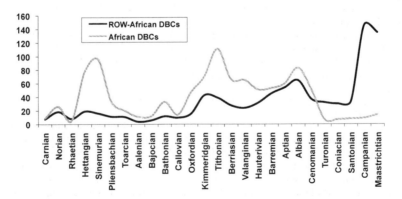

Fig. 5. Graphs of 'rest of the world' (minus Africa) DBCs and African DBCs. N.B. 'Rest of the world' DBCs have been divided by 10 so that they can be plotted on the same *y*-axis as African DBCs.

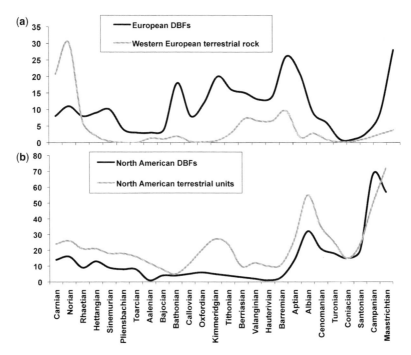

Fig. 6. Graphs of: (**a**) European DBFs and western European terrestrial rock outcrop area; and (**b**) North American DBFs and North American terrestrial units. N.B. western European terrestrial rock outcrop area has been divided by 10 so that it can be plotted on the same *y*-axis as European DBFs.

American DBFs v. the North American terrestrial units of Peters & Heim (2010)) (Fig. 6b).

Regional sampling metrics and observed diversity

Regional sampling v. global diversity. Analyses G1–G13 compare various regional sampling metrics (e.g. North American DBFs) with global dinosaur diversity. Six of these analyses (G1, G4–6, G12, G13) support significant positive correlations, and G2, G3 and G10 fail narrowly. Application of the Bonferroni correction results in the failure of analyses G4–G6, but the statistical significance of analyses G1, G12 and G13 persists. Thus, there is strong support for positive correlations between sampling metrics comprising North American DBCs, African DBCs, or African DBFs and global dinosaur diversity (Fig. 7).

Regional sampling metrics v. regional observed diversity. Analyses H1–H14 compare regional dinosaur diversity (e.g. European dinosaur diversity) with regional sampling metrics taken from the same region (in the case of the current example, these are European DBCs, European DBFs, and western European rock outcrop area). Nine of these

analyses (H1, H2, H4–H6, H11–H14) demonstrate the presence of significant positive correlations (see Fig. 8a, b for exemplar plots of the untransformed data series), and four others (H3, H7, H9 and H10) fail the tests only narrowly. With the exception of analysis H2, the significance of these correlations persists even after the application of the Bonferroni correction. These results therefore suggest that in all regions, except perhaps South America, regional sampling metrics are strongly positively correlated with regional diversity counts.

Regional sampling metrics and regional diversity have also been compared for each of the three major clades of dinosaurs (Fig. 9a–c): sauropodomorphs (analyses I1–I14); theropods (analyses J1–J14); and ornithischians (analyses K1–K14). For sauropodomorphs, eight analyses (I1, I4–I8, I13, I14) demonstrate significant positive correlations between regional sampling and regional diversity, and analyses I2 and I9–11 fail the tests only narrowly. Support for three of these positive correlations (I1, I5, I6) persists even when the Bonferroni correction is applied, and three of the new failures (I4, I8, I13) fail only narrowly. Nine of the theropod analyses (J1, J2, J4–J6, J10–J14) produce significant positive correlations, and one analysis (J7) fails the statistical tests narrowly.

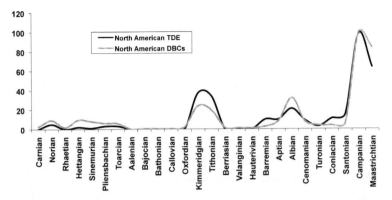

Fig. 7. Graphs of North American DBCs and observed dinosaur diversity. N.B. North American DBCs have been divided by 10 so that they can be plotted on the same *y*-axis as North American observed dinosaur diversity.

Application of the Bonferroni correction produces seven remaining significant results (J1, J5, J6, J10–J14). For ornithischians, eight analyses yield significant positive correlations (K1–4, K6, K11, K13, K14) and one (K5) fails the tests narrowly. Three of these analyses (K1, K4, K13) continue to pass the statistical tests (and K2 and K14 only fail narrowly) even after the application of the Bonferroni correction. In general, these results demonstrate the existence of strong positive correlations between sampling and the diversity of sauropodomorphs, theropods and ornithischians at the regional scale. The strongest correlations typically occur in North America, Europe and Africa, with comparisons involving Asia and South America failing more frequently.

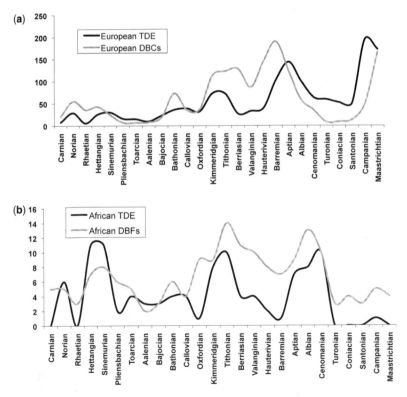

Fig. 8. Graphs of: (**a**) European observed dinosaur diversity and European DBCs; and (**b**) African observed dinosaur diversity and African DBFs.

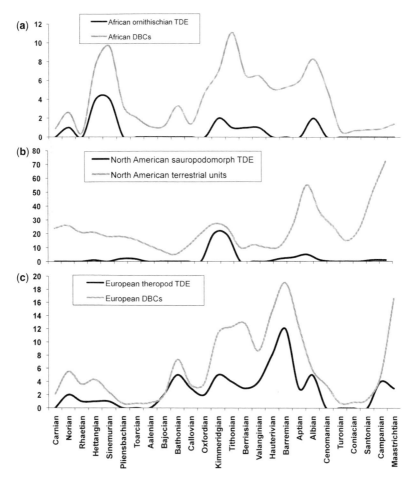

Fig. 9. Graphs of: (**a**) African ornithischian observed diversity and African DBCs; (**b**) North American sauropodomorph observed diversity v. North American terrestrial units; and (**c**) European theropod observed diversity and European DBCs.

Global sampling and regional diversity

Analyses L1–L10 compare global sampling (DBCs and DBFs) with dinosaur diversity in each of the five regions. Four of these analyses (L1, L4, L8, L9) demonstrate the presence of positive correlations between global sampling and regional diversity (Fig. 10), and two others (L3, L6) fail the tests only narrowly. With the exception of analysis L1, all of these significant results disappear when the Bonferroni correction is applied.

Residual diversity at global and regional scales

Global residuals for dinosaurs and three major clades. Figures 11 and 12a–c show the residual diversity estimates (RDEs) for dinosaurs, theropods,

sauropodomorphs and ornithischians, calculated using DBCs. Analyses M1–M3 make pairwise comparisons between the RDEs of the three sub-clades: those of theropods and ornithischians are positively correlated, whereas there are no correlations with the RDEs of sauropodomorphs. The statistical significance of the correlation between theropod and ornithischian RDEs disappears when the Bonferroni correction is applied.

Figure 13a–e show the residual diversity estimates for dinosaur diversity in each of the five geographical regions. Analyses N1–N5 compare the global (rest of the world) dinosaur RDEs with the RDEs for each of these regions. All of these analyses fail the correlation tests; one fails marginally (N5).

Analyses O1–O10 are a series of pairwise comparisons between the RDEs of the five geographical

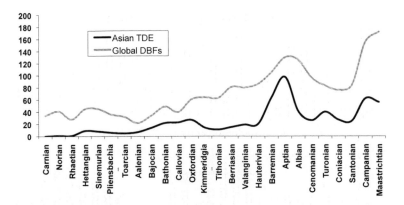

Fig. 10. Graphs of Asian observed dinosaur diversity and global DBFs.

regions (e.g. O1, North American RDE v. European RDE). All of these analyses fail the correlation tests.

In order to enable direct comparisons of our results with those of Barrett *et al.* (2009), we have also calculated the RDEs for theropods, sauropodomorphs and ornithischians using DBFs (see Fig. 14a–c and the supplementary material). Analyses P1–P3 carry out pairwise correlation tests between these three RDEs. The results indicate that theropod and ornithischian RDEs, and theropod and sauropodomorph RDEs, are strongly positively correlated, but a correlation between sauropodomorph and ornithischian RDEs cannot be detected.

Long-term diversity trends

In order to examine the possible presence of long-term trends in dinosaur diversity during the

Mesozoic, we have applied a runs test to the global DBC-based RDEs for dinosaurs (analyses Q1–Q3), DBC-based RDEs of sauropodomorphs (Q4–Q6), theropods (Q7–Q9) and ornithischians (Q10–Q12), and DBF-based RDEs for these three clades (Q13–Q21). For each set of analyses, long-term trends have been investigated for the Mesozoic as a whole, the Late Triassic–Jurassic inclusive, and the Cretaceous. Analysis Q1 suggests that there is a long-term trend of increasing RDE values through the Mesozoic, but Q2 and Q3 indicate that this results from the combination of a weakly supported upward trend during the Triassic and Jurassic, and no discernible trend during the Cretaceous. The DBC-based residuals produce different results for each of the three clades considered. Ornithischians display no detectable long-term trend in RDEs, whereas the increase in RDEs for theropods during

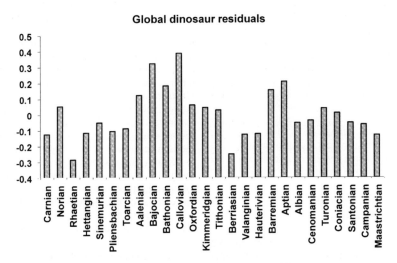

Fig. 11. Residual diversity estimates of global dinosaur diversity based on DBCs.

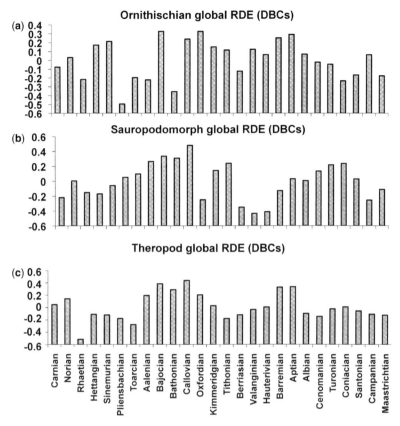

Fig. 12. Residual diversity estimates based on DBCs for: (**a**) global ornithischian diversity; (**b**) global sauropodomorph diversity; and (**c**) global theropod diversity.

the Mesozoic is apparently generated by an upward trend during the Triassic–Jurassic and no trend in the Cretaceous. Sauropodomorphs are interesting because they display evidence for upward trends in RDE values for both the Triassic–Jurassic and Cretaceous, but these combine together to produce only a weakly supported trend for the Mesozoic as a whole. This can be explained as the result of combining two temporally distinct upward trends that are separated by the decrease in sauropod diversity at the Jurassic/Cretaceous boundary (see below). None of the DBF-based RDEs show any evidence for an upward diversity trend during the Mesozoic or in either of the shorter time slices.

Discussion

Diversity and sampling

Several previous studies (e.g. Wang & Dodson 2006; Lloyd *et al.* 2008; Barrett *et al.* 2009; Butler *et al.* 2011; Mannion *et al.* 2011), have

demonstrated that global dinosaur diversity (or that of subclades such as Theropoda) correlates positively with global sampling metrics such as the numbers of DBCs, DBFs and/or dinosaur-bearing localities. This conclusion is supported by the results of analyses A1–A8 here. One exception to this is the diversity of sauropodomorphs v. the number of DBFs (analysis A4). Barrett *et al.* (2009) found a weak, non-significant, negative correlation (based on Spearman's rank and Kendal's tau tests) between phylogenetic diversity estimates for sauropodomorphs and the number of DBFs, whereas the same study found a non-significant positive correlation between observed sauropodomorph taxic diversity and DBFs. Mannion *et al.* (2011) and the current study, found no statistical support for either a negative or positive correlation for the Mesozoic as a whole, but Mannion *et al.* (2011) did find a positive correlation for the Cretaceous by itself.

Approximately half of our other analyses that have compared sampling with diversity also

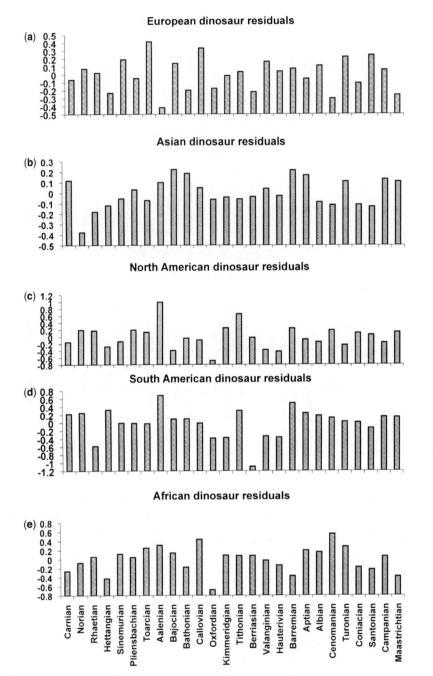

Fig. 13. Regional residual diversity estimates for dinosaurs based on DBCs: (**a**) Europe; (**b**) Asia; (**c**) North America; (**d**) South America; and (**e**) Africa.

demonstrate the presence of strong positive correlations between these parameters. These analyses include comparisons of global dinosaur diversity and regional sampling (G analyses – six passes out of 13 comparisons), regional dinosaur diversity with regional sampling (H analyses – nine passes out of 14 comparisons), regional sauropodomorph, theropod and ornithischian diversity v. regional

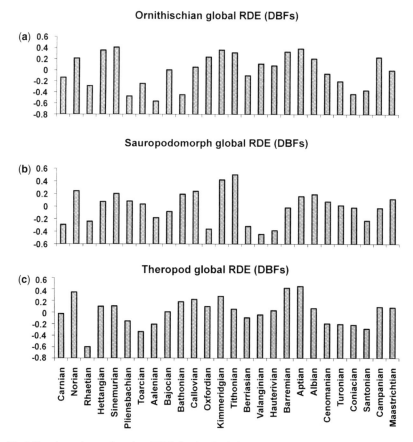

Fig. 14. Residual diversity estimates based on DBFs for: (**a**) Ornithischia; (**b**) Sauropodomorpha; and (**c**) Theropoda.

sampling (I, J and K analyses – 25 passes out of 42 comparisons), and regional dinosaur diversity v. global sampling (L analyses – four passes out of 10 comparisons). There are three principle explanations for these strong positive correlations between the sampling metrics and observed diversity at the various geographical and phylogenetic scales:

(i) *Common cause.* A third factor, such as sea-level, might control both sampling (e.g. by determining how much fossiliferous sediment is preserved or destroyed during each time bin) and genuine diversity (e.g. via the effects of sea-level on the fragmentation/connection of land areas and/or increases/decreases in land area) (Sepkoski 1976; Peters & Foote 2001; Smith *et al.* 2001; Peters 2005, 2006; Benton & Emerson 2007; Lloyd *et al.* 2008; Wall *et al.* 2009; Butler *et al.* 2011; Mannion *et al.* 2011). Butler *et al.* (2011) presented a series of analyses that demonstrate that

apparent correlations between sea-level or land area and observed dinosaur diversity disappear once data series are detrended and the effects of temporal autocorrelation are taken into account. In contrast, positive correlations between dinosaur diversity and sampling metrics persist even after data transformation has removed the effects of long-term trends. Here, therefore, we do not regard sea-level common cause as a convincing explanation for positive correlations between the sampling metrics and observed dinosaur diversity. At present, we are not aware of any other hypothesized common cause mechanisms that could explain positive correlations between sampling metrics and observed dinosaur diversity.

(ii) *Circular reasoning or redundancy.* A second possibility is that DBCs and DBFs correlate with dinosaur diversity because of circular reasoning. If dinosaurian taxa have played a significant role in the recognition and definition of geological units (e.g. formations), then the discovery of new dinosaurs might prompt the naming of new geological formations, and a tendency for DBFs and dinosaur diversity to be positively correlated would be automatically built into these data series. Similarly, it could be argued that those time periods when dinosaurs were genuinely more diverse and abundant will tend to produce outcrops with richer fossil assemblages. Such outcrops might in turn tend to attract disproportionate attention from palaeontological expeditions because of the perceived increased probability of a successful and productive field season. Moreover, a genuine decrease in diversity is likely to place constraints on the number of DBCs: palaeontologists may still search intensively, but must report few DBCs if specimens are rare. If these phenomena occur, then a tendency for observed dinosaur diversity to be positively correlated with the number of DBCs might also be generated automatically. This problem has also been termed 'redundancy' (see Benton *et al.* 2011).

In previous analyses and those carried out here, however, the dangers of circular reasoning have been minimized by counting all DBCs and DBFs irrespective of the quality or quantity of material recovered from each collection or formation. This means that a formation that has only produced a few scraps of indeterminate dinosaur material is still considered to be 'an opportunity to observe' dinosaur diversity (see Upchurch & Barrett 2005), and carries as much weight in the analyses as a geological unit (such as the Late Jurassic Morrison Formation of the USA) that has produced hundreds of skeletons belonging to tens of diagnosable dinosaurian species (Weishampel *et al.* 2004). At present, quantitative data on the extent to which dinosaur discoveries influence the naming of geological units is not available. However, we suspect that the decision to divide sedimentary sequences into separate geological formations is usually based on abiotic evidence (e.g. boundaries between formations are recognized on the basis of changes in facies), and the recognition and naming of geological units often precedes the discovery of dinosaurs in them. When biotic evidence is used to distinguish between formations, this may include the presence/absence of certain dinosaurian taxa, but is frequently based on non-dinosaurs such as ostracods, shelled invertebrates, fish and other tetrapods. For example, the recognition of the Dinosaur Park Formation (Alberta, Canada) and the 'Dinosaur beds' of Malawi, as distinct geological units, seems to result from the presence of dinosaur fossils. On the other hand, the Cabao Formation in Libya was named prior (approximately in 1963) to the discovery of dinosaur remains in the 1980s' (Le Loeuff *et al.* 2010): this formation therefore seems to have been distinguished from other geological units on the basis of sedimentological differences rather than the presence of certain types of dinosaur. Moreover, any circular relationships between DBCs/DBFs and observed diversity should be at least partially disrupted when comparisons are made between these metrics and clades within Dinosauria: after all, a formation that has been recognized on the basis of the occurrence of a distinctive ornithischian dinosaur is not linked (in terms of circular reasoning at least) to the observed diversity of sauropodomorphs. Yet, analyses A3–A8 demonstrate that the observed diversities of the three main clades of dinosaurs also correlate with numbers of DBCs and DBFs (with the exception of sauropodomorphs v. DBFs – see above), and evidence for these positive correlations also occurs when these clades are compared with regional sampling metrics. These results suggest that there is more to the relationships between these data series than a simple circularity based on counts of geological units or collections defined by the presence of dinosaurs.

Finally, it should be noted that the results of analyses F1–F4 indicate that there are positive correlations between different sampling metrics in North America (i.e. DBFs and DBCs v. the terrestrial rock units and collections of Peters & Heim 2010) and in Europe (i.e. DBFs and DBCs v. the western European outcrop area estimates of Smith & McGowan 2007). Importantly, the Peters & Heim (2010), and especially the Smith & McGowan (2007) sampling metrics are not defined on the basis that they are 'opportunities to observe dinosaur diversity' (e.g. sedimentary rock outcrop area is counted in square km). Thus, aside from an impact on sampling rates, there is no reason to expect a correlation between these sampling metrics and the number of dinosaur taxa present in these sediments. Agreement among sampling metrics, each based on a different criterion, suggests that they are collectively 'homing in' on some form of sampling signal, and correlations cannot be explained merely in terms of their definition based on the presence/absence of dinosaurs. We suggest therefore that while circular reasoning cannot be completely ruled out as a contributory factor to the occurrence of positive correlations between sampling metrics and observed dinosaur diversity, it is unlikely that this phenomenon entirely (or even largely) accounts for these correlations.

(iii) *Sampling influences observed diversity.* In agreement with previous analyses, such as Fastovsky *et al.* (2004), Wang & Dodson (2006), Barrett *et al.* (2009), Butler *et al.* (2011) and Mannion *et al.* (2011), current evidence indicates that observed dinosaur diversity is strongly controlled by sampling, suggesting that sampling regimes in the terrestrial realm have an important impact on observed diversity patterns (as is also the case in the marine realm (e.g. Smith & McGowan 2007; Benson *et al.* 2010)). The fact that positive correlations between sampling metrics and observed diversity persist even when data are transformed and detrended, indicates that short-term fluctuations in sampling are probably responsible for short-term changes in observed diversity (Butler *et al.* 2011).

Common cause, circular reasoning and sampling are not mutually exclusive phenomena, and it is possible that all three factors have played some role in shaping observed dinosaur diversity. At present, however, the evidence supports a dominant role for sampling, and indicates that it is legitimate to attempt to reduce or remove the effects of uneven sampling via techniques such as subsampling or residual diversity estimation.

Dinosaur diversity at global and regional scales

Previous studies of dinosaur diversity have typically focused on patterns at the global scale (Carrano 2005; Upchurch & Barrett 2005; Wang & Dodson 2006; Lloyd *et al.* 2008; Barrett *et al.* 2009; Butler *et al.* 2011; Mannion *et al.* 2011) or have examined a single region such as North America or Africa (e.g. Carrano 2008*a*; Mannion 2009). At present, therefore, little is known about how dinosaur diversity varied from region to region, and there has been virtually no work on how such regional comparisons might be affected by sampling (though see Fastovsky *et al.* 2004; Mannion & Upchurch 2011). Analyses B1–B5 and C1–C10 examine the relationships of observed dinosaur diversity at global and regional scales. The results (one pass out of 15 comparisons) indicate that there is very little support for positive correlations between either global ('rest of the world') observed diversity and that of a given region, or between the diversities of pairs of regions. Thus, although individual peaks or troughs in observed diversity in two or more regions may coincide occasionally (e.g. Late Jurassic diversity peaks in North America and Africa associated with

the rich dinosaur faunas of the Morrison Formation and Tendaguru beds respectively), there is no compelling evidence that fluctuations in observed dinosaur diversity occurred in a uniform way across the globe during the Mesozoic (though see discussion of the Jurassic/Cretaceous boundary extinction below). However, this lack of uniformity or similarity might be an artefact created by the distorting effects of uneven sampling of the fossil record in each region. For example, simultaneous peaks in genuine diversity might have occurred in all geographical regions, but poor sampling in some regions might obscure the truly global nature of this evolutionary radiation. This issue is addressed by analyses N1–N5 and O1–O10, which examine correlations between residual dinosaur diversities at global and regional scales and between pairs of regions. The results demonstrate that there is little evidence for a truly uniform global pattern of evolutionary radiations among dinosaurs, even after the removal of the effects of sampling.

These results have two important implications. First, they suggest that sampling regimes have not imposed either artificially uniform or artificially disparate patterns on diversity across our five geographical regions. Second, it seems that dinosaur evolution (at least in terms of the radiation and extinction of species) followed different patterns in different regions (or at least that genuine differences outweigh similarities sufficiently often to preclude significant passes of correlation tests). This is perhaps not a particularly surprising result. Given regional differences in both biotic and abiotic conditions that inevitably occur as a result of climatic factors and evolutionary history, there is generally no *a priori* expectation that dinosaur diversity should have changed in a similar manner, at similar rates, in each region. Regional differences in dinosaur evolution could have occurred for many reasons. For example, each continental region may have had distinctive climatic, environmental and/or biotic conditions that provoked unique evolutionary responses among their particular dinosaurian faunas. Furthermore, it is conceivable that certain clades originated in one particular region, generating a spike in diversity in that region alone. Whether or not these regional radiations resulted in subsequent changes in diversity elsewhere in the world would then depend on the complex interactions between palaeogeography (e.g. the creation/destruction of dispersal routes) and palaeoecology (e.g. the ability of dispersing taxa to occupy niches in newly invaded areas). Similarly, there may have been region-specific extinctions: for example, diplodocoids in North America apparently died out at the Jurassic/Cretaceous boundary, whereas this clade continued to flourish in South America during the Early Cretaceous. However, one obvious exception to this 'lack of uniformity' might occur when mass extinction events result in a simultaneous global reduction in diversity, as may have occurred at the Triassic/Jurassic, Jurassic/Cretaceous and Cretaceous/Palaeogene boundaries. These extinction events are discussed at both global and regional scales in 'Dinosaur radiations and extinctions revisited' below.

Sampling metrics at regional and global scales

As noted above, some sampling metrics such as numbers of North American terrestrial units/ collections (Peters & Heim 2010) and estimates of sedimentary rock outcrop area (Smith & McGowan 2007; McGowan & Smith 2008), circumvent the potential problem of 'circular reasoning' because they estimate our opportunities to find fossils using units of measurement that are not defined by the presence of the taxa in which we are interested. One limitation with these sampling metrics, however, is that their temporal, and particularly their geographical extents are often limited by the quality and quantity of geological data available. For example, the intensive investigation of European stratigraphy during the past two centuries has led to a generally well-understood, detailed and accurate knowledge of the ages and distributions of sediments in this area, which in turn means that estimates of rock outcrop area per time bin can be made relatively easily and accurately. However, it is no coincidence that such regional sampling metrics have been produced for North America and Europe, but rarely for other parts of the world. In Asia, for example, uncertainties about the ages of rock outcrops are far more prevalent (e.g. the Mengyin Formation, that has yielded the sauropod *Euhelopus*, could be either Late Jurassic or Early Cretaceous in age – see Wilson & Upchurch 2009 and references therein). Moreover, the combination of extensive remote areas and a shorter history of geological research in parts of Asia, Africa and South America, means that many sediments remain to be mapped and/or dated accurately. These circumstances raise the question as to whether detailed and accurate sampling metrics based on data for North America and/or Europe are suitable as sampling metrics for studies of diversity at global scales. A similar issue was raised by Smith & McGowan (2007) in their study of global marine invertebrate diversity and western European rock outcrop area. These authors demonstrated the existence of positive correlations between their regional sampling metric and the global diversity of marine invertebrates. Smith & McGowan (2007) proposed two hypotheses to account for this correlation. One possibility is that a regional sampling metric

captures information on global sampling regimes: for example, fluctuations in rock outcrop area in western Europe could reflect the impact of factors that might control sedimentation at a global scale (such as the eustatic component of sea-level). Alternatively, the marine invertebrate data set might actually be dominated by taxa from North America and western Europe because of the longer history of palaeontological research in these regions. If the latter is correct, then the positive correlation between data series might occur because the analysis is actually comparing a regional sampling metric with a regional (or at least regionally biased) taxonomic count.

Inevitably, there is some regional skew in our dinosaur data set, which is dominated by Northern Hemisphere taxa in general and Asian forms in particular (Table 1). Nevertheless, the data set is truly global in terms of its taxonomic coverage and two of the sampling metrics (DBCs and DBFs). In addition, the dominance of any given region fluctuates through time, so that no one region is uniformly dominant in the sample throughout the Mesozoic. This means that evidence for positive correlations between regional sampling metrics and global diversity cannot be easily explained as a result of a regional taxonomic bias (especially for those analyses that do not use Asian DBCs and DBFs as the regional sampling metric). Thus, if positive correlations occur, they probably indicate that the regional sampling metric has captured at least part of the sampling signal at the global scale. These issues are examined here via the series of analyses that compare global and regional sampling (D analyses), inter-regional sampling (E and F analyses), regional sampling and global observed diversity (G analyses), global sampling and regional observed diversity (L analyses), and regional sampling v. regional observed diversity (H–K analyses).

Analyses D1–D14 and E1–E20 compare global ('rest of the world') and regional sampling metrics, or carry out pairwise comparisons between sampling metrics for individual regions. Most of these analyses (27 out of 34 comparisons) fail the statistical tests. However, there are some striking examples of positive correlations between data series. Interestingly, most such examples involve African DBCs or DBFs, suggesting that sampling regimes in this region are not independent of those in other regions. In general, however, the results of analyses D and E suggest that regional sampling metrics are not particularly effective at capturing global sampling signals. This may be because there is no single uniform regime that has been imposed on the sampling of the dinosaurian fossil record by an external factor such as sea-level. Even though sea-level will rise and fall in a coherent manner across the globe, this may have different effects on the sampling of the terrestrial fossil record in different regions, perhaps depending on local conditions such as climate, continental size and configuration, continental shelf slope and area, and also biotic factors such as variations in the extent to which the terrestrial faunas tended to occupy coastal, fluviolacustrine and/or arid inland environments. Similarly, the lack of correlations between 'rest of the world' and regional DBCs also indicates that there is no evidence for a uniform global pattern in collecting effort. This could mean that palaeontologists are attempting to find dinosaur fossils throughout the known stratigraphical and geographical range of this group, without marked biases towards particular time periods (e.g. the Late Cretaceous) or regions (e.g. North America). It is also conceivable, however, that palaeontologists do display biases in their collecting effort, but these are constrained and overwhelmed by geological controls on rock availability.

Comparison of regional sampling with global observed diversity (G analyses) and global sampling with regional diversity (L analyses), yield a surprising number of statistical passes (10 out of 23 comparisons, although only four passes remain after the application of the Bonferroni correction). Not surprisingly, however, significant positive correlations occur far more often (34 out of 56 comparisons) when the observed diversity of a region is compared with a sampling metric for that same region (H–K analyses). The general lack of correlation between global sampling and regional observed diversity, and regional sampling and global observed diversity, may stem from the combination of both regional variation in genuine diversity patterns and regional variation in sampling rates (see above). These results have at least two important implications. First, although regional sampling metrics can be used as proxies for global sampling under some conditions, in general it is better to employ global sampling metrics when dealing with global observed diversity and a regional sampling metric when investigating the observed diversity of that region. Second, global diversity patterns and global sampling regimes should be viewed as generalized patterns that represent summaries across regions and therefore tend to obscure important regional variations. There are thus increased risks of error when extrapolating from a regional diversity pattern to the global scale and when using a regional sampling metric as a proxy for global sampling.

Dinosaur radiations and extinctions revisited

The following sections discuss the evolutionary history of dinosaurs based on our residual diversity estimations and comparisons between these RDEs

and raw taxic diversity. Before doing so, however, some caveats regarding the interpretation of RDEs are warranted. Some deviations from the model's predictions will occur by chance, but the greater the distance between the straight line representing modelled diversity and a given data point representing observed diversity, the higher the probability that the observed diversity value genuinely departs from the model. The assumption that every fluctuation in RDE, however small, is in some way meaningful in macro-evolutionary terms, runs a severe risk of producing over-interpretations of the results. Barrett *et al.* (2009) addressed this issue by generating 95% confidence intervals, implemented based on the distribution of residual diversity, so that only approximately 5% of data points fell outside of the intervals. Thus, by definition, the vast majority of fluctuations in RDEs lie within the 95% confidence limits (*sensu* Barrett *et al.* 2009). If the confidence intervals generated by this approach are interpreted literally, very few deviations from the sampling model can be interpreted in terms of diversity change: however, this approach does not take account of the goodness of model fit, and is probably not appropriate. Moreover, it ignores the size of relative changes between stages. For example, one of the most dramatic changes in the RDEs presented by Barrett *et al.* (2009, fig. 2c) occurs when sauropodomorph diversity crosses the Jurassic/Cretaceous (J/K) boundary, yet this deflection lies entirely within the 95% confidence intervals. Given that a major decrease in sauropodomorph diversity occurs at the J/K boundary according to raw taxic data, phylogenetic diversity estimates, RDEs and rarefaction (Mannion *et al.* 2011), this would seem to be a genuine macroevolutionary event that would be overlooked if the 95% confidence intervals (*sensu* Barrett *et al.* 2009) were strictly enforced. A more appropriate method for calculating confidence intervals is being formulated (G. T. Lloyd, pers. comm. 2011). For the present, however, we take the view that qualitative shifts in RDEs can still provide useful information on genuine diversity patterns and the impact of sampling, especially when RDEs are compared with raw taxic diversity and 'other sources of sampling-corrected' diversity data (e.g. subsampling curves). Nevertheless, the reader should be aware that many of the fluctuations in RDEs reported below are unlikely to qualify as statistically significant (at least in terms of Barrett *et al.*'s (2009) confidence intervals).

Late Triassic–Early Jurassic. The earliest unequivocal dinosaurs are known from the late Carnian (*c.* 230 Ma) Ischigualasto Formation of Argentina (e.g. *Herrerasaurus*, *Eoraptor*, *Panphagia*) and the Santa Maria Formation of Brazil (e.g.

Staurikosaurus, Saturnalia) (Brusatte *et al.* 2010; Langer *et al.* 2010), although footprints may extend the stratigraphical range of the group into pre-Carnian time (Langer *et al.* 2010, and references therein). Benton (1983, 1994) argued for an end-Carnian extinction event (now likely to be dated within the Norian, because of revisions to the Triassic time-scale (Mundill *et al.* 2010; Martinez *et al.* 2011)), in which the dominant herbivorous groups (rhynchosaurs and dicynodonts) were dramatically depleted in abundance/diversity. Benton suggested that this extinction was more significant for terrestrial vertebrates than the subsequent end-Triassic event (see below). He also hypothesized that the Carnian–Norian event opened ecological space that allowed sauropodomorphs to radiate opportunistically (see also Brusatte *et al.* 2008*a, b*), in contrast to the classical scenario involving a long-term competitive replacement of synapsids, rhynchosaurs, and crurotarsans by dinosaurs (e.g. Charig 1984). Dinosaurs became widespread in the Norian and Rhaetian, particularly sauropodomorphs (e.g. lower Elliot Formation, southern Africa; Los Colorados Formation, Argentina; Löwenstein Formation, Germany). One exception is North America (Chinle Group) where sauropodomorphs are absent, and theropods are the only dinosaur group known (Nesbitt *et al.* 2007; Brusatte *et al.* 2010; Langer *et al.* 2010) (N.B., several Late Triassic body fossil remains (Long & Murray 1995; Hunt *et al.* 1998; Harris *et al.* 2002) and tracks (Wilson 2005) from North America have been attributed to basal sauropodomorphs, but all of these were rejected by Nesbitt *et al.* (2007) on the basis of a lack of sauropodomorph synapomorphies, with most reinterpreted as indeterminate archosauriforms). Ornithischians are extremely scarce throughout the Late Triassic and may have been geographically restricted to southern Gondwana (Butler *et al.* 2007; Irmis *et al.* 2007; Nesbitt *et al.* 2007). A peak in dinosaurian diversity during the Late Triassic, especially the Norian, has been reported by most analyses, including those based on raw taxic data (e.g. Dodson 1990; Haubold 1990; and the current study – see Fig. 1), phylogenetic diversity estimates (PDEs) (e.g. Sereno 1997, 1999) and those that have employed more sophisticated subsampling and residual methods (e.g. Lloyd *et al.* 2008; Barrett *et al.* 2009; Mannion *et al.* 2011).

Throughout the Late Triassic and Early Jurassic, observed taxic diversity seems to track sampling very closely (Fig. 1), suggesting that the Norian peak is at least partly an artefact. However, global RDEs for dinosaurs (Fig. 11) also show a peak in the Norian relative to the Carnian and Rhaetian (N.B., it should be noted that this results from lower than expected diversity in the latter two stages rather than elevated levels of diversity in

the Norian: the latter has a RDE score just above zero). Examination of RDEs for the five geographical regions (Fig. 13a–e) indicates that this pattern of a Norian peak bracketed by Carnian and Rhaetian troughs is far from being a global phenomenon. European RDEs come closest to the 'global pattern', whereas Norian Asian diversity is much lower than expected and is considerably lower than that of either the Carnian or Rhaetian. This Asian 'pattern', however, is spurious because it is based on inadequate data (only one out of 37 Late Triassic dinosaur species is known from Asia). The 'global' pattern is apparently created by adding together disparate regional patterns in which Norian diversity is relatively higher everywhere except Asia; Carnian diversity is lower in North America, Africa and Europe; and Rhaetian diversity is lower in Asia and South America. In terms of the three main dinosaurian clades (Figs 12 & 14), the sauropodomorph RDEs conform most closely to the 'general pattern', although Norian peaks also occur in most RDEs for theropods and ornithischians as well. Given that sauropodomorphs contribute 59% of Late Triassic dinosaurian species, it is not surprising that a peak in the diversity of this clade should produce a peak in Norian dinosaur RDEs as a whole. Thus, the Norian peak in sauropodomorph diversity is consistent with Benton's (1983, 1994) proposal that this clade radiated opportunistically after a Carnian/Norian boundary extinction of non-dinosaurian herbivores. However, claims regarding the severity of this extinction event and the dynamics of subsequent ecological replacements should be viewed with caution given new data on the stratigraphical ranges of many groups (Langer *et al.* 2010), the observation that this peak is perhaps exaggerated by sampling biases, and the apparent regional variation in the timing and magnitude of the Late Triassic peak in RDEs. The recent recalibration of the Triassic time-scale has made the Norian the longest stage in the Mesozoic, with a greater disparity in temporal length relative to its neighbours than almost any other. It has also resulted in repositioning many formerly 'Carnian' species as Norian ones, and enhanced the existence of a diversity 'peak'. It is likely that the ability to place taxa more precisely within the Norian would reveal that many are not contemporaneous with one another (as is the case for many Campanian dinosaurs, Carrano 2008*b*). Finally, the drop in diversity during the Rhaetian should be treated with caution because of the uncertainties surrounding the dating of many Late Triassic deposits.

The Triassic/Jurassic (T/J) boundary mass extinction event has been identified as one of the 'big five' Phanerozoic mass extinctions, with estimates suggesting that up to 30% of 'families' became extinct (Benton 1995). Extinctions occurred in both terrestrial and marine realms across a breadth of taxonomic groups, and included the final loss of several primitive groups of crurotarsan or more basal archosauriforms (e.g. phytosaurs, rauisuchians, aetosaurs). Several previous studies (e.g. Lloyd *et al.* 2008; Barrett *et al.* 2009) have suggested that dinosaur diversity increased in the wake of this end-Triassic extinction. For example, Lloyd *et al.* (2008, fig. 2b) found evidence for an increase in dinosaur taxonomic diversity across the T/J boundary when subsampling approaches were used. They also found that the Early and Middle Jurassic marked a period of significantly elevated levels of diversification. Here, a Hettangian–Sinemurian peak in diversity, similar in magnitude to that witnessed in the Norian, occurs in the raw taxic diversity data (Fig. 1). This apparent peak in diversity corresponds closely with a peak in the number of DBCs (Fig. 1), suggesting that it may be artefactual. Genuine increases in diversity during the earliest Jurassic are supported by some sampling-corrected diversity estimates, but the details vary from clade to clade and depend on the analytical method utilized (see below). Our global RDEs for dinosaurs (Fig. 11) indicate a drop in diversity in the Rhaetian (but see above), followed by a relative recovery in the Early Jurassic (although diversity remains depressed relative to many other time bins). This pattern occurs in Asia and South America, whereas there are extinction events among dinosaurs at the T/J boundary in Africa, Europe and North America (Fig. 13a–e). This complex pattern may partly reflect the diversity of sauropodomorphs in each area during the Late Triassic–Early Jurassic transition. In regions where sauropodomorphs were scarce in the Late Triassic (e.g. Asia), Early Jurassic radiations among theropods and/or ornithischians would produce an apparent increase in dinosaur diversity after the T/J boundary. In contrast, regions where sauropodomorphs were diverse in the Late Triassic (i.e. Africa and Europe) may give the impression of an overall dinosaur diversity decrease at the T/J boundary because the losses experienced by this clade (see below) were not entirely compensated for by new Early Jurassic theropods and ornithischians. However, it is not clear why North America, which has no Late Triassic sauropodomorphs, displays an apparent decrease in diversity across the T/J boundary.

Much of the putative Early Jurassic dinosaurian radiation has been attributed to diversification events among theropods and ornithischians, with sauropodomorphs supposedly displaying few negative or positive effects of the T/J extinction event. For example, the immediate aftermath of the end-Triassic extinction in eastern North America is marked by a notable increase in theropod body size (based upon the ichnological record: Olsen *et al.* 2002), which may reflect a global theropod radiation following the extinction of carnivorous crurotarsan lineages (Brusatte *et al.* 2010). However, Tanner *et al.* (2004, p. 113) noted that larger-bodied Late Triassic theropod body fossils are known from Europe (*Liliensternus*) and even North America (*Gojirasaurus*), and Barrett *et al.* (2009) argued for a decline in theropod diversity, from the Late Triassic to Early Jurassic, based on declining RDE values. Both our DBC-based RDEs (Fig. 12c) and DBF-based RDEs (Fig. 14c) suggest that theropods underwent an increase in diversity (relative to the Rhaetian) in the two earliest stages of the Jurassic, although diversity remained relatively depressed during the Pliensbachian and Toarcian and did not start to increase markedly until the Aalenian. The fact that our DBF-based RDEs support this pattern, in contrast to Barrett *et al.*'s (2009) DBF-based RDEs, suggests that the disagreement between the latter authors and ourselves (plus Lloyd *et al.* 2008) stems from differences in data sets rather than methodological approach. One possibility is that the Lloyd *et al.* (2008) data set and the updated taxic data set used here are more similar to each other than they are to the data set of Barrett *et al.* (2009), which was based on data in Weishampel *et al.* (2004): some of the same gaps in the dinosaurian fossil record may have been filled since 2004, either by ghost range reconstruction (Lloyd *et al.* 2008) or via the addition of taxa discovered during the past 6–8 years (the current data set; see Fig. 15). It should also be noted that, outside of Europe, it is difficult to date most Early Jurassic dinosaurian taxa accurately to the stage level, so interpretations of diversity changes at this time should be treated with caution.

An Early Jurassic global radiation in the distribution, diversity and abundance of ornithischian dinosaurs has been attributed to an expansion into vacant ecological space (Butler *et al.* 2007). Barrett *et al.* (2009) also found that although residual diversity for ornithischians is negative in the Late Triassic, it displays a small positive peak in the earliest Jurassic. Both our DBC-based (Fig. 12a) and DBF-based (Fig. 14a) RDEs support the conclusion that ornithischians increased in diversity during at least the first two stages of the Jurassic, relative to the Late Triassic.

Brusatte *et al.* (2010) and Langer *et al.* (2010) have argued that the end-Triassic extinction appears to have had little impact on sauropodomorphs. Moreover, Barrett *et al.* (2009) found that there is substantial positive residual diversity for sauropodomorphs from the Norian through the Early Jurassic. Our DBC-based RDEs (Fig. 12b)

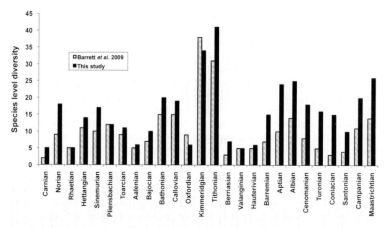

Fig. 15. Comparison of the number of sauropodomorph species per Standard European Stage in the Barrett *et al.* (2009) data set (open columns) and the data set used here (filled columns). N.B. the Barrett *et al.* data on sauropodomorph diversity was obtained from Upchurch *et al.* (2004), the latter being accurate only for taxa named prior to 2003.

support the view that sauropodomorphs experienced little diversity change at the T/J boundary. In the Early and Middle Jurassic, however, sauropodomorph diversity increases steadily to reach a peak in the Callovian: this could reflect the initial radiation of Eusauropoda and the origination of a number of neosauropod lineages (see below). DBF-based RDEs (Fig. 14b) also suggest that sauropodomorphs experienced an increase in diversity during the early stages of the Jurassic, but like theropods and ornithischians, Pliensbachian and Toarcian diversity is depressed relative to this peak and again in the Bathonian–Callovian. Thus, accurate reconstruction of diversity change among sauropodomorphs across the T/J boundary is difficult because the results depend on which analytical approach and/or sampling proxy are used. Sauropodomorph diversity was no doubt negatively affected by the loss of some basal forms at the end of the Triassic, but this seems to have been more than compensated for by their invasion of areas such as North America in the Early Jurassic, coupled with the onset of the eusauropod radiation.

Middle Jurassic to the J/K boundary. Raw taxic dinosaur diversity increases steadily from the beginning of the Middle Jurassic (Aalenian) to the end of the Jurassic, with a dip in the Oxfordian and dramatic decrease at the Jurassic/Cretaceous (J/K) boundary (Fig. 1). Sampling (in the form of DBCs) shows approximately the same trends during the Middle and Late Jurassic, suggesting that the observed diversity pattern may be artefactual. However, the various RDEs also confirm the occurrence of Middle and Late Jurassic diversity increases and a decrease in the earliest Cretaceous (Figs 11–14). The precise pattern of increases and

decreases varies depending on which RDEs are examined. Global DBC-based RDEs (Fig. 11) indicate that dinosaur diversity peaked in the Middle Jurassic (Bajocian–Callovian) and then gradually declined in the Late Jurassic, before a more dramatic extinction at the J/K boundary. The J/K boundary drop in RDEs is the third largest of 12 stage-to-stage decreases in our data set (surpassed in magnitude only by the drops in diversity during the Rhaetian and Oxfordian). This extinction is most clearly observed in Europe, North America and South America (Fig. 13a, c, d). In contrast, Asia and Africa show the Late Jurassic decline and Oxfordian extinction respectively, but there is little evidence for an extinction at the J/K boundary (Fig. 13b, e). DBC-based RDEs suggest that theropods and ornithischians underwent gradual reductions in diversity during the Late Jurassic followed by extinctions among the latter clade at the J/K boundary (Fig. 12a, c), whereas sauropodomorphs experienced peaks in diversity during the late Middle Jurassic (Bathonian and Callovian) and late Late Jurassic (Kimmeridgian and Tithonian), with more dramatic decreases in the Oxfordian and especially at the J/K boundary (Fig. 12b: as also noted by Upchurch & Barrett 2005; Mannion *et al.* 2011). DBF-based RDEs for theropods, sauropodomorphs and ornithischians suggest a more uniform increase in diversity towards the end of the Jurassic, and all display losses at the J/K boundary (Fig. 14) (although the losses among theropods are relatively small). Analysis of the theropod data without Mesozoic birds (Fig. 16) indicates a more profound decrease in diversity during the Late Jurassic and across the J/K boundary (e.g. the drop in DBC-based RDE values for non-avian dinosaurs becomes the largest of 11 stage-to-stage decreases).

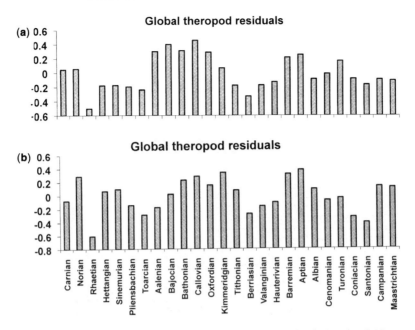

Fig. 16. Residual diversity estimates for non-avian theropods: (**a**) based on DBCs; (**b**) based on DBFs.

This implies some selectivity in the extinction whereby larger bodied non-avian theropods (and other large dinosaurs such as sauropods) were adversely affected and smaller volant forms remained unscathed or even diversified. In general, these results are in broad agreement with the DBF-based RDEs of Barrett *et al.* (2009), which identified marked drops in diversity in all three clades in the Oxfordian, decreases in sauropodomorph and ornithischian diversity at the J/K boundary, and a more gradual decrease in theropod diversity through the Late Jurassic and into the Early Cretaceous.

Raup & Sepkoski (1986) identified the J/K boundary event as one of eight major extinctions that have occurred during the last 250 million years. Subsequently, however, most studies have regarded this event as only a minor extinction (e.g. Hallam & Wignall 1997; Bambach 2006; Arens & West 2008). Recent studies of dinosaur diversity have either not commented on a J/K extinction, or have argued that this event strongly affected sauropods (Upchurch & Barrett 2005; Mannion *et al.* 2011), but probably had little impact on the diversity of ornithischians and theropods (e.g. Barrett *et al.* 2009). However, Orcutt *et al.*'s (2007) work on terrestrial tetrapods, and Benson *et al.*'s (2010) study of marine reptiles (see also Bakker 1993; Bardet 1994), suggests that the J/K boundary extinction event may have been more significant and widespread than previously realized. Both our raw taxic

data and RDEs for dinosaurs support this view and indicate that, while sauropodomorphs and stegosaurs (Bakker 1978; Galton & Upchurch 2004; Maidment *et al.* 2008) might have been particularly strongly affected, many other dinosaur clades probably also declined in diversity at this time.

Cretaceous diversity and the end-Cretaceous extinction. After a period of apparent lowered diversity in the Early Cretaceous, raw taxic diversity of dinosaurs displays prominent peaks in the Aptian–Albian and the Campanian–Maastrichtian, with an apparent trough in the Cenomanian–Santonian (Fig. 1). However, this pattern of peak-trough-peak follows the sampling of DBFs and DBCs very closely. The various RDEs generally agree that diversity increased throughout much of the Early Cretaceous, with peaks in the Barremian–Albian (Figs 11, 12a, c, 13 & 14), although in Africa the mid-Cretaceous peak occurs in the Cenomanian (Fig. 13e). One exception to this is the DBC-based RDEs for sauropodomorphs (Fig. 12b), which shows a steady increase in diversity from the beginning of the Cretaceous to a peak in the Coniacian (see also the results of runs test Q6), followed by a decline into the Campanian and a rise in the Maastrichtian.

Patterns in dinosaur diversity during the early Late Cretaceous are more difficult to elucidate. In general, DBC-based RDEs display moderately lower diversity during this time compared to the

Barremian–Aptian, with some results suggesting that a peak in diversity may have occurred during the Turonian (e.g. global dinosaur RDEs, European RDEs, Asian RDEs). DBF-based RDEs show a similar pattern, but lack the Turonian peak and indicate a more profound decrease in diversity in the Cenomanian–Santonian, especially during the Coniacian–Santonian (Fig. 14). Thus, the very low raw taxic diversity levels throughout the early Late Cretaceous are mainly the result of poorer sampling (particularly in the case of the Cenomanian and Turonian), but a genuine drop in diversity seems to have occurred in the Coniacian and Santonian.

The end-Cretaceous or Cretaceous/Palaeogene (K/Pg) mass extinction has received more attention from palaeobiologists than almost any other event in Earth history. Whether non-avian dinosaurs went extinct abruptly at the K/Pg boundary, or the final extinction was preceded by a slow decline during the Campanian and/or Maastrichtian, remains controversial. Many studies based on raw taxic data have concluded that dinosaurs underwent a gradual terminal decline prior to the extinction (e.g. Sloan *et al.* 1986; Sarjeant & Currie 2001; Sullivan 2006), although Sheehan *et al.* (1991) and Pearson *et al.* (2002) argued for a sudden extinction. Most recent studies, in which sampling biases have been addressed, have concluded that there is little evidence for a gradual decline (e.g. Fastovsky *et al.* 2004; Wang & Dodson 2006; Carrano 2008*a*; Lloyd *et al.* 2008; Mannion *et al.* 2011). For example, Carrano (2008*a*) carried out a regional study based on North American dinosaurs: he concluded that fluctuations in observed diversity during the Campanian and Maastrichtian largely reflect ecological differences between geological units and sampling issues, rather than long-term macro-evolutionary changes. He also noted that lumping all Campanian dinosaurs together into a single diversity count obscured the fact that many species were not contemporaneous, but members of successive faunas. One exception to this recent trend is the work of Barrett *et al.* (2009): these authors argued that their DBF-based RDEs indicated gradual declines in dinosaur diversity in the lead up to the K/Pg boundary (although detailed inspection of their results indicates that sauropodomorph diversity does not display a clear decrease, and theropods display a far less dramatic decline relative to ornithischians, across the Campanian/ Maastrichtian boundary).

Our global RDEs for dinosaurs suggest that the diversity of this group in the latest Cretaceous was generally similar to that seen in the Late Jurassic (i.e. lower than in the Middle Jurassic and Barremian–Aptian, but higher than in the earliest Cretaceous), with a small decrease from the Campanian to the Maastrichtian (Fig. 11). This decrease

from the Campanian to the Maastrichtian is consistent with the conclusions of Barrett *et al.* (2009). However, the RDEs of the major clades within Dinosauria indicate a more complex pattern. Both DBC-based and DBF-based RDEs display a marked decrease for ornithischians (Figs 12a & 14a), virtually no decrease for theropods (Figs 12c & 14c), and an increase for sauropodomorphs (Figs 12b & 14b), from the Campanian to the Maastrichtian. Moreover, the fluctuations in RDEs for dinosaurs as a whole (and for the three major clades) in the latest Cretaceous are no greater than those observed during other periods of their evolutionary history, and indeed are often smaller. For example, the decrease in DBC-based RDEs from the Campanian to the Maastrichtian is only the sixth largest out of 12 stage-to-stage decreases during the Mesozoic.

The final extinction of the non-avian dinosaurs is also more complicated than expected when examined at the regional scale. The RDEs for Europe and Africa display marked declines from the Campanian to the Maastrichtian (Fig. 13a, e). However, in South America and Asia, dinosaurian RDEs remain fairly stable from the Campanian to the Maastrichtian (Fig. 13b, d), and the RDE for North America actually displays an increase in the Maastrichtian (Fig. 13c). The latter pattern might be regarded as particularly significant, given the fact that North America possesses the best sampled record of latest Cretaceous dinosaurs.

In summary, analyses of Late Cretaceous dinosaur diversity that take sampling biases into account, generally agree that there is little support for a gradual decline leading up to the K/Pg boundary. This conclusion has been reached by both regional and global studies, and through the application of rarefaction (Fastovsky *et al.* 2004; Mannion *et al.* 2011), abundance-based coverage estimation (Wang & Dodson 2006), and sampling metrics based on DBCs, DBFs and the numbers of dinosaur-bearing localities (Lloyd *et al.* 2008; the current study). Until the current work, the disagreement between previous studies and Barrett *et al.* (2009) could have been attributed to the latter's unique use of DBF-based residuals. However, the same approach has been used here and failed to support clear pre-K/Pg declines in theropods and sauropodomorphs (although Ornithischia may still conform to the gradualistic extinction scenario). Failure to find support for consistent Campanian–Maastrichtian drops in DBF-based RDEs (as well as our DBC-based RDEs) may reflect the effects of an influx of new data on the Late Cretaceous since 2004. It is interesting to note, for example, that both DBC-based and DBF-based RDEs for sauropodomorphs (Figs 12 & 14) indicate an increase in the diversity of this clade from the

Campanian to the Maastrichtian, which potentially reflects the discovery of numerous new Late Cretaceous sauropods in the past eight years: compare the numbers of sauropod taxa in the Campanian and Maastrichtian available to Barrett *et al.* (2009) and the current study (Fig. 15). Finally, one caveat should be noted: the current study has examined dinosaur diversity using stage-level time bins, rather than the substages employed by Barrett *et al.* (2009). As such, we can comment on diversity fluctuations from the Campanian to the Maastrichtian, but have no information on changes within the Maastrichtian. If, for example, dinosaur diversity was relatively high in the early Maastrichtian, and then underwent a gradual decline during the middle and late Maastrichtian, our data would fail to capture this pattern and would produce a diversity estimate that is averaged across this stage. Thus, our results contradict a long-term gradual decline in dinosaur diversity during the 15 million years of the Campanian–Maastrichtian, but this still leaves open the possibility that a gradual decline occurred globally over the final 3–4 million years prior to the K/Pg boundary (but see Fastovsky *et al.* 2004).

Long-term trends during the Mesozoic. The raw taxic diversity of dinosaurs as a whole, and theropods and ornithischians separately, generally increases throughout the Jurassic and Cretaceous, culminating in a large peak in the Campanian–Maastrichtian (Figs 1 & 2). Sauropodomorphs follow a slightly different pattern: essentially, their observed diversity increases during the Jurassic, suffers a serious set-back at the J/K boundary (see above), and then recovers during the Cretaceous. This apparent general increase in raw taxic diversity raises the question as to whether this represents a genuine evolutionary phenomenon (i.e. dinosaur diversification rates increased through time, perhaps conforming to an exponential growth curve), or is an artefact of sampling (e.g. the tendency for younger sedimentary deposits to be preserved in greater abundance than older ones). The analyses of Fastovsky *et al.* (2004) and Wang & Dodson (2006) concluded that a general increase in diversity can still be detected even after uneven sampling rates are taken into account. However, this pattern of long-term growth in diversity is not supported by other recent studies and the analyses presented here. Lloyd *et al.* (2008) noted that the Campanian peak in observed diversity is somewhat reduced when ghost range data and sampling metrics are considered. The results of our runs tests (analyses Q) provide very little support for the existence of a persistent upward trend in dinosaur diversity. For example, none of the DBF-based RDEs, or any of the RDEs for ornithischians, demonstrate

the presence of a trend of increasing diversity during the Mesozoic or the Triassic–Jurassic and Cretaceous time slices (see runs tests Q10–Q21). Although the runs test for DBC-based RDEs for all Mesozoic dinosaurs (Q1) does support the presence of such an upward trend, time-sliced analyses (Q2, Q3) show that this result is the product of combining increasing diversity through the Triassic–Jurassic with no detectable trend in the Cretaceous. It appears that theropods and sauropodomorphs may have experienced an upward trend in diversity during the former time slice, whereas in the Cretaceous only sauropods show such a trend (Q4–Q9). Importantly, the Triassic–Jurassic and Cretaceous trends in sauropodomorph diversity are not part of one continuous pattern, rather they are two distinct trends separated by the J/K boundary extinction that reset their diversity to low levels during the Early Cretaceous.

Changes in RDEs, and the results of the runs tests, suggest that the long-term history of dinosaur diversity is better characterised as a series of radiations punctuated by 'set-backs', rather than a steady (perhaps exponential) rise towards an end-Cretaceous peak (contrast this model, for example, with Fastovsky *et al.* 2004, fig. 1). Key phases in dinosaur evolution would thus include:

(i) Divergences into the three main clades, and major radiation of sauropodomorphs, during the Late Triassic (especially the Carnian–Norian).

(ii) A marked decrease in diversity during the Rhaetian (although this might have been exaggerated by uncertainties in the dating of Late Triassic rocks, perhaps 'back-smearing' the effects of a T/J boundary mass extinction event).

(iii) Major radiations of all three clades during the Early and Middle Jurassic, with theropods and sauropodomorphs displaying statistically detectable upward trends in diversity (see below).

(iv) A gradual decline in diversity through the Late Jurassic (theropods and ornithischians) combined with a more severe extinction event at the J/K boundary.

(v) A recovery phase in the Early Cretaceous, culminating in peaks in diversity during the Barremian–Albian.

(vi) A moderate decline in diversity, reaching a low point in the Coniacian–Santonian.

(vii) Moderately increased diversity in the Campanian–Maastrichtian, driven especially by the radiation of titanosaur sauropods (only sauropods display a statistically significant upward trend in diversity during the Cretaceous). Despite the appearance of many new forms of theropod (e.g. Cretaceous birds) and ornithischians (radiations of ceratopsians, hadrosaurs and ankylosaurs), these were apparently insufficient to result in a persistent upward trend in diversity, perhaps because originations were offset by high extinction rates. Consequently, although sauropod diversity during the Campanian–Maastrichtian recovers to levels comparable to those of the Late Jurassic (see also Mannion *et al.* 2011), the end of the Cretaceous does not have abnormally elevated levels of diversity when compared to earlier peaks, such as those that occur during the Middle Jurassic and mid-Cretaceous (Figs 11–14).

(viii) The extinction of all non-avian dinosaurs at the Cretaceous/Palaeogene boundary.

The radiation of dinosaurs during the Early and Middle Jurassic could be regarded as one of the most important phases in dinosaur evolution, with significant upward trends in theropod and sauropod diversity, and the first clear radiations of theropods and ornithischians. The proposal that the Early and Middle Jurassic represents a key phase in dinosaurian diversification is consistent with several other recent studies and lines of evidence. For example, ghost ranges indicate that many of the distinct dinosaurian clades originated prior to the late Middle Jurassic (Upchurch & Barrett 2005) and this is reflected in the discovery of early members of clades that were originally found in the Cretaceous but are now known from the Jurassic (e.g. tyrannosaurs (Xu *et al.* 2006; Benson 2008), deinonychosaurs (Hu *et al.* 2009), titanosaurs (Day *et al.* 2002, 2004), and ankylosaurs (Carpenter *et al.* 1998)). Lloyd *et al.* (2008) also used a time-calibrated dinosaurian supertree to argue that diversification rates peaked during the Middle Jurassic but were no greater than background rates during most of the Cretaceous. Finally, Upchurch *et al.* (2002) suggested that the presence of continent-scale vicariance patterns among Late Jurassic and Cretaceous dinosaurs is consistent with the dispersal of members of most clades across Pangaea prior to the onset of its fragmentation in the Middle Jurassic. In short, it seems probable that the true pattern of dinosaurian diversification is 'bottom heavy' (i.e. cladogenetic events are concentrated in the early phases of dinosaur evolution), but the impression of a more 'top heavy' history (i.e. high diversity concentrated towards the middle and Late Cretaceous) is an artefact generated by better preservation of younger fossiliferous deposits, coupled with the more intensive sampling of Campanian–Maastrichtian rocks because of disproportionate interest in the K/Pg boundary extinction.

Conclusions, caveats and prospects

The results of this study have the following implications for our understanding of dinosaur evolutionary history and the effects of sampling in the terrestrial realm on palaeodiversity patterns:

(i) The observed taxic diversity of dinosaurs is positively correlated with the numbers of dinosaur-bearing collections and dinosaur-bearing formations. Correlations persist even when the data set is partitioned into the clades Theropoda, Sauropodomorpha and Ornithischia, or into continental regions (i.e. Africa, Asia, Europe, and North and South America). It is therefore difficult to escape the conclusion that many of the fluctuations in observed diversity actually represent artefacts generated by uneven sampling.

(ii) Sampling regimes in the terrestrial realm did not cause the observed diversity patterns of separate regions to be artificially more or less similar to each other.

(iii) The diversity histories of dinosaur faunas on different continents apparently varied from region to region. Global environmental change may have imposed some congruence in terms of the timing and magnitude of radiations and extinctions, but this is not reflected in significant positive correlations between the diversity patterns in separate regions.

(iv) Regional sampling metrics are not particularly effective at capturing global sampling signals in the terrestrial realm. This might reflect regional variations in sampling rates, so that the 'global signal' is in fact a summary of disparate regional patterns. In contrast, regional sampling metrics are typically strongly positively correlated with regional observed diversity. These results suggest that, if a regional sampling metric is found to correlate with supposed global diversity, it is important to rule out the possibility that the 'global' taxonomic data set is biased in terms of over-representing taxa from the region, which has yielded the sampling metric.

(v) Different sampling metrics for the same region (i.e. DBFs, DBCs, rock units and collections for North America, western European terrestrial sedimentary rock outcrop area) typically correlate with each other, suggesting that they are 'homing in' on approximately the same sampling signal. Nevertheless, it should also be noted that choice of sampling metric could affect the details of residual diversity estimates (e.g. the extent to which sauropod diversity was relatively depressed during the Cenomanian–Coniacian) depends on whether DBFs or DBCs are used to produce RDEs (compare Figs 12b & 14b).

(vi) Concerns that correlations between observed diversity and sampling metrics result from circularity, rather than the controlling effects of sampling, are addressed by the observation that numbers of DBFs and DBCs correlate with other sampling metrics such as rock outcrop area and gap-bound packages. Moreover, correlations between observed diversity and sampling metrics persist even when the criterion for recognising a unit of sampling (e.g. counting all formations containing any evidence of the presence of dinosaurs, even trackways or fragmentary body fossils) is far broader than the taxonomic group under investigation (e.g. Theropoda).

(vii) The three main dinosaurian clades appear to have experienced early bursts in diversification rates at different times. The initial sauropodomorph diversification occurred during the Late Triassic (Norian), followed by strong radiations of theropods and ornithischians in the Early Jurassic.

(viii) Both the Rhaetian and Oxfordian appear to be times of genuinely low diversity for most dinosaurian clades according to raw TDEs and RDEs. However, both of these stages are affected by dating issues whereby deposits that should be assigned to them may have been incorrectly dated as Norian in the case of the Rhaetian and Kimmeridgian in the case of the Oxfordian (e.g. see Mannion et al. 2011).

(ix) An extinction among dinosaurs apparently occurred at the Jurassic/Cretaceous boundary. Previously this was thought to have mainly affected sauropodomorphs, but all three of the main dinosaurian clades display either a decline in diversity throughout the Late Jurassic and into the Early Cretaceous (non-avian theropods) or a more dramatic decline at the J/K boundary itself (sauropodomorphs and ornithischians). Combined with recent work on marine reptile diversity (Bakker 1993; Bardet 1994; Benson et al. 2010), these results suggest that the J/K boundary extinction event was probably more significant than previously acknowledged.

(x) There is little evidence for a significant gradualistic decline in dinosaur diversity during the latest Cretaceous. Where a decrease in diversity from the Campanian to the Maastrichtian does occur, such declines are no larger (and are typically much smaller) than declines in diversity that were experienced earlier in dinosaur evolution. Regional residual diversity estimates, and the RDEs for clades within Dinosauria, often display either no decrease, or even an increase in diversity, from the Campanian to Maastrichtian. These observations do not rule out a gradual decline in dinosaur diversity prior to the K/Pg boundary event, but if such a decline did occur, it must have happened during the middle and/or late Maastrichtian, with relatively high diversity levels in the early Maastrichtian.

(xi) Dinosaur diversity does not display a gradual increase throughout the Mesozoic, culminating in a Campanian–Maastrichtian peak, as has been claimed by several previous studies (e.g. Sereno 1997, 1999; Fastovsky et al. 2004; Wang & Dodson 2006). This apparent long-term trend is probably an artefact generated by better sampling available in younger rocks, combined with disproportionate sampling effort regarding Campanian–Maastrichtian rocks by workers interested in the causes of the K/Pg mass extinction. When sampling biases are taken into account, the history of dinosaur diversity is characterised by a series of growth phases (i.e. Carnian–Norian, Early and Middle Jurassic, Early Cretaceous to Barremian–Albian and Campanian) punctuated by gradual decreases and occasional severe extinction events (i.e. the Rhaetian, Oxfordian, J/K boundary and Coniacian–Santonian). This pattern is consistent with the work of Lloyd et al. (2008), who noted a statistically significant increase in diversity rates during the Early and Middle Jurassic, followed by background rates of diversification throughout the rest of dinosaurian evolutionary history.

There are a number of caveats that should be borne in mind when analysing data sets of terrestrial vertebrates and when applying methods such as residual diversity estimation. Although statistical comparisons are a vital component of quantitative palaeobiology, they can obscure important details as well as reveal patterns. One reason for this is that which patterns are found or not found depends heavily on how the researcher partitions their data for analysis. For example, here we have used five continental areas in order to examine regional patterns in diversity: it is hoped that these five regions represent genuine geographical units that had some kind of biological reality during the Mesozoic. However, palaeogeographical events mean that, for example, Africa and South America might be considered one area during the Early Jurassic but are clearly two distinct regions during the Late Cretaceous.

Similarly, the search for 'signal' in large complex data sets creates a dilemma: analysis of all or most of the data means that 'patterns' can be identified with greater statistical rigour, but the 'general pattern' may obscure many important secondary signals (e.g. separate clades showing different responses to the same environmental events). However, attempts to locate these secondary signals by partitioning the data set results in fewer data points for analysis, which may make it more difficult to resolve true signal against the backdrop of 'noise' (e.g. see the 'Discussion' in Mannion & Upchurch 2010*b*). Thus, our choice to look at diversity change among Dinosauria, Theropoda, Sauropodomorpha and Ornithischia, has probably obscured important events that affected smaller clades (e.g. the extinction of many stegosaurs at the J/K boundary and the radiation of ankylosaurs during the Cretaceous).

Another problem is that, while RDEs offer one of the best ways to remove the effects of uneven sampling from raw diversity data, the resulting fluctuations in RDEs may be over-interpreted: many such fluctuations lie within the bounds of statistical 'noise'. Much of our interpretation and discussion has hinged on whether two data series pass or fail statistical tests. However, two data series may correlate well over part of the time range and not correlate at all over the remainder (as was found by Benson *et al.* (2010) and Mannion *et al.* 2011). Thus, while we have found that there is little evidence for a common global pattern of dinosaurian diversification, this is a general statement and really means that there is not enough agreement among the diversity patterns from each region to allow a statistical pass. In reality, there may still be many peaks and troughs (in either observed diversity or RDEs) that coincide, but these are not quite sufficient to produce positive correlations.

Finally, in this paper, we have discussed increases and decreases in diversity in terms of diversification and extinction events respectively. However, the majority of dinosaur genera and species are point occurrences in terms of their stratigraphical ranges, or have almost certainly had their ranges artificially truncated by poor sampling: thus, it is not possible to calculate meaningful origination and extinction rates. Consequently, the true dynamics of diversity fluctuations during dinosaur evolution are difficult to elucidate using the currently available data. For example, the J/K boundary reduction in standing diversity could reflect a dramatic increase in extinction rates, or it might stem from a decrease in origination rates (with extinction rates remaining largely unchanged) after the Early–Middle Jurassic 'burst' of diversification.

Our results and conclusions suggest a number of lines of future study that may reveal important insights into dinosaur evolution, terrestrial sampling, and the methods we use to deal with sampling and diversity. First, there are several other ways in which the data set could be partitioned (e.g. comparisons of Northern and Southern Hemisphere patterns, time-slicing into Jurassic and Cretaceous subsets, etc.). Second, although the current data set is the largest available for terrestrial Mesozoic organisms, it remains narrow relative to the full spectrum of available tetrapods. This raises questions such as: do any of the observed patterns in dinosaur diversity also occur in other tetrapods? Does the addition of other non-dinosaurian groups reinforce or contradict apparent regional differences in diversity patterns? Does the observed diversity of terrestrial vertebrates as a whole correlate with sampling of the terrestrial fossil record, rock outcrop area, etc.? Finally, the estimation, and especially the correction, of the effects of uneven sampling on observed diversity patterns, has yet to overcome several methodological challenges. In particular, debate still rages over whether or not sampling has produced significant distortions of observed diversity, and even if this has occurred, there is the possibility that our methods for creating sampling-corrected diversity curves may produce additional distortions that result from data transformations. At present, many workers seem to fall into one of two camps – subsampling (e.g. Alroy *et al.* 2008) v. residuals (e.g. Smith & McGowan 2007) – but it is often feasible to apply both methods to the same data set. Indeed, on one of the few occasions when this has been implemented (Mannion *et al.* 2011), subsampling and residuals produced very similar reconstructions of diversity fluctuations in sauropodomorphs, which may be reassuring for those palaeobiologists who are more concerned with understanding the evolution of their study organisms than they are with methodological issues.

This paper presents the results of just one case study of diversity and sampling in the terrestrial realm, and it is obviously dangerous to overstate the generality of its conclusions. Nevertheless, it demonstrates the need to take sampling into account when reconstructing the diversity history of terrestrial Mesozoic organisms, and shows that some intriguing, unexpected and thought-provoking conclusions can result. We hope that this study will prompt further analyses by those wishing to either build upon our results or to overturn them. Such studies will play a key role in capturing the complexity of evolutionary patterns at regional and global scales.

This paper is a Paleobiology Database official publication No. 133. PU gratefully acknowledges support from the Abbey International Collaboration Grant scheme (funded by Santander Bank and administered by University College London), and a Palaeontological Association Research Grant, which enabled collection of data on sauropodomorph dinosaurs. PDM was supported by a University College London NERC studentship (NER/S/A/ 2006/14347) and a grant from the Jurassic Foundation. RJB is supported by an Alexander von Humboldt Research Fellowship. MTC received support from the Evolution of Terrestrial Ecosystems (ETE) program at the Smithsonian Institution; this is ETE publication 231. We also thank A. Smith and A. J. McGowan for their invitation to talk at the Lyell meeting (part of the 3rd International Paleontological Congress, London, 2010) and to participate in this edited volume.

References

ALROY, J. 2000. Successive approximations of diversity curves: ten more years in the library. *Geology*, **28**, 1023–1026, doi: http://dx.doi.org/10.1130/0091-7613(2000)28<1023:SAODCT>2.0.CO;2.

ALROY, J., MARSHALL, C. R. ET AL. 2001. Effects of sampling standardization on estimates of Phanerozoic marine diversification. *Proceedings of the National Academy of Sciences, USA*, **98**, 6261–6266.

ALROY, J., ABERHAN, M. ET AL. 2008. Phanerozoic trends in the global diversity of marine invertebrates. *Science*, **321**, 97–100, doi: 10.1126/science.1156963.

ARENS, N. C. & WEST, I. D. 2008. Press-pulse: a general theory of mass extinction? *Paleobiology*, **34**, 456–471.

BAKKER, R. T. 1978. Dinosaur feeding behaviour and the origin of flowering plants. *Nature*, **274**, 661–663.

BAKKER, R. T. 1993. Plesiosaur extinction cycles – events that mark the beginning, middle and end of the Cretaceous. *Geological Association of Canada, Special Papers*, **39**, 641–664.

BAMBACH, R. K. 2006. Phanerozoic biodiversity: mass extinctions. *Annual Review of Earth and Planetary Science*, **34**, 127–155.

BARDET, N. 1994. Extinction events among Mesozoic marine reptiles. *Historical Biology*, **7**, 313–324.

BARRETT, P. M., McGOWAN, A. J. & PAGE, V. 2009. Dinosaur diversity and the rock record. *Proceedings of the Royal Society, B*, **276**, 2667–2674, doi: 10.1098/rspb.2009.0352

BENSON, R. B. J. 2008. New information on *Stokesosaurus*, a tyrannosauroid (Dinosauria: Theropoda) from North America and the United Kingdom. *Journal of Vertebrate Paleontology*, **28**, 732–750.

BENSON, R. B. J., BUTLER, R. J., LINDGREN, J. & SMITH, A. S. 2010. Mesozoic marine tetrapod diversity: mass extinctions and temporal heterogeneity in geological megabiases affecting vertebrates. *Proceedings of the Royal Society, B*, **277**, 829–834, doi: 10.1098/rspb.2009.1845

BENSON, R. B. J. & BUTLER, R. J. 2011. Uncovering the diversification history of marine tetrapods: ecology influences the effect of geological sampling biases. *In*: McGOWAN, A. J. & SMITH, A. B. (eds) *Comparing the Geological and Fossil Records: Implications for Biodiversity Studies*. Geological Society, London, Special Publications, **358**, 191–208.

BENTON, M. J. 1983. Dinosaur success in the Triassic: a noncompetitive ecological model. *Quarterly Review of Biology*, **58**, 29–55.

BENTON, M. J. 1994. Late Triassic to Middle Jurassic extinctions among continental tetrapods: testing the pattern. *In*: FRASER, N. C. & SUES, H.-D. (eds) *In the Shadow of the Dinosaurs*. Cambridge University Press, Cambridge, 366–397.

BENTON, M. J. 1995. Diversification and extinction in the history of life. *Science*, **268**, 52–58.

BENTON, M. J. & EMERSON, B. J. 2007. How did life become so diverse? The dynamics of diversification according to the fossil record and molecular phylogenetics. *Palaeontology*, **50**, 23–40, doi: 10.1111/j.1475-4983.2006.00612.x.

BENTON, M. J., TVERDOKHLEBOV, V. P. & SURKOV, M. V. 2004. Ecosystem remodelling among vertebrates at the Permian–Triassic boundary in Russia. *Nature*, **432**, 97–100.

BENTON, M. J., DUNHILL, A. M., LLOYD, G. T. & MARX, F. G. 2011. Assessing the quality of the fossil record: insights from vertebrates. *In*: McGOWAN, A. J. & SMITH, A. B. (eds) *Comparing the Geological and Fossil Records: Implications for Biodiversity Studies*. Geological Society, London, Special Publications, **358**, 63–94.

BRUSATTE, S. L., BENTON, M. J., RUTA, M. & LLOYD, G. T. 2008a. Superiority, competition, and opportunism in the evolutionary radiation of dinosaurs. *Science*, **321**, 1485–1488.

BRUSATTE, S. L., BENTON, M. J., RUTA, M. & LLOYD, G. T. 2008b. The first 50 mya of dinosaur evolution: macroevolutionary pattern and morphological disparity. *Biology Letters*, **4**, 733–736.

BRUSATTE, S. L., NESBITT, S. J., IRMIS, R. B., BUTLER, R. J., BENTON, M. J. & NORELL, M. A. 2010. The origin and early radiation of dinosaurs. *Earth-Science Reviews*, **101**, 68–100.

BUTLER, R. J., BARRETT, P. M., NOWBATH, S. & UPCHURCH, P. 2009. Estimating the effects of sampling biases on pterosaur diversity patterns: implications for hypotheses of bird/pterosaur competitive replacement. *Paleobiology*, **35**, 432–446, doi: 10.1666/0094-8373-35.3.432.

BUTLER, R. J., SMITH, R. M. H. & NORMAN, D. B. 2007. A primitive ornithischian dinosaur from the Late Triassic of South Africa, and the early evolution and diversification of Ornithischia. *Proceedings of the Royal Society, B*, **274**, 2041–2046.

BUTLER, R. J., BENSON, R. B. J., CARRANO, M. T., MANNION, P. D. & UPCHURCH, P. 2011. Sea level, dinosaur diversity, and sampling biases: investigating the 'common cause' hypothesis in the terrestrial realm. *Proceedings of the Royal Society, B*, **278**, 1165–1170; doi: 10.1098/rspb.2010.1754

CARPENTER, K., MILES, C. & CLOWARD, K. 1998. Skull of a Jurassic ankylosaur (Dinosauria). *Nature*, **393**, 782–783.

CARRANO, M. T. 2005. The dinosaur fossil record. *Journal of Vertebrate Paleontology*, **25** (Suppl. 3), 42A.

CARRANO, M. T. 2008a. Patterns of diversity among latest Cretaceous dinosaurs in North America. *Journal of Vertebrate Paleontology*, **28** (Suppl. 3), 61A.

CARRANO, M. T. 2008b. Taxonomy and classification of non-avian Dinosauria. *Paleobiology Database Online Systematics Archive*, **4**, www.paleodb.org

CHARIG, A. J. 1984. Competition between therapsids and archosaurs during the Triassic Period: a review and synthesis of current theories. *Symposium of the Zoological Society of London*, **52**, 597–628.

CRAMPTON, J. S., BEU, A. G., COOPER, R. A., JONES, C. M., MARSHALL, B. & MAXWELL, P. A. 2003. Estimating the rock volume bias in paleobiodiversity studies. *Science*, **301**, 358–360, doi: 10.1126/science.1085075.

DAY, J. J., UPCHURCH, P., NORMAN, D. B., GALE, A. S. & POWELL, H. P. 2002. Sauropod Trackways: evolution and behavior. *Science*, **296**, 1659.

DAY, J. J., NORMAN, D. B., GALE, A. S., UPCHURCH, P. & POWELL, H. P. 2004. New Middle Jurassic dinosaur trackways from Oxfordshire, U.K. *Palaeontology*, **47**, 319–348.

DODSON, P. 1990. Counting dinosaurs: how many kinds were there? *Proceedings of the National Academy of Science, USA*, **87**, 7608–7612.

DYKE, G. J., MCGOWAN, A. J., NUDDS, R. L. & SMITH, D. 2009. The shape of pterosaur evolution: evidence from the fossil record. *Journal of Evolutionary Biology*, **22**, 890–898.

FASTOVSKY, D. E., HUANG, Y., HSU, J., MARTIN-MCNAUGHTON, J., SHEEHAN, P. M. & WEISHAMPEL, D. B. 2004. Shape of Mesozoic dinosaur richness. *Geology*, **32**, 877–880.

FRÖBISCH, J. 2008. Global taxonomic diversity of anomodonts (Tetrapoda, Therapsida) and the terrestrial rock record across the Permian–Triassic boundary. *PLoS ONE*, **3**, 1–14, doi: 10.1371/journal.pone.0003733

GALTON, P. M. & UPCHURCH, P. 2004. Stegosauria. *In*: WEISHAMPEL, D. B., DODSON, P. & OSMOLSKA, H. (eds) *The Dinosauria II*. University of California Press, Berkeley and Los Angeles, 343–362.

GOSWAMI, A. & UPCHURCH, P. 2010. The dating game: a reply to Heads (2010). *Zoologica Scripta*, **39**, 406–409.

GRADSTEIN, F. M., OGG, J. G. & SMITH, A. G. 2005. *A Geologic Time Scale 2004*. Cambridge University Press, Cambridge.

HALLAM, A. & WIGNALL, P. B. 1997. *Mass Extinctions and their Aftermath*. Oxford University Press, Oxford.

HAMMER, Ø. & HARPER, D. A. T. 2006. *Paleontological Data Analysis*. Blackwell, Oxford.

HAMMER, Ø., HARPER, D. A. T. & RYAN, P. D. 2001. PAST: Palaeontological Statistics software package for education and data analysis. *Palaeontologica Electronica*, **4** (1).

HARRIS, S. K., HECKERT, A. B., LUCAS, S. G. & HUNT, A. P. 2002. The oldest North American prosauropod, from the Upper Triassic Tecovas Formation of the Chinle Group (Adamanian: Latest Carnian), West Texas. *New Mexico Museum of Natural History & Science Bulletin*, **21**, 249–252.

HAUBOLD, H. 1990. Dinosaurs and fluctuating sea levels during the Mesozoic. *Historical Biology*, **4**, 75–106.

HEADS, M. 2010. Evolution and biogeography of primates: a new model based on molecular phylogenetics, vicariance and plate tectonics. *Zoologica Scripta*, **39**, 107–127.

HU, D., HOU, L., ZHANG, L. & XU, X. 2009. A pre-*Archaeopteryx* troodontid theropod from China with ling feathers on the metatarsus. *Nature*, **461**, 640–643.

HUNT, A. P., LUCAS, S. G., HECKERT, A. B., SULLIVAN, R. M. & LOCKLEY, M. G. 1998. Late Triassic dinosaurs from the western United States. *Geobios*, **31**, 511–531.

IRMIS, R. B., PARKER, W. G., NESBITT, S. J. & LIU, J. 2007. Early ornithischian dinosaurs: the Triassic record. *Historical Biology*, **19**, 3–22.

JACKSON, J. B. C. & JOHNSON, K. G. 2001. Measuring past biodiversity. *Science*, **293**, 2401–2404.

LANE, A., JANIS, C. M. & SEPKOSKI, J. J., JR. 2005. Estimating paleodiversities: a test of the taxic and phylogenetic methods. *Paleobiology*, **31**, 21–34.

LANGER, M. C., EZCURRA, M. D., BITTENCOURT, J. S. & NOVAS, F. E. 2010. The origin and early evolution of dinosaurs. *Biological Reviews*, **85**, 55–110.

LE LOEUFF, J., MÉTAIS, E. *ET AL*. 2010. An Early Cretaceous vertebrate assemblage from the Cabao Formation of NW Libya. *Geological Magazine*, **147**, 750–759.

LLOYD, G. T., DAVIS, K. E. *ET AL*. 2008. Dinosaurs and the Cretaceous terrestrial revolution. *Proceedings of the Royal Society, B*, **275**, 2483–2490, doi: 10.1098/rspb.2008.0715.

LONG, R. A. & MURRY, P. A. 1995. Late Triassic (Carnian and Norian) tetrapods from the Southwestern United States. *New Mexico Museum of Natural History & Science Bulletin*, **4**, 1–254.

MAIDMENT, S. C. R., NORMAN, D. B., BARRETT, P. M. & UPCHURCH, P. 2008. Systematics and phylogeny of Stegosauria (Dinosauria, Ornithischia). *Journal of Systematic Palaeontology*, **6**, 367–407.

MANNION, P. D. 2009. Review and analysis of African sauropodomorph dinosaur diversity. *Palaeontologia Africana*, **44**, 108–111.

MANNION, P. D. & UPCHURCH, P. 2010*a*. Completeness metrics and the quality of the sauropodomorph fossil record through geological and historical time. *Paleobiology*, **36**, 283–302, doi: 10.1666/09008.1

MANNION, P. D. & UPCHURCH, P. 2010*b*. A quantitative analysis of environmental associations in sauropod dinosaurs. *Paleobiology*, **36**, 253–282, doi: 10.1666/08085.1

MANNION, P. D. & UPCHURCH, P. 2011. A re-evaluation of the 'mid-Cretaceous sauropod hiatus', and the impact of uneven sampling of the fossil record on patterns of regional dinosaur extinction. *Palaeogeography, Palaeoclimatology, Palaeoecology*, **299**, 529–540.

MANNION, P. D., UPCHURCH, P., CARRANO, M. T. & BARRETT, P. M. 2011. Testing the effect of the rock record on diversity: a multidisciplinary approach to elucidating the generic richness of sauropodomorph dinosaurs through time. *Biological Reviews*, **86**, 157–181, doi: 10.1111/j.1469-185X.2010.00139.x

MARTINEZ, R. N., SERENO, P. C., ALCOBER, O. A., COLOMBI, C. E., RENNE, P. R., MONTAÑEZ, I. P. & CURRIE, B. S. 2011. A basal dinosaur from the dawn of the dinosaur era in southwestern Pangaea. *Science*, **331**, 206–210.

MARX, F. G. 2009. Marine mammals through time: when less is more in studying palaeodiversity. *Proceedings of the Royal Society, B*, **276**, 887–892.

MARX, F. G. & UHEN, M. D. 2010. Climate, critters, and cetaceans: Cenozoic drivers of the evolution of modern whales. *Science*, **327**, 993–996.

MCGOWAN, A. J. & SMITH, A. B. 2008. Are global Phanerozoic marine diversity curves truly global? A study of the relationship between regional rock records and global Phanerozoic marine diversity. *Paleobiology*, **34**, 80–103.

McKINNEY, M. L. 1990. Classifying and analysing evolutionary trends. *In*: McNAMARA, K. J. (ed.) *Evolutionary Trends*. University of Arizona Press, Tucson, 28–58.

MUNDILL, R., PALFY, J., RENNE, P. R. & BRACK, P. 2010. The Triassic timescale: new constraints and a review of geochronological data. *In*: LUCAS, S. G. (ed.) *The Triassic Timescale*. Geological Society, London, Special Publications, **334**, 41–60.

NESBITT, S. J., IRMIS, R. B. & PARKER, W. G. 2007. A critical re-evaluation of the Late Triassic dinosaur taxa of North America. *Journal of Systematic Palaeontology*, **5**, 209–243.

NORELL, M. A. & NOVACEK, M. J. 1992*a*. Congruence between superpositional and phylogenetic patterns comparing cladistic patterns with fossil evidence. *Cladistics*, **8**, 319–337.

NORELL, M. A. & NOVACEK, M. J. 1992*b*. The fossil record and evolution: comparing cladistic and paleontologic evidence for vertebrate history. *Science*, **255**, 1690–1693.

OLSEN, P. E., KENT, D. V. *ET AL.* 2002. Ascent of dinosaurs linked to an iridium anomaly at the Triassic–Jurassic boundary. *Science*, **296**, 1305–1307.

ORCUTT, J., SAHNEY, S. & LLOYD, G. T. 2007. Tetrapod extinction across the Jurassic–Cretaceous boundary. *Journal of Vertebrate Paleontology*, **27** (Suppl. 3), 126A.

PEARSON, D. A., SCHAEFER, T., JOHNSON, K. R., NICHOLS, D. J. & HUNTER, J. P. 2002. Vertebrate biostratigraphy of the Hell Creek Formation in southwestern North Dakota and northwestern South Dakota. *Geological Society of America Special Paper*, **361**, 145–167.

PETERS, S. E. 2005. Geologic constraints on the macroevolutionary history of marine animals. *Proceedings of the National Academy of Science, USA*, **102**, 12326–12331, doi: 10.1073/pnas.0502616102

PETERS, S. E. 2006. Genus extinction, origination, and the durations of sedimentary hiatuses. *Paleobiology*, **32**, 387–407, doi: 10.1666/05081.1

PETERS, S. E. & FOOTE, M. 2001. Biodiversity in the Phanerozoic: a reinterpretation. *Paleobiology*, **27**, 583–601, doi: http://dx.doi.org/10.1666/0094-8373 (2001)027<0583:BITPAR>2.0.CO;2

PETERS, S. E. & HEIM, N. A. 2010. The geological completeness of paleontological sampling in North America. *Paleobiology*, **36**, 61–79.

RAUP, D. M. 1972. Taxonomic diversity during the Phanerozoic. *Science*, **177**, 1065–1071, doi: 10.1126/science.177.4054.1065

RAUP, D. M. 1976. Species diversity in the Phanerozoic: an interpretation. *Paleobiology*, **2**, 289–297.

RAUP, D. M. & SEPKOSKI, J. J., JR. 1986. Periodic extinction of families and genera. *Science*, **231**, 833–836.

RICE, W. R. 1989. Analyzing tables of statistical tests. *Evolution*, **43**, 223–225.

SAHNEY, S., BENTON, M. J. & FERRY, P. A. 2010. Links between global taxonomic diversity, ecological diversity and the expansion of vertebrates on land. *Biology Letters*, **6**, 544–547, doi: 10.1098/rsbl.2009. 1024

SARJEANT, W. A. S. & CURRIE, P. J. 2001. The Great Extinction that never happened: the demise of the dinosaurs considered. *Canadian Journal of Earth Sciences*, **38**, 239–247.

SEPKOSKI, J. J., JR. 1976. Species diversity in the Phanerozoic: species-area effects. *Paleobiology*, **2**, 298–303.

SERENO, P. C. 1997. The origin and evolution of dinosaurs. *Annual Review of Earth and Planetary Science*, **25**, 435–489, doi: 10.1146/annurev.earth.25.1.435

SERENO, P. C. 1999. The evolution of dinosaurs. *Science*, **284**, 2137–2147.

SHEEHAN, P. M., FASTOVSKY, P. M., HOFFMANN, R. G., BERGHAUS, C. B. & GABRIEL, D. L. 1991. Sudden extinction of the dinosaurs: Latest Cretaceous, Upper Great Plains, U.S.A. *Science*, **254**, 835–839.

SLOAN, R. E., RIGBY, J. K., JR., VAN VALEN, L. M. & GABRIEL, D. L. 1986. Gradual dinosaur extinction and simultaneous ungulate radiation in the Hell Creek Formation. *Science*, **234**, 1173–1175.

SMITH, A. B. 2001. Large-scale heterogeneity of the fossil record: implications for Phanerozoic biodiversity studies. *Philosophical Transactions of the Royal Society of London, B Biological Sciences*, **356**, 351–367, doi: 10.1098/rstb.2000.0768

SMITH, A. B. & McGOWAN, A. J. 2007. The shape of the Phanerozoic marine palaeodiversity curve: how much can be predicted from the sedimentary rock record of western Europe? *Palaeontology*, **50**, 765–774, doi: 10.1111/j.1475-4983.2007.00693.x

SMITH, A. B., GALE, A. S. & MONKS, N. E. A. 2001. Sea-level change and rock-record bias in the Cretaceous: a problem for extinction and biodiversity studies. *Paleobiology*, **27**, 241–253, doi: http://dx.doi.org/10.1666/ 0094-8373(2001)027<0241:SLCARR>2.0.CO;2

SULLIVAN, R. M. 2006. The shape of Mesozoic dinosaur richness: a reassessment. *New Mexico Museum of Natural History and Science Bulletin*, **35**, 403–405.

TANNER, L. H., LUCAS, S. G. & CHAPMAN, M. G. 2004. Assessing the record and causes of Late Triassic extinctions. *Earth Science Reviews*, **65**, 103–139.

UHEN, M. D. & PYENSON, N. D. 2007. Diversity estimates, biases, and historiographic effects: resolving cetacean diversity in the Tertiary. *Palaeontologia Electronica*, **10** (2).

UPCHURCH, P. & BARRETT, P. M. 2005. A taxic and phylogenetic perspective on sauropod diversity. *In*: CURRY ROGERS, K. & WILSON, J. A. (eds) *The Sauropods: Evolution and Paleobiology*. University of California Press, Berkeley, 104–124.

UPCHURCH, P., HUNN, C. A. & NORMAN, D. B. 2002. An analysis of dinosaurian biogeography: evidence for the existence of vicariance and dispersal patterns caused by geological events. *Proceedings of the Royal Society, B*, **269**, 613–622.

UPCHURCH, P., BARRETT, P. M. & DODSON, P. 2004. Sauropoda. *In*: WEISHAMPEL, D. B., DODSON, P. & OSMOLSKA, H. (eds) *The Dinosauria II*. University of California Press, Berkeley and Los Angeles, 259–322.

WAITE, S. 2000. *Statistical Ecology in Practice: A Guide to Analyzing Environmental and Ecological Field Data*. Pearson Education Limited, Harlow.

WALL, P. D., IVANY, L. C. & WILKINSON, B. H. 2009. Revisiting Raup: exploring the influence of outcrop area on diversity in light of modern sample-standardisation techniques. *Paleobiology*, **35**, 146–167, doi: 10.1666/ 07069.1

WANG, S. C. & DODSON, P. 2006. Estimating the diversity of dinosaurs. *Proceedings of the National Academy of Science, USA*, **103**, 13601–13605, doi: 10.1073/pnas.0606028103

WEISHAMPEL, D. B. & JIANU, C.-M. 2000. Planteaters and ghost lineages: dinosaurian herbivory revisited. *In*: SUES, H.-D. (ed.) *The Evolution of Herbivory in Terrestrial Vertebrates. Perspectives from the Fossil Record.* Cambridge University Press, Cambridge, 123–143.

WEISHAMPEL, D. B., BARRETT, P. M. *ET AL.* 2004. Dinosaur distribution. *In*: WEISHAMPEL, D. B., DODSON, P. & OSMOLSKA, H. (eds) *The Dinosauria II.* University of California Press, Berkeley and Los Angeles, 517–606.

WILSON, J. A. 2005. Integrating ichnofossil and body fossil records to estimate locomotor posture and spatiotemporal distribution of early sauropod dinosaurs: a stratocladistic approach. *Paleobiology*, **31**, 400–423.

WILSON, J. A. & UPCHURCH, P. 2009. Redescription and reassessment of the phylogenetic affinities of *Euhelopus zdanskyi* (Dinosauria: Sauropoda) from the Early Cretaceous of China. *Journal of Systematic Palaeontology*, **7**, 199–239.

XU, X., CLARK, J. M. *ET AL.* 2006. A basal tyrannosauroid dinosaur from the Late Jurassic of China. *Nature*, **439**, 715–718.

Index

Page numbers in *italics* refer to Figures. Page numbers in **bold** refer to Tables.